X-RAY MICROANALYSIS
IN BIOLOGY

X-RAY
MICROANALYSIS
IN BIOLOGY

edited by
M. A. Hayat, Ph.D.
Professor of Biology
Kean College of New Jersey
Union, New Jersey

First published in the USA 1980 by University Park Press, Baltimore.

Published in the UK 1981 by Scientific and Medical Division
MACMILLAN PUBLISHERS LTD
London and Basingstoke
Companies and representatives throughout the world.

Printed in the USA.

ISBN 0-333-32355-6

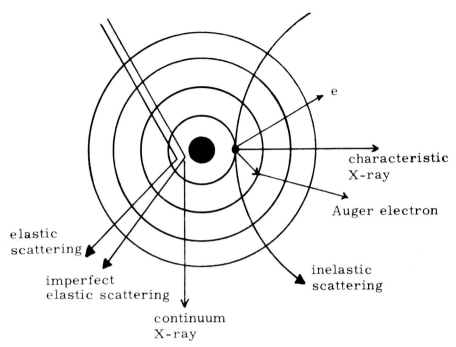

Figure 1.   Interactions of primary electrons with target atoms.

Outer-shell electrons lose energy and occupy the vacancy created in the inner shell and in successive outer shells. The energy loss is in the form of an emitted x-ray. Thus an x-ray spectrum is produced that is related to the number of orbital shells. The spectral lines have energy values that are specific for the element concerned. Not all of these characteristic x-rays escape from the atom. In low atomic number elements particularly, an internal interaction may occur. An x-ray may give up its energy to an orbital shell electron, which is then itself emitted as an Auger electron. The fraction of electron interactions (or ionizations) that results in emission of characteristic x-rays is known as the *fluorescent yield*. The fluorescent yield is strongly dependent upon atomic number (Figure 2).

The energy values of characteristic x-rays are related to the atomic number of the element from which they are emitted (Figure 3). This was first defined by Moseley (1913).

Inelastic Scattering

The majority of primary electrons interact with the weakly bound electrons in the outer shell of the sample atoms. This results in small-angle scattering and dissipation of energy by the primary electrons. Such interactions are the principal factor in defining the size of the volume through which the primary electrons diffuse in the sample.

Continuum Radiation

If a primary electron passes close to the nucleus of a sample atom, it may experience a large change in angular direction and in addition give up some of its energy in the form of an x-ray. In this case the scattering of the electron is said to be not perfectly elastic. The x-radiation produced is known as the *continuum radiation* (also known as background radiation, Bremsstrahlung, and white radiation). Theoretically, continuum radiation has a distribution from infinity up to the energy of the primary electron. The intensity variation is defined by the Kramers equation (Kramers, 1923):

$$I_b \simeq \overline{Z} \frac{E_o - E}{E}$$

(where I_b = continuum intensity, Z = atomic number, E_o = accelerating voltage, and E = x-ray energy). At the low-energy end of the spectrum, continuum radiation is absorbed within the sample and by the window of the x-ray detector and therefore does not reach the theoretical intensity value (Figure 4).

Elastic Scattering

When a primary electron passes close to the nucleus of a sample atom, it may experience a large change in angular direction as before, but without loss of energy. This is referred to as *elastic or Rutherford scattering* and is

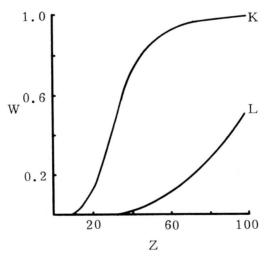

Figure 2.   Relationship between fluorescent yield (W) and atomic number (Z).

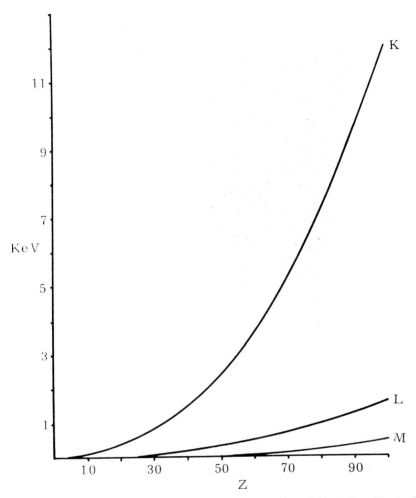

Figure 3. Variation in critical excitation potentials for K, L, and M shells, with atomic number (Z).

the principal factor defining the shape of the volume through which primary electrons diffuse in a sample before losing all their energies.

INTENSITY OF CHARACTERISTIC X-RAYS

The intensity (I) of characteristic x-ray emission produced by bombardment of a sample with primary electrons in an electron microscope is linearly proportional to the beam current (i) (Birks, 1963):

$$I \simeq i$$

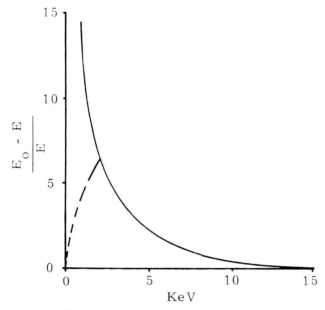

Figure 4.   Distribution of continuum radiation.

There are, however, a number of other factors that influence intensity, and these vary depending upon whether the sample is in the form of a *thin section* or a *bulk specimen*.

Bulk Specimens

In bulk specimens, intensity is proportional to the number of atoms encountered by a primary electron (concentration), fluorescent yield, and the ionization cross-section—the probability that ionization of the inner shell will occur.

The *ionization cross-section* (Q) is a function of the energy of the primary electrons (E_o) and the critical excitation potential (E_c) of the particular shell of the element concerned:

$$Q \simeq \frac{1}{E_c^2}\left(\frac{\ln U}{U}\right) \text{where } U = E_o/E_c$$

or

$$Q \simeq \frac{1}{E_o E_c}\left(\frac{\ln E_o}{E_c}\right)$$

If Q is plotted as a function of overvoltage ratio U (Figure 5), it can be seen that Q (and therefore intensity) should reach a maximum when $U \simeq 3$, i.e., when the accelerating voltage is approximately three times the critical excitation potential.

However, the efficiency of x-ray production in a bulk specimen is determined by the rate of decrease in energy of the primary electrons as they diffuse through the sample. Taking this into account, Green and Cosslett (1961) and Green (1963a) have shown theoretically and experimentally that:

$$I \simeq (U - 1)^{1.63}$$

This relationship indicates that in order to increase intensity U should be made large. A consideration that imposes an upper limit on accelerating voltage (E_o) is the decline in intensity with increasing E_o due to *absorption*. As E_o increases, the electrons penetrate further into the specimen and the path lengths of x-rays that emerge from the specimen are correspondingly longer. Consequently the possibility of absorption occurring is increased.

Absorption is a consequence of x-ray interaction with atomic electrons in the sample. The x-rays lose all their energies and photoelectrons are ejected from the atoms. The attenuation of x-ray intensity is described by Beer's law:

$$I = I^{\circ} \exp [(u/\rho)\rho x]$$

(where I is the observed intensity, I° is the original intensity, u/ρ is the mass absorption coefficient, ρ is density, and x is the distance the x-rays travel through the sample).

Absorption is particularly acute for light elements (Andersen, 1966) (Figure 6). There is a maximum value of accelerating voltage for light elements at which absorption balances generation and thereafter exceeds generation, resulting in a decline in the observed intensity. Some values for

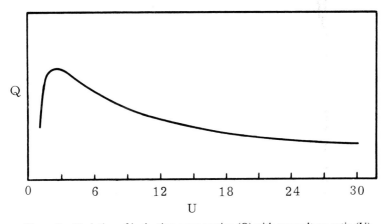

Figure 5.   Variation of ionization cross-section (Q) with overvoltage ratio (U).

these maxima have been experimentally determined by Andersen (1967a) and are given in a graph.

The mass absorption coefficients of absorbing elements vary continuously with respect to the energy of the absorbed x-rays, except for discrete steps termed *absorption edges* (Figure 7), which correspond to the critical ionization potentials of the various energy shells of the absorbing element. X-rays with energy levels slightly higher than the critical excitation potentials are strongly absorbed and may interact with inner orbital electrons to produce a secondary x-ray of lower energy. This phenomenon is known as *fluorescence*.

Absorption and fluorescence have important implications for both qualitative and quantitative analysis.

Thin Sections

In thin sections it is assumed that electrons pass through the section without loss of energy and that absorption is negligible. It therefore follows that x-ray intensity (I) should be simply proportional to the number of atoms encountered by a primary electron, fluorescent yield, ionization cross-section (Q), and beam current (i).

$$I \simeq i\,Q$$

The relationship between intensity and accelerating voltage in thin foils at constant beam current has been investigated by Kyser and Geiss (1977)

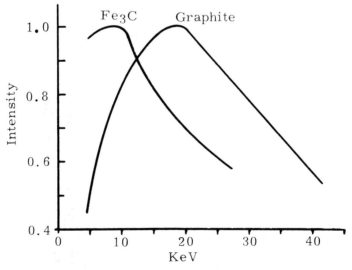

Figure 6.   Graph showing decline in intensity of C Kα radiation in graphite and Fe_3C due to absorption (after Andersen, 1966).

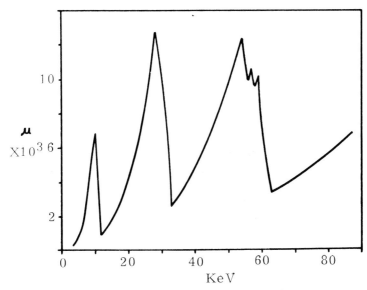

Figure 7. Variation in mass absorption coefficient (μ) for Na Kα radiation in relation to critical excitation potentials of absorbing elements, showing absorption edges.

up to 45 kV and overvoltage ratios of 20 to 21. Their data are in accord with predicted variations in relation to the ionization cross-section.

INTENSITY OF CONTINUUM RADIATION

Bulk Specimens

The fundamental equation defining the generation of continuum radiation in bulk specimens is that of Kramers (1923). Although this equation has been modified in various ways (see Fiori et al., 1976), the dependence of continuum intensity on accelerating voltage can be expressed simply as:

$$I_b \simeq \frac{E_o - E}{E}$$

(where I_b is continuum intensity, E_o is accelerating voltage, and E is the energy of the continuum radiation).

Thin Sections

The intensity of continuum radiation emitted from a thin section is inversely proportional to accelerating voltage (Hall, 1968; Reed, 1975):

$$I_b \simeq \frac{1}{E_o E}$$

(where the symbols have the same meaning as above).

Unlike the situation in bulk specimens, where the continuum radiation is emitted isotropically, the emission of continuum radiation from thin sections is strongly anisotropic (Reed, 1975). Most of the radiation is directed downward from the section in the direction of exiting electrons. This has some practical consequences for analysis that are discussed later.

PEAK-TO-BACKGROUND RATIOS

For the purposes of both qualitative and quantitative analysis, the discrimination of characteristic x-ray peaks in the spectrum from background is of obvious importance. The intensities of characteristic x-ray and continuum generation have been defined in terms of voltage dependency. It is therefore possible to predict the *theoretical* change in peak-to-background (P/B) ratio with overvoltage ratio at constant beam current (Figures 8 and 9). It must be stressed that these are theoretical predictions.

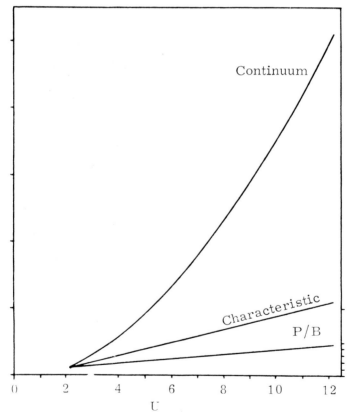

Figure 8.   Graph showing calculated relationship between overvoltage ratio and intensities of characteristic radiation, continuum radiation, and peak-to-background ratio (P/B) in bulk specimens.

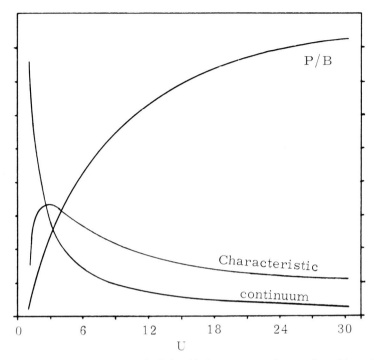

Figure 9.   Graph showing calculated relationship between overvoltage ratio and intensities of characteristic radiations, continuum radiation, and peak-to-background ratio in thin sections.

In the case of bulk specimens, the P/B ratio should continue to rise as accelerating voltage increases because of the difference in the rates of generation of characteristic and continuum radiations; however, this does not occur, as was shown by Andersen (1966) and Russ (1976b). Russ found that for several elements P/B ratios rose until the overvoltage ratio was about 10 to 15, when they leveled off and then fell. It was suggested by Russ that this phenomenon is due to increased background, which is possibly attributable to scattered electrons. It is concluded that in practice there is no advantage to be gained by using overvoltage ratios greater than 10. Andersen found that, at constant current, P/B ratios for Fe $K\alpha$ radiation reached a maximum when the overvoltage ratio was 3.

The situation with respect to thin specimens or sections is controversial. Hall (1968) predicted, on a theoretical basis, that, if maximum available current were put into a spot of fixed size, counting rate should rise with increasing E_o with no maximum, as should P/B ratio. This conclusion has also been arrived at on theoretical grounds by Russ (1977c). However, some authors have found experimentally that P/B ratios reach a maximum at relatively low accelerating voltages (Fowler and Goyer, 1975; Mizuhira, 1976). Russ (1977c) shows that the drop-off in P/B ratios is probably attributable to

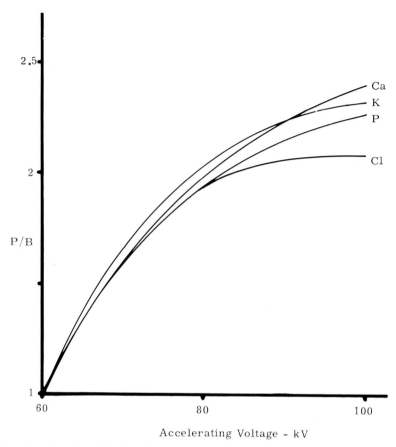

Figure 10.   Experimentally determined peak-to-background ratios (P/B) in a TEM at various accelerating voltages.

an increase in extraneous background as accelerating voltage increases. This background probably varies considerably from one make of instrument to another and will also depend upon various operating parameters (such as the diameter of the condenser lens variable aperture—see section on "Extraneous X-Ray Radiation"), which are seldom specified. Another possible factor is the position of the detector crystal. The further the detector penetrates into the polepiece, the greater will be the likelihood that it will "see" extraneous radiation. This is closely associated with the degree of collimation (see the section on "Collimation").

Marshall and Forrest (1977) found that P/B ratios continued to increase with increasing accelerating voltage (Figure 10), following the predicted variation with the product of beam current and ionization cross-section. It was suggested that this may be due to different microscope and detector geometries. In these experiments the detector did not penetrate into the polepiece, and this may account for some of the differences between these results and those of other investigators.

SPATIAL RESOLUTION

Spatial resolution in bulk samples depends upon the volume through which electrons will diffuse in a given sample. This in turn depends upon the accelerating voltage and the density of the specimen. It is important to realize that the volume of excitation will be different for every element at any particular accelerating voltage. This is due to the differences in critical excitation potential of different elements. As electrons diffuse through a sample they lose energy; clearly, electrons having sufficient energy to excite sodium will go considerably farther than electrons capable of exciting zinc, for example. This is of little importance in a homogeneous sample and particularly so in samples of high atomic number and high density, where diffusion will be limited in extent. However, in samples of low atomic number and low density, such as biological samples, the diffusion paths will be very long. In addition, a biological sample is highly heterogeneous (Figure 11). The investigator must therefore exercise considerable caution in interpreting analyses of bulk biological specimens. This applies particularly to freeze-dried specimens (Marshall, 1974) (Figure 12).

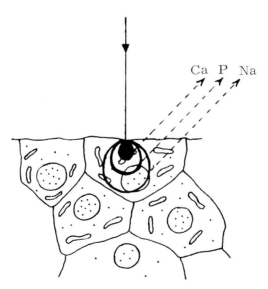

Figure 11.   Diagram showing increase in excitation volume for x-rays of increasingly lower energies within a heterogeneous specimen.

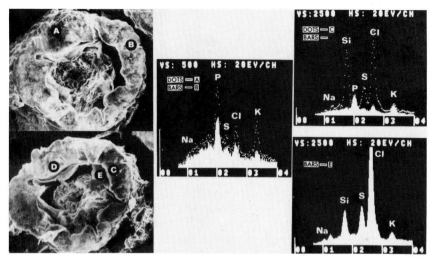

Figure 12.   Freeze-dried biological specimen (hindgut of insect larva). The effect of tilting the specimen is to greatly alter the spectra, due to excessive beam penetration in the underlying tissues. Spectrum A is not markedly different from D, but B is considerably different from C. The beam has penetrated to the lumen contents at C, as shown by the similarity of spectra C and E.

The shape of the volume of excitation varies with the average atomic number of the sample. In low atomic number samples, such as biological specimens, the volume is pear-shaped, whereas in samples of high atomic number it is hemispherical. The shape and extent of the excited volume can be determined by calculation of electron trajectories by the Monte Carlo method (Green, 1963b) or experimentally using plastic "resists" (Everhart et al., 1972). Estimates of spatial resolution in terms of depth of electron penetration and lateral spread have been made for biological specimens by Andersen (1967a,b), Hall (1971), and Hall et al. (1972). These data are presented in graphic form and allow estimates of resolution to be made very quickly for a number of biologically important elements in a matrix of any density.

The present author's unpublished observations suggest that calculated estimates for spatial resolution in dried biological specimens tend to be underestimates. Analyses of differing thicknesses of tissue placed on different substrates give values for electron penetration that are substantially higher than calculated values. Tissue sections of varying thickness were prepared from fixed specimens (toad gastrocnemius muscle and liver) sectioned in paraffin wax, placed on different substrates (glass, NaCl, $CaCl_2$), dewaxed, and air dried. The accelerating voltages at which signals from the substrates were obtained were noted and are plotted in graphic form in Figure 13. This figure gives a comparison with penetration calculated from a range equation developed by Reed (1975) and based on data for copper. It can be seen that there is a considerable difference. A comparison with estimates made from Hall's graph also indicates large differences. The density of gastrocnemius muscle was calculated from precisely cut blocks of muscle that were measured and weighed before and after air drying from amyl acetate or critical point drying from amyl acetate. The value for density (0.30–0.31 g cm^{-3}) was used to calculate the beam penetration from Hall's graph for silicon radiation at an accelerating voltage of 7 kV. This gave a figure of 4.4 μm, compared to a measured penetration of 15 μm.

The possible differences in excitation volume for elements with different critical excitation potentials should be carefully assessed before any attempt is made to compare concentrations of different elements at the same point of analysis. If the excitation volumes are sufficiently similar, then comparisons can be made, but if they are not, then comparisons should not be made when the sample is heterogeneous.

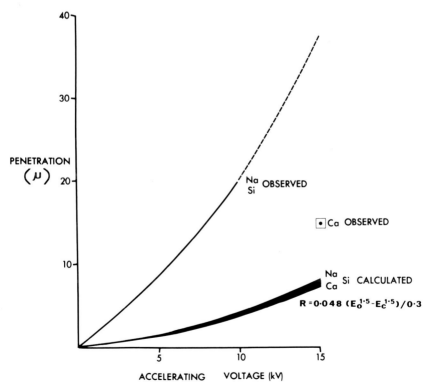

Figure 13. Experimentally determined penetration depth in dried biological tissue compared to depth calculated by typical range equation.

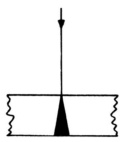

Figure 14.   Electron scatter in a thin section.

In order to obtain the best spatial resolution, the accelerating voltage should be only slightly higher than the critical excitation potential of the analyte. However, as accelerating voltage is reduced, so is x-ray intensity, and it is therefore necessary to arrive at a compromise between the spatial resolution desired and the lowest tolerable x-ray intensity. This decision is in the hands of the investigator and must be made largely on an empirical basis.

In samples in the form of sections, the majority of electrons pass through the section with relatively little scatter (Figure 14). Therefore, the excited volume is smaller than in a bulk specimen and the spatial resolution is higher.

The resolution has been calculated for rather thick sections by Hall et al. (1972), who show that for accelerating voltages above 30 kV the lateral resolution will be between 1 and 2 μm in sections of embedded, frozen, or freeze-dried tissue that are 5 μm thick.

Since the maximum beam diameter in the SEM[1] and TSEM will be of the order of 0.5 to 1.0 μm, it is advisable to increase beam diameter during analysis until beam current has to be limited because of specimen damage. Spatial resolution will not be impaired and the count rate will be maximized.

In thin sections (< 1 μm) that are analyzed in the TSEM, STEM, and TEM, spatial resolution can be of the order of 0.25–0.5 μm for resin-embedded sections (Marshall, unpublished observations). In order to obtain maximum resolution it is advisable to limit beam diameter to 200 nm. In sections that are less than 250 nm thick, spatial resolution is limited by beam diameter. Spatial resolutions of a few hundred Ångstroms can be obtained if beam diameter is small enough. Beam diameters of this magnitude are obtained in the TSEM and STEM. Russ (1972) has estimated the degree of scatter for a beam 20 nm in diameter for embedded and freeze-dried sections, and these data are summarized in Figure 15.

---

[1] For definitions of abbreviations, see the section of this chapter entitled "Microscopes, Available Systems."

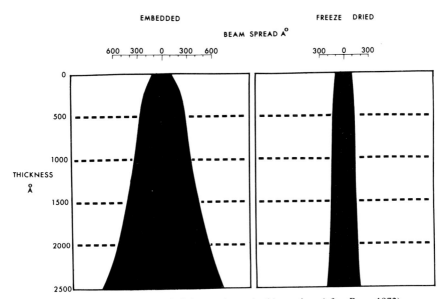

Figure 15.   Lateral spread of electron beam in thin sections (after Russ, 1972).

As an indication of the spatial resolution that is possible, mention is made of the experiments on metal films by Kyser and Geiss (1977), who achieved a spatial resolution of 19 nm with a beam 15 nm in diameter at 100 kV. It should also be noted that Kyser and Geiss comment that the current in beams less than 50 nm in diameter is not enough for quantitative analysis, although semiquantitative analysis can be achieved. This comment applies only to beams produced by hairpin tungsten filaments (see section on "Beam Current").

It should be realized when considering the selection of optimum beam diameter for the analysis of sections in the TSEM and STEM that beam diameter has a profound influence on optical resolution. It may well happen, particularly in an unstained section, that optimizing beam diameter for analysis is limited by the degradation of image quality. It is difficult to analyze what cannot be seen!

ENERGY-DISPERSIVE DETECTOR

The detector is in essence a silicon wafer some 3.5 to 6 mm in diameter. When x-ray photons enter the detector, ionizations of the constituent silicon atoms occur, which results in a cascade of photoelectrons. These electrons finally lose their energy by raising the energy of the valence electrons to that of the conduction band. The silicon may thus be thought of as becoming more conductive, containing negative charges in the form of mobile electrons and "holes" in the valence band that can behave as free positive charges. The total negative charge that is produced in this way can be measured, and since the formation of an electron hole pair requires 3.8 eV of energy, the total charge is a measure of the energy of the ionizing x-ray:

$$E_c = N_t \, e$$

(where E_c is x-ray energy, N_t is total charge, and e = 3.8 eV).

The silicon wafer is produced from a single crystal of p-type silicon. Since it is impossible to prepare silicon free of lattice imperfections and impurities, lithium ions (which are of small radius and are efficient electron donors) are "drifted" into the crystal. The lithium neutralizes the electron acceptors present in the wafer. Thus a large, sensitive zone of high resistivity is created.

The lithium-drifting procedure is vital in ensuring that the conversion of x-rays into electron charge is a linear process. If regions of high conductivity exist in the silicon, due to impurities or lattice defects, then charge carriers (electrons and holes) may be trapped. Trapping will occur to some extent even in the best detectors. Most of the trapped charge carriers will be released by thermal excitation, but time is required for this to occur. The practical consequence is that resolution will be affected by the time constant of the main amplifier. Thus a large time constant will allow more charge carriers, which have been momentarily trapped, to be included in the detector output signal. This will improve the statistics of the charge collection process and will thereby improve resolution.

The lithium-drifted detector crystal is maintained at low temperature in a cryostat. This is partly due to the necessity of preventing redistribution of the lithium within the crystal and, perhaps more importantly, is due to the requirement for low levels of electronic noise. In front of the crystal is a thin beryllium window (typically 7.5 μm) that seals the cryostat vacuum and protects the crystal from contamination.

On the x-ray receiving side the crystal is covered by a layer of gold. To this is connected a negative bias voltage of about -750 V. The other side of the crystal is also coated with metal and is connected directly to a field effect transistor (FET), which represents the first stage of the signal amplification process. When incident x-rays create electron hole pairs, the negative bias voltage ensures that the electrons are swept out of the crystal to the gate of the FET while the positive holes move to the negative supply voltage electrode (Figure 16).

The voltage output from the preamplifier is a ramp or staircase function of the charge collected by the FET. Each step is linearly proportional to the charge. In order to maintain this linearity the voltage has to be periodically reset to zero. This is accomplished with minimal electronic noise by

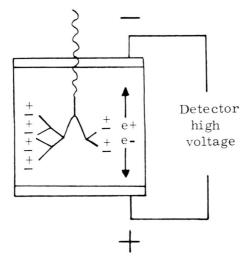

Figure 16.   Ionization within the detector crystal.

making use of a photoelectric effect to "short out" the gate-drain junction in the FET. This junction is photosensitive and light is directed onto it from a light-emitting diode (Figure 17). During the "shorting out" period, processing of signals from the preamplifier is inhibited, thereby further reducing noise. This process is termed *pulsed optoelectronic feedback*. The details of the detector are summarized in Figure 17.

Other systems of resetting the FET voltage exist. A method that does not use a light-emitting diode is found in ORTEC detectors. This method is referred to as "dynamic charge restoration" or "drain feedback" (Elad, 1972). An advantage of this method is that it is not necessary to stop pulse processing while the resetting of the FET voltage is being carried out. This means that the slight increase in dead-time introduced by pulsed optoelectronic feedback is avoided.

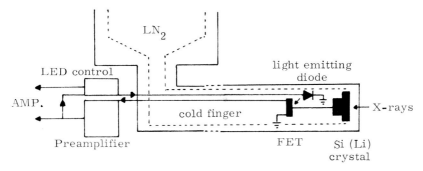

Figure 17.   Schematic diagram of energy-dispersive detector.

DETECTOR RESOLUTION

The ionizations that occur when an x-ray photon enters the silicon crystal are subject to statistical variation, i.e., not all of the photon energy is converted into electron hole pairs. If the formation of electron hole pairs were a completely random process, then the standard deviation could be expressed as:

$$SD = \sqrt{\frac{E_c}{e}}$$

(where SD is standard deviation, E_c is x-ray energy, and e = 3.8 eV). Since the process is not a completely random factor, the Fano factor (Fano, 1947) is introduced into the equation:

$$SD = \sqrt{\frac{FE_c}{e}}$$

(where F is the Fano factor).

The pulses emerging from the preamplifier have an approximately Gaussian distribution and detector resolution is usually stipulated as the full peak width at half maximum height of the peak (FWHM). Since the FWHM is 2.35 times the standard deviation of a Gaussian distribution, the resolution is defined as:

$$E = 2.35 \sqrt{FE_c e}$$

(where E is the resolution in eV at FWHM).

Added to this statistical variation is electronic noise from the detector and amplifier, which results in further peak broadening.

The change in resolution with energy for any detector can be predicted from the observed resolution at a specified energy. The energy usually used is that of the Mn $K\alpha$ peak. Thus FWHM at energy E is:

$$FWHM_E = [R^2 + 2.74 (E - 5890)]^{\frac{1}{2}}$$

(where R is the FWHM for Mn $K\alpha$). The 2.74 value includes the Fano factor and is derived from $[2.35 (Fe)^{\frac{1}{2}}]^2$.

It has been shown (Aitken and Woo, 1971) that, because of statistical and noise limitations, resolution cannot be expected to be reduced below 100 eV (measured at 5.9 keV). Recent improvements in the manufacture of silicon crystals and improved amplifier electronics have resulted in the availability of resolutions of the order 140 eV. This represents a dramatic improvement since the first reports of the application of solid state detectors appeared, when resolutions of the order of 1,000 eV were usual (Fitzgerald et al., 1968).

From the foregoing it is apparent that resolution or peak width varies with x-ray energy (Figure 18). The practical importance of good resolution

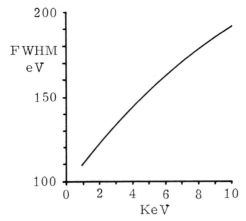

Figure 18.   Variation of resolution with x-ray energy for a detector having a resolution for Mn Kα of 160 eV FWHM.

is related to peak separation in complex spectra and to peak-to-background (P/B) ratios. P/B ratios are improved as FWHM decreases, since the same numbers of pulses are compressed into higher peaks.

Resolution is also a function of crystal area. The larger the area, the wider will be the FWHM. Typical values at the present time (stipulated at the Mn Kα peak and at 3,000 cps) are 149 eV for a 10-mm² crystal and 154 eV for a 30-mm² crystal.

DETECTOR EFFICIENCY

Because x-rays must pass through a beryllium window to enter the detector crystal, they suffer attenuation, due to absorption, if they are of low energy. Absorption also occurs, but to a much lesser extent, in the gold layer and in the dead layer of silicon at the crystal face. The degree of attenuation depends principally on the window thickness and on the angle of incidence of the x-rays. Most manufacturers provide curves purporting to show detector efficiency varying with window thickness and atomic number. However, these may be calculated curves and may not represent the efficiencies as found under conditions of electron beam excitation. Moreover, there is often a variation in window thickness from one detector to another. In other words, the specified window thickness tends to be a mean value. It is advisable to test the efficiency of the detector by use of isoatomic standards in the form of a thin film. A method for obtaining standards for this purpose, which is most suitable for biological x-ray microanalysis (i.e., the standards cover the range of elements most frequently encountered in biological samples), has been described (Morgan et al., 1975; Marshall, 1977). Figure 19

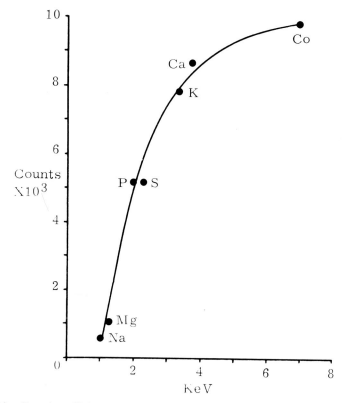

Figure 19. Detection efficiency curve determined experimentally from dried microdroplets containing isoatomic ratios of elements.

shows an efficiency curve derived in this way. This is a most important parameter when investigations of sodium and magnesium distributions are contemplated.

DETECTOR GEOMETRY

Window Thickness

In order to minimize absorption by the detector of x-rays from elements of low atomic number, the detector should be arranged so that the angle of incidence for an x-ray travelling from the specimen to the center of the detector window is 90° (Figure 20).

Absorption by the beryllium window is defined by Beer's law:

$$\frac{I}{I_o} = e^{-\mu\rho x}$$

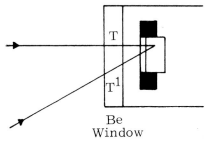

Figure 20. Absorption path lengths of x-rays through detector window and dead layer $T^1 > T$.

(where I/I_o is the fraction of transmitted x-rays, e = 2.714, μ is the mass absorption coefficient of beryllium for the particular x-ray energy, ρ is the density of beryllium (1.844), and x is the path length through the window). The mass absorption coefficient is the principal factor; Table 1 gives values for elements of biological interest. However, if the path length x is increased by reducing the angle of incidence (Figure 20), the transmission of sodium and magnesium x-rays can be significantly affected, as is shown in Table 2 for a 7.5-μm window. Thus, for a change in angle of incidence from 90° to 30°, the transmission of sodium x-rays falls from 56 to 32%.

Take-off Angle

Since x-rays are absorbed as they travel through the sample, the *take-off angle* (θ) should be as large as possible. The take-off angle is the angle that the emergent x-rays make (in line of sight with the detector) with respect to the sample surface (Figure 21). For reasons that are dealt with later, it is advantageous with biological specimens to have the sample surface horizontal, i.e., at zero tilt. Consequently a high take-off angle must be achieved by means of an inclined detector cold finger. In a SEM, an inclined detector can give take-off angles as high as 40° when the untilted sample is at a long working-distance position (Figure 21).

If for any reason it is undesirable or impossible (as in the case of some of the newer, low-cost SEMs) to place the sample at a long working distance, then a number of manufacturers offer a compromise solution. For example, Princeton Gamma-Tech offers a detector with variable Z motion,

Table 1. Mass absorption coefficients for Kα radiation

			Emitter			
Na	Mg	P	S	Cl	K	Ca
418	246	63	43	30	15	11

Table 2.

Angle of incidence	Path length	I/I_o
90°	7.5 μ	0.56
50°	9.79 μ	0.49
30°	15.0 μ	0.32

which means that the detector cold finger can be inclined through a modest angle. This means that at a short working distance, take-off angle can be increased by tilting the specimen while retaining a 90° incidence angle. This not only improves light element transmission but also increases the detector solid angle, which also depends on the angle of incidence (Figure 22).

A novel detector with a tilted window is manufactured by United Scientific and is called the Hypersense detector. This detector gives a substantial increase in take-off angle at short working distances with an untilted sample (Figure 23).

In the STEM and TEM it may be advantageous to incline the detector and not to tilt the specimen. This is not done in order to reduce absorption (which will be negligible in a thin section) but is done in order to take full advantage of the anisotropic emission of the continuum radiation. Such a geometry should minimize background.

Solid Angle

The efficiency of collection of x-rays by the detector is directly proportional to the solid angle. The solid angle is defined as the three-dimensional angle subtended at the sample by the detector, and may be expressed as:

$$\Omega \simeq \frac{A}{S^2} \sin \alpha$$

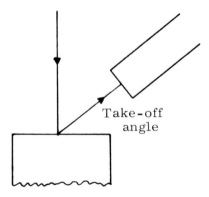

Figure 21.   Take-off angle with inclined detector having a 90° angle of incidence.

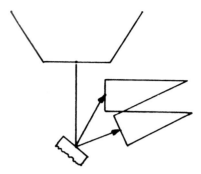

Figure 22.   Variable tilt detector.

where Ω is the solid angle in steradians, A is crystal area in mm², S is sample-to-crystal distance, and α is the angle of incidence of x-rays to the crystal. If it is assumed that α will be 90° in the ideal detector configuration, then it can be seen that the solid angle (and therefore the efficiency of x-ray collection or count rate) will increase with increased crystal area and a reduction in sample-to-crystal distance (Figure 24).

In the analysis of biological specimens, count rates are inherently low; it is therefore usually desirable to increase the solid angle as much as possible in order to increase count rate. The crystal areas commonly obtainable at the present time are 10 mm², 12.5 mm², and 30 mm². It will be apparent that the count rate obtainable with a 30-mm² crystal will be three times higher than that obtainable with a 10-mm² crystal. There is, however, a reduction in resolution with a 30-min² crystal compared with a 10- or 12.5-mm² crystal.

Count rate will increase as the detector is moved closer to the sample by virtue of the inverse square relationship between solid angle and sample-to-detector distance. It follows, therefore, that the detector should be retractable, so that the sample-to-detector distance can be optimized. The

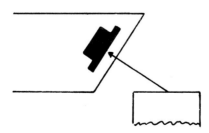

Figure 23.   Hypersense detector.

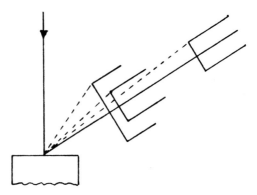

Figure 24.   Variation in solid angle with sample-to-detector distance and crystal area.

influence of the distance from sample to detector crystal and of crystal size on solid angle is shown in Table 3.

It should be noted that while as a general rule for biological analyses the solid angle should be maximized, there are occasions and circumstances when positioning the detector close to the specimen is disadvantageous.

Collimation

In the majority of electron microscopes, if not in all, the electron beam will produce x-ray radiation not only from the specimen but also from components of the microscope. This may occur either directly via scattered primary electrons or indirectly via backscattered electrons or fluorescing x-rays. Microscopes differ greatly in their peculiarities in this respect, and the subject is further discussed in the section dealing with microscopes. The detection of extraneous radiation produced in this way may be restricted quite substantially by proper collimation.

Table 3.   Influence of crystal size and sample-to-detector distance on solid angle

Crystal area (mm²)	Sample-to-detector distance (mm)	Solid angle (steradians)
10	50	0.0040
	40	0.0062
	30	0.0111
	20	0.0250
	10	0.1000
30	50	0.0120
	40	0.0187
	30	0.0333
	20	0.0750
	10	0.3000

The collimator should be lined with spectroscopically pure carbon and constructed so that the aperture next to the window does not occlude the crystal. In some cases the crystal may not be positioned at the center of the beryllium window (Marshall, 1975b). The optimum entrance aperture diameter (d') is defined by:

$$d' = db/a$$

(where d = detector diameter, a = detector-to-specimen distance, and b = aperture-to-specimen distance) (Reed, 1975).

The area of specimen from which x-rays are accepted depends on the distance between entrance aperture and specimen and between entrance aperture and detector crystal (Geiss and Huang, 1975).

This has been formulated (Reed, 1975) as:

$$d'' = 2d\,b/(a - b)$$

(where d'' is the diameter of the "penumbra" and other symbols are as before). Values for d'' for different collimator sizes are listed in Table 4.

It is obvious that the best collimation is obtained by moving the entrance aperture as close to the specimen as possible and by having a long collimator. The cost of doing this is a decrease in solid angle and consequently in count rate. It can also be seen that a detector with a large crystal area cannot be as well collimated as one with a small area. These factors should all be considered when selecting a detector. For example, if extraneous radiation is a problem, approximately the same collimation is achieved with a 12.5-mm^2 detector at a distance of 40 mm as with a 30-mm^2 detector at 60 mm. In this case the solid angle for each detector is approximately the same (0.0078 and 0.0083 steradians, respectively) and thus the count rate will also be the same. Consequently, there would be no advantage to be gained in such a situation by purchasing a 30-mm^2 detector; indeed, the lower resolution of the 30-mm^2 detector would be a disadvantage.

The question of collimation should be considered together with

Table 4.   Values for d'' for different collimator sizes

Crystal diameter (mm)	Detector-specimen distance (mm)	Aperture-specimen distance (mm)	d'' (mm)
4 (12.5 mm^2)	20	10	8.0
	40	10	2.7
	60	10	1.6
6 (30 mm^2)	20	10	12.0
	40	10	4.0
	60	10	2.4

methods for decreasing extraneous radiation in electron microscopes; the latter is dealt with in the section on microscopes.

BACKSCATTERED ELECTRONS

If high-energy backscattered electrons are able to pass through the beryllium window and enter the detector, a substantial increase in background and dead-time results. This can readily occur in a scanning electron microscope that is being operated at high (>40 kV) accelerating voltages. Accelerating voltages of this magnitude could be used for the analysis of sections. It has been shown by Hall (1977) that background produced in this way can be eliminated by deflecting magnets mounted in front of the detector window.

WINDOWLESS DETECTORS

Windowless detectors are available from several manufacturers. However, only the Edax Econ detector (Russ et al., 1976; Sandborg and Lichtinger, 1977) appears to be suitable for routine use on standard scanning electron microscopes. Windowless detectors are either without a window of any type or they utilize a window made from an organic film. Some mechanism must be present to seal the detector when the microscope column is brought to atmospheric pressure. In the Econ detector, backscattered electrons are prevented from entering the detector by a magnetic electron trap, which is effective for accelerating voltages up to approximately 12 kV, and contamination of the detector crystal is prevented by a cryogenic trap. The presence of these traps in front of the silicon crystal places restrictions on the detector-to-sample distance and gives a relatively poor solid angle.

The analysis of carbon, oxygen, and nitrogen is practical with the Econ detector, but there are limitations, in that peak position is not accurately defined and pulse pile-up will occur at count rates in excess of 100 cps. These factors mean that quantitation is difficult.

AMPLIFIER AND PULSE PROCESSOR

The main amplifier or pulse processor is the heart of the analytical system and considerable attention should be paid both to the initial selection of the amplifier and to its function. It is concerned not only with linear amplification of the preamplifier output but also with pulse shaping or noise filtering, which affects resolution, pulse pile-up rejection, which affects sensitivity and quantitative accuracy, base-line restoration or peak shift, which affects peak deconvolution and quantitation, and live-time correction, which is essential for quantitative analysis. The output from the preamplifier is in the form of a step signal (Figure 25) in the millivolt range. This is amplified in the main amplifier to produce positive pulses up to 10 V in amplitude (Figure 25).

During the amplification process the signals are passed through a pulse-shaping network, which is essentially a process of differentiation and integration via a capacitance-resistance circuit. This circuit acts as a filter for preamplifier noise. The initial differentiation filters out low-frequency noise and the subsequent integrations filter out high-frequency noise. The resulting pulse is semi-Gaussian in form and the duration of the pulse is determined by the time constant (product of resistance and capacitance) of the pulse-shaping network. The larger the time constant, the better is the noise filtering. This means that the pulse height will be subject to less variation; consequently, the energy spread of the pulses produced by x-rays of a given energy will be narrower. In other words, the resolution of the system will be improved. However, the increased duration of the pulse means that the interval that must occur between recognition of successive pulses will be longer, i.e., the throughput rate of pulses by the amplifier will be lower.

Semi-Gaussian pulse shaping by capacitance-resistance circuits is employed by the majority of manufacturers. However, the Link System 290 and System 860 analyzers employ a pulse processor that uses time-variant filters (Kandiah, 1966). The advantage of this system is that for the same degree of noise filtering provided by a Gaussian amplifier a much higher throughput can be achieved.

In any amplifier it will be an advantage if the operator has control over the selection of the pulse-shaping time constant. Thus a long time constant

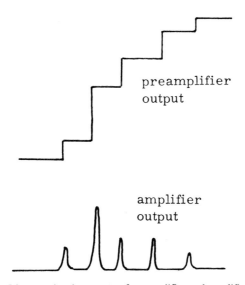

Figure 25.   Diagram showing nature of preamplifier and amplifier outputs.

can be selected to give maximum resolution at low count rate, whereas a short time constant will facilitate the acquisition of pulses at very high count rates.

Pulse Pile-up Rejection

If a pulse is accepted by the main amplifier while it is still processing the previous one, the energies will add, thereby distorting the spectral information. The nature of the distortion will be seen as increased background and sum peaks (Figure 26). In order to prevent this, the amplifier contains a pile-up rejection circuit, which uses a second channel in the amplifier that is parallel to the measurement channel and is usually called a fast inspection channel (Figure 27). Whereas the pulses in the measurement channel are shaped with a long time constant, those in the inspection channel are shaped with a very short time constant (less than 1 μs). The pulse amplitude is therefore extremely variable. However, this is not important, since the fast pulses are only used to establish the presence or absence of pulses entering the amplifier while a pulse is still being processed in the measurement channel. If the arrival of a second pulse during the processing of the first is recognized via the inspection channel, then one or both pulses are rejected (Figure 28). In some amplifiers both pulses are always rejected. However, in better amplifiers only the second pulse is rejected when it occurs at the trailing edge of the first pulse.

The threshold of the inspection channel is usually set just above the noise level. If it is set too high, it will miss low-energy pulses and pile-up protection will be inadequate for the low-energy region of the spectrum. If

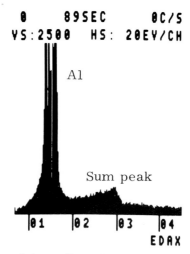

Figure 26.   Sum peak due to pile-up of aluminum x-rays at high count rate.

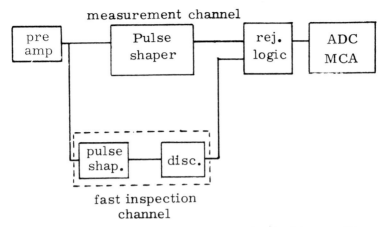

measurement channel

fast inspection
channel

Figure 27.   Block diagram of a pulse pile-up rejection circuit in an amplifier.

the threshold is set too low, noise will be recognized as pulses and the throughput will be reduced.

The time constant for the inspection channel pulse determines the pulse height variation for a given energy. The longer the time constant, the lower the variation. Thus a pulse duration of 1.0 μs will ensure that low-energy

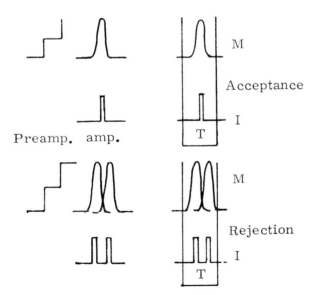

Preamp.   amp.

Acceptance

Rejection

Figure 28.   Pulse pile-up rejection diagram showing that if two pulses in the inspection channel (I) fall within time interval T, the rejection logic rejects the overlapped pulses in the measurement channel (M).

pulses down to 1 keV will be above threshold at reasonable noise rates. On the other hand, a pulse duration of 0.1 μs will not provide pulse pile-up protection below 3 keV because the pulse height variation will severely lower the pulse detection efficiency in the inspection channel (Reed, 1972; Statham, 1977). In this case the throughput rate will be higher. In the Nuclear Semiconductor 513 amplifier, two inspection channels are used. One channel uses pulses of 0.18-μs duration for pile-up rejection at high energies and the other channel uses 1-μs pulses for pulse pile-up rejection at low energies. A similar system, but with an even faster "fast" inspection channel, is used in the Link 2010 pulse processor.

It should be noted that none of the presently available commercial systems is capable of providing pulse pile-up protection below 1 keV. Consequently the problem of pile-up imposes a severe limitation on the usefulness of windowless or thin window detectors. In order to avoid pulse pile-up, these detectors must be operated under conditions in which the count rate does not exceed approximately 100 cps (Russ et al., 1976). A means of providing pulse pile-protection that would be effective at low energies has been proposed by Statham (1976a). This involves a technique in which "beam switching" is employed. In "beam switching" the detection of a pulse in the amplifier results in the blanking-off of the electron beam for the time that it takes to measure that pulse. The beam is switched on when the electronics are ready to deal with another pulse.

Base Line Restoration

The shaped pulse that emerges from the measuring channel of the amplifier exhibits undershoot. At high count rates there is insufficient time for the base line to reach its original level before the next pulse arrives. This means that the apparent energy of the pulse is less than its real energy and that, over all, there is a shift in peak position toward low energy values and a deterioration in resolution (Figure 29).

A base-line restorer in the amplifier attempts to attenuate the duration of the undershoot so that peak shift does not occur until very high count rates are achieved. In most commercial instruments this is usually well above the count rate that would normally be obtained from a biological sample.

For quantitative analyses it is essential that peak shift does not occur or is at least very limited. Consequently, if analyses with high count rates are envisaged, some attention should be paid to the selection of an amplifier that has good base-line restoration characteristics. Specifications for peak shift can vary widely, e.g., the Kevex 4530 amplifier is stable up to 20–30,000 cps, Nuclear Semiconductor 513 up to almost 10,000 cps, Ortec 739 up to 10,000 cps, and the Link 2010 pulse processor is stable to a count

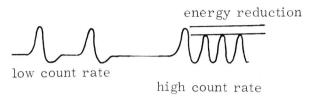

Figure 29.   Energy shift at high count rate due to lack of base-line restoration.

rate well in excess of 60,000 cps. There is also a wide variation in specifications for resolution degradation with count rate.

Live-Time Correction

Differing concentrations of an element in a sample or samples will yield x-rays at different count rates. To compare these concentrations, counting must be carried out for a set time. There are, however, finite time periods during which the system is not recording pulses and these periods vary as count rate increases. This loss in counting time is referred to as dead-time and must be corrected for by a proportionate increase in the overall counting time or live-time. The dead-time arises principally from the inhibitory period during which the FET voltage is reset, losses during pulse pile-up rejection (which is highly variable), and the time taken by the logic circuits in the multichannel analyzer to process the pulses.

The amplifier should have a means of accurately measuring the dead-time and of extending the live-time by the same period. Many live-time correction circuits use the difference between input rate and output rate of the amplifier for measuring dead-time. There is an inherent danger in this technique (Statham, 1977) of which the operator should be aware.

If the discriminator threshold for the inspection channel is set low to collect more low-energy pulses, it will trigger on noise, and the input rate will be artificially high as seen by the live-time correction circuit. If the discriminator threshold is set high, then the input rate will be artificially low.

Consider a situation where the true input rate is 5,000 cps. On the one hand, if the discriminator is set very low, so that the noise rate is 1,000 cps, then the apparent input rate will be 6,000 cps, and in a preset live-time of 100 s the total counts will be 600,000 instead of 500,000. On the other hand, if the discriminator is set very high, so that 50% of the pulses are missed, the apparent input rate will be 2,500 cps and the total counts in 100 s live-time will be 250,000.

Thus with this system the discriminator threshold in the inspection channel must be set so that the noise rate is minimal; otherwise live-time correction is not accurate. Consequently, in quantitative analyses the threshold should not normally be lowered to improve pile-up rejection at low energies when counting on a preset live-time basis. Even if these precautions are taken, it is still possible to have incorrect live-time correction if the spectrum contains an unusually large number of low-energy pulses. This will occur when frozen-hydrated bulk specimens are being analyzed. The very high continuum count from the low-energy end of the spectrum leads to an output rate that is greater than input rate (Marshall, unpublished observations).

MULTICHANNEL ANALYZER

The pulses from the amplifier are sorted on the basis of amplitude, and are stored in the multichannel analyzer (MCA) so that successive points in the digitized spectrum are stored in successive channels. Each channel contains a record of the number of x-ray counts from a specific region of the spectrum.

The spectrum is digitized in the analogue-to-digital converter (ADC), which is the first part of the MCA. Digitizing is done via a high-frequency clock, which measures the time taken for a capacitor to discharge after being charged up by a pulse from the amplifier. Each pulse in the spectrum is given a value in terms of time or, more specifically, clock pulses. This value corresponds to the pulse height or energy of the spectral pulse. The number of clock pulses is determined by a scaler, and this number is used as the memory channel address. The total number of counts in the channel at this address is then increased by one. The data stored in the memory channels are continuously accessed for display on a video terminal or cathode ray tube.

At the present time there are three types of MCA systems:

1.   Hardwired MCAs, in which all spectral manipulation and reduction are accomplished by logic circuits in the MCA. The more sophisticated hardwired MCAs are frequently interfaced with a minicomputer. Low-priced models include the Tracor Northern TN 1705 and the Ortec 5200 M, while higher-priced models include the Ortec 6230 and Edax 711 MCAs.

2.   MCAs that use part of the standard memory of a minicomputer for pulse height analysis. These MCAs represent a logical step from the combination of MCA and minicomputer described above and are a more elegant approach. This system is used in the Ino-tech Ultima II and in the Link 290. In both cases the computer can also be used as a stand-alone computer. All data manipulation is controlled by software in these systems. In the Link 860, part of the memory of a microcomputer is used as the MCA and part is in the form of PROM memory, which controls the basic analyzer functions.

3.   MCAs in which a microcomputer is used to control the data

manipulation. Examples of this system are seen in the Tracor Northern TN-1710 and the Kevex μX analyzers. In the TN-1710, data manipulation is controlled by firmware in the form of ROM modules and the microcomputer can be operated as a stand-alone computer. In the μX, data control is largely by means of software. Integration of analyzer and computer has been developed even further in the Ortec EEDS II, the Tracor Northern TN-2000, and the Edax 9100 systems.

It seems inevitable that large-scale integration (LSI) will be used increasingly in the form of microprocessors and microcomputers in MCAs. This will certainly lead to faster, more stable, and much more automated systems.

A LOW-COST SYSTEM

For the biologist who requires qualitative or semi-quantitative analyses it is possible to put together a very low-cost but excellent system for analyses that have inherently low count rates. A system that this author has used in his laboratory consists of an Edax detector with model 183 preamplifier/amplifier interfaced to a Tracor Northern TN 1705 MCA. The Edax detector has a combined preamplifier/amplifier and consequently no extra cost is involved in the purchase of a separate amplifier, as is the case with most other makes of detector. A power supply can be easily constructed. A low-voltage ($\pm$12 V and $\pm$24 V) power supply for the preamplifier/amplifier can be built using the widely known voltage regulator IC 723. Suitable circuits employing this IC are described in *Linear Data Book* [National Semiconductor Corporation, 1976 (Figure 3, p. 1–71, and Figure 5, p. 1–72)]. The high-voltage (750 V) supply for the detector is a typical zener diode stabilizer. Its sole peculiarity is a high time constant at the output (22 MΩ and 0.5 μF), needed for protection of the detector against transients.

The signal from the measuring channel is led to the MCA input and the inhibiting signal during resetting of the FET can be led to the anticoincidence gate. It is not possible to incorporate pulse pile-up rejection and the system can be effectively used only at modest count rates (up to 1,000 cps).

MICROSCOPES

This section reviews the features of electron microscopes that affect x-ray microanalysis and the modifications that may be made to commercial instruments for the purpose of improving x-ray microanalysis.

Available Systems

Analyses may be carried out in a variety of microscopes with a variety of detector geometries. The *scanning electron microscope* (SEM) is best suited to the analysis of bulk specimens, although sections can also be analyzed with or without the aid of a transmitted electron detector. In the latter case the microscope is probably best referred to as a *transmission scanning electron microscope* (TSEM). Section analysis is best carried out in the *transmission electron microscope* (TEM) or *scanning transmission electron microscope* (STEM), although it is also possible to analyze small bulk samples in the latter instrument. A STEM with an x-ray analyzer is commonly called an *analytical electron microscope*. In addition there are a number of instruments that have been specifically designed as analytical instruments, e.g., the AEI Corinth (Anderson et al., 1976).

In the SEM, the detector may enter the specimen chamber horizontally or it may be inclined, depending on the working distance required, and the specimen can usually be tilted toward the detector if necessary. In the TEM and STEM, a side entry goniometer stage must be used if the detector enters horizontally. This permits the specimen to be tilted toward the detector.

Beam Current

The usual electron source in an electron microscope is a heated tungsten wire in the form of a hairpin filament. The emission of electrons from the filament is controlled by four factors: temperature, filament-to-grid distance, gun bias voltage, and accelerating voltage (Haine and Einstein, 1952). All of these parameters are under the control of the operator.

When the filament is heated, electron emission rises rapidly and is stabilized by means of negative feedback via a variable resistor to the grid or Wehnelt cylinder (Figure 30). The function of the grid is to act as an electrostatic lens, making the electrons converge with a crossover diameter

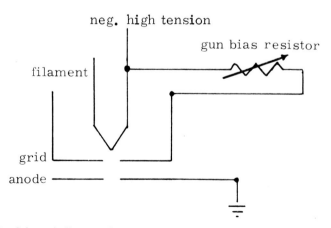

Figure 30.   Schematic diagram of a typical electron gun utilizing a hairpin tungsten filament.

of 10 to 100 μm. Negative bias applied to the grid reduces the area of the grid aperture through which electrons can travel, thus determining the area of the filament from which electrons can pass to the anode. A high bias resistance or voltage tends to give a beam with a lower but more stable beam current and also a more coherent beam.

The position of the filament can be changed with respect to the grid aperture. If it is moved toward the aperture, then beam current will be increased (Revell, 1970), but at the expense of some degradation in resolution, since the energy spread of the electrons increases (Boersh effect) (Broers, 1973). It should be noted that filament life is considerably shortened and the stability is more susceptible to filament movement (Reed, 1975).

The shape of the Wehnelt cap or grid has a marked influence on the beam current and beam coherence. The conical Johansen Wehnelt grid gives improved performance, in terms of a brighter gun (higher beam current), more coherent beam, low contamination rate, and longer life (Johansen, 1973). This type of Wehnelt is standard on all JEOL SEMs and TEMs.

Beam current can be increased by the use of a pointed filament. However, this increase is marginal compared with a hairpin filament with a conical Wehnelt but is substantial compared to a hairpin filament used with a re-entrant grid. Pointed filaments are usually used with flat Wehnelt grids. They suffer from the disadvantage of being relatively expensive and have a very short lifetime.

Lanthanum hexaboride tips have a very high electron emission at relatively low temperatures and give a considerably increased gun brightness. However, unless the vacuum in the gun chamber is better than 10^{-6} torr and the tip is mounted in a well-designed holder, the beam current is unstable. With a well-designed holder, good stability and long life are attainable (Nakagawa and Yanaka, 1975).

Single crystal tungsten wire can be etched to radii of the order of 20 nm. When placed in a positive electric field the crystal emits electrons that are converged by a second electrode held at negative potential. A very small cross-over can be achieved with a very high current density. A current of 10^{-8} A in a 5-nm beam is claimed for one system (Bovey et al., 1977). This type of gun requires an ultra-high vacuum (10^{-10} torr). However, this requirement can be relaxed to 10^{-8} to 10^{-9} torr if the filament tip is heated. Very small beam diameters (1 nm) can be obtained with beam currents of the order of 6×10^{-11} A (Harada et al., 1976), which are probably adequate for section analysis. Unfortunately, field emission guns do not have a very high degree of stability, which renders quantitative analysis difficult unless some system of beam current integration is used (e.g., Vadimsky, 1976).

Beam current is also increased by increasing accelerating voltage and the angular aperture of the beam. Thus beam current varies with the square of the diameter of the final lens aperture in a SEM and the variable condenser aperture in a TEM or STEM.

The spherical aberration of the lens system also affects beam current according to the equation:

$$i \simeq \frac{d^{8/3} \times B}{C_s}$$

(where i = beam current, d = beam diameter, B = gun brightness, and C_s = spherical aberration). In a STEM, C_s is much less than in a SEM, consequently, beam current is increased for the same beam diameter.

Beam Current Stability

Since x-ray intensity is directly proportional to beam current, the stability of the beam current should be of the order of 0.1% (Fitzgerald, 1964) and manufacturers generally specify values of this order for SEMs (e.g., ± 0.2% per hour for JEOL JSM 35) (Figure 31). For TEMs and STEMs, data on

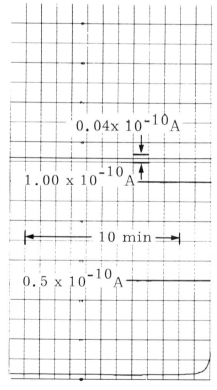

Figure 31.   Beam current stability in JSM 35.

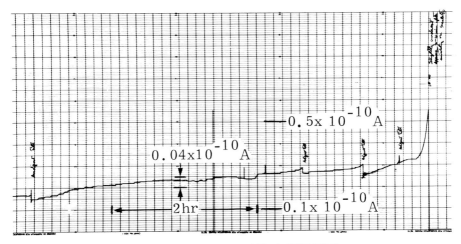

Figure 32.   Beam current stability in JEM 100 B.

beam current stability are not usually found in the manufacturer's specifications. If data are available they are given for very short periods of time.

The author is not aware of any published data on beam stability in the TEM or STEM. His own unpublished observations on the beam current in a JEOL JEM 100 B STEM are shown in Figure 32.

A stable beam current is mandatory in any form of quantitative analysis based on standards, and measuring absolute concentrations as mass per unit volume is required. It should be noted that the average electron microscope will require 1 to 3 h to reach electronic and thermal stability after all lenses, high tension, and filament emission have been switched on.

Measuring Beam Current

Beam current can be measured with the aid of a Faraday cup, placed at the position of the specimen, and a sensitive electrometer or specimen current amplifier. The beam is positioned in the cup in SEM or STEM mode. The cup can be a narrow-bore hole (1 mm diameter) in which the length is four to five times the diameter. In this case the electron capture will be slightly less than 100%. An even simpler Faraday cup can be constructed from a grid cemented onto graphite. The current measured when the beam is positioned in the center of a grid hole is within 1 or 2% of the total current (Wiesner, 1976). A Faraday cup that can be used to measure beam current to within 0.1% is described by Joy (1974).

In addition to an absolute knowledge of beam current, a knowledge of beam current stability is essential for some forms of quantitative analysis. Beam current during an analysis can be monitored by measuring the current from an aperture. Reed (1975) recommends a double aperture in which the second or lower aperture is used to measure the current. The author has found a single insulated final lens aperture to be satisfactory. If the aperture is not repositioned there is a good degree of linearity between the current measured at a Faraday cup and at the aperture.

In a STEM the best system devised so far permits monitoring stability only before and after an analysis without moving the specimen. This is achieved in the JEOL JEM 100 CX STEM by deflecting the beam above the level of the specimen onto a current collecting aperture (actually the anticontamination aperture). In a TEM, beam current stability can be monitored at the level of the viewing screen by means of a Faraday cup.

Correcting Beam Current

If the beam current can be monitored, drift can be corrected by adjusting condenser lens current or gun bias voltage.

Beam current is proportional to beam diameter:

$$i \simeq \frac{d^{8/3} \times B}{C_s}$$

(where i = beam current, d = beam diameter, B = gun brightness, and C_s = spherical aberration). Beam diameter depends upon the demagnification of the source diameter, and this is under the control of the condenser lens.

In the SEM, beam diameter can be changed substantially without affecting the spatial resolution of a bulk analysis. Thus the condenser lens current can be adjusted to correct beam current by changing beam diameter.

Changes in beam diameter beyond certain narrow limits cannot be made in section analysis without affecting the spatial resolution, and in most TEMs and STEMs the current in the first condenser lens can be adjusted only in large steps. Consequently the only way in which beam current can be corrected in the TSEM, TEM, and STEM is to adjust the gun bias. However, this procedure is not straightforward, since in the majority of microscopes gun bias resistance is not continuously variable, but is stepped. This is illustrated in Figure 33. In this situation all that can be done is to adjust gun bias and filament heating current to achieve correction in the TEM and STEM. This means that the filament may be running in the undersaturated condition, with the disadvantageous consequence that beam stability is reduced. In the TSEM it should be possible to make minor adjustments to condenser lens current, together with gun bias and filament heating current, without a great sacrifice of spatial resolution.

In summary, beam current in the SEM can be corrected fairly easily by hand, and it is not difficult to devise a negative feedback system to the condenser lens supply to provide automatic correction if desired. In the TEM and STEM a continuously variable gun bias is necessary to provide a

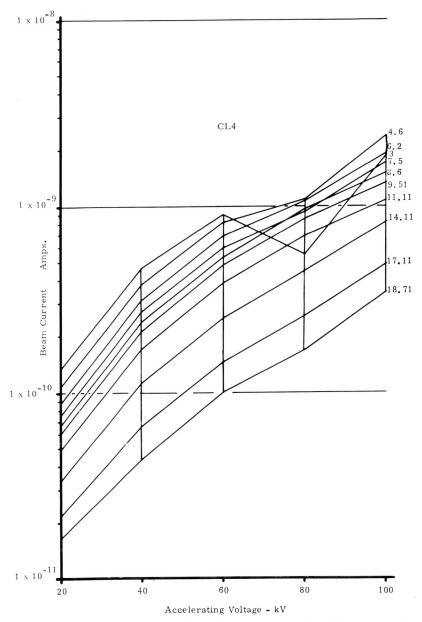

Figure 33.   Variation in beam current with accelerating voltage and gun bias resistance in a JEM 100 B STEM.

reasonable facility for correcting beam current. This facility is not commonly available on commercial instruments at the present time.

Beam Diameter

It is important to know the upper limit of beam diameter for analyses of bulk specimens so that maximum current can be applied to the specimen without a decline in spatial resolution. Since this upper limit would be of the order of 1 μm, it is rarely of importance in the SEM because beam diameters would rarely exceed this value, even with fully weakened condenser lens excitation. However, for section analysis in the TSEM and in the STEM, beam diameter can be the principal factor that influences spatial resolution.

Beam diameter can be measured directly from the fluorescent screen in the TEM, but is rather more difficult in the SEM, TSEM, and STEM. The basis of all methods is to estimate beam diameter by observing the signal change at the sharp edge of a specimen. This can be done either by measuring Fresnel fringes at the borders of holes in collodion films using a transmitted electron detector or by measuring the "fuzziness" at the edge of a specimen photographed using the secondary electron signal (Wiesner, 1976). If a densitometer is available, the latter can be quantified. The distance on the scan profile between the 10 to 90% limits (Figure 34) represents the diameter, which contains 56% of the beam current, assuming that the distribution of current across the beam is Gaussian.

Beam diameter can also be measured directly by observing the change in signal intensity in a line scan, either in the secondary electron signal if the sharp edge specimen is placed on a carbon support (Wiesner, 1976), or in the transmitted electron signal (Joy, 1974). Again, if the 10 to 90% limits

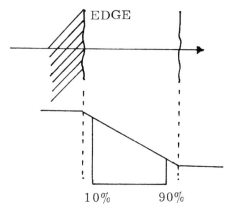

Figure 34.   Densitometer or line scan profile for determining beam diameter, which contains 56% of beam current.

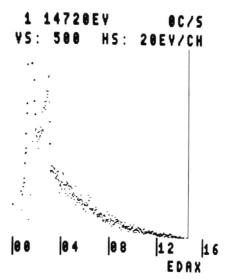

Figure 35. Determination of accelerating voltage by observing the energy at which the continuum falls to zero (indicated by cursor).

are used, the distance between them gives the beam diameter, which contains 56% of the current.

Accelerating Voltage

For bulk specimen analysis in the SEM, it is desirable to be able to select accelerating voltage with 1-kV increments. This is essential for optimizing spatial resolution and count rate for light element analysis in biological specimens. For the analysis of sections, very high accelerating voltages are desirable. The accelerating voltage should be highly stable, since x-ray intensity is proportional to accelerating voltage. The stability requirements for scanning and transmission electron microscopy are usually in excess of the requirements for quantitative x-ray microanalysis. However, for quantitative analysis an accurate knowledge of the value of the accelerating voltage is required. This may not coincide with the instrument-designated value and so the true accelerating voltage should be measured.

The easiest method of measuring accelerating voltage is to observe the energy level at which the continuum x-ray radiation in a recorded spectrum falls to zero, ignoring the high-energy tail that may occur due to the resolution limit of the detector (Beaman and Solosky, 1972) (Figure 35). Methods for more direct measurement are given by Wiesner (1976).

Extraneous X-Ray Radiation

Extraneous radiation can arise from many sources, including the specimen holder, the condenser or final lens apertures, the condenser or final lens polepiece, and the objective lens polepiece. The problem is perhaps greater in a SEM than in a TEM or STEM, since electrons in the specimen chamber are not confined by a magnetic field as they are in the latter microscopes and may more readily produce x-rays from microscope components.

In the SEM and TSEM, x-rays may be produced by backscattered electrons from the specimen that strike the polepiece or components surrounding the specimen. This problem may be overcome by the use of shielding, e.g., a beryllium plate attached to the bottom of the polepiece, or by coating components with colloidal carbon. It should be noted, however, that although this procedure will remove unwanted characteristic x-rays, it will not entirely eliminate unwanted continuum radiation and is not a substitute for good collimation (Statham, 1976b). Specimen holders may also be constructed of carbon or beryllium.

Electrons may be scattered around the final lens aperture (Figure 36). This can be reduced by the use of a limiting aperture placed immediately above the final aperture (Bolon and McConnell, 1976). Electron scattering in a Philips EM 300 was eliminated by Williams and Goldstein (1977). They achieved this by placing a thick (>1 mm) lead aperture below the variable aperture of the second condenser lens. A dirty final lens aperture can also lead to electron scatter, which will similarly result in extraneous x-rays. If the final lens aperture is a thin foil, x-rays may also be produced that will cause x-ray fluorescence in the sample. Substituting thick apertures reduces this problem but does not completely eliminate it, since x-rays can still be produced at the thin edge of the annulus (Bolon and McConnell, 1976).

In the TEM and STEM, the production of extraneous x-rays by scat-

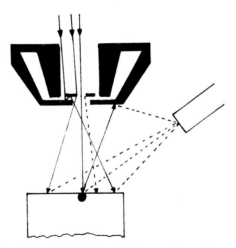

Figure 36.   Scattering of electrons (solid lines) and x-rays (dotted lines) in a SEM (after Bolon and McConnell, 1976).

tered electrons is less severe than in the SEM, but nevertheless it is significant. Moreover, the contribution from x-rays generated in the illuminating system (Kenik and Bentley, 1977) and at apertures is perhaps greater than in the SEM. The severity of the effect varies with make of microscope and no universal solution to the problem can be offered.

The effect of backscattered electrons in the specimen chamber can be reduced by withdrawing the objective lens aperture (Marshall, 1975b) and by using Nylon or beryllium grids and grid holders made from carbon or beryllium. Murray (1973) devised a shield that entirely covered the specimen and specimen holder. It was constructed of aluminum coated with colloidal carbon and had a small hole for the exit of x-rays to the detector.

X-rays fluoresced from the variable aperture in the condenser lens are a major problem for quantitative analysis (Kenik and Bentley, 1977). This source of x-rays in a Philips EM 300 was reduced by Goldstein and Williams (1978) by using a double-thickness variable aperture in the final condenser lens. JEOL recommend the use of thick tantalum apertures as the condenser lens variable aperture and an additional aperture at the base of the upper polepiece in the specimen chamber. Nicholson et al. (1977) have also used a thick carbon aperture below the upper polepiece in the specimen chamber of a JEOL JEM 100 C. These measures appear to provide a substantial reduction both in continuum radiation and in the appearance of peaks from microscope components.

Vacuum System

In order to reduce contamination, all possible steps should be taken to ensure a clean vacuum system. If the instrument is not an ultra-high vacuum instrument, the vacuum can be improved by simple measures, such as the use of diffusion pump oils with low vapor pressures, the use of molecular sieve traps on the roughing pump to prevent backstreaming of rotary pump oil, and dry nitrogen backfilling. If nitrogen backfilling is used, steps should be taken to ensure that gas pressure in the column cannnot rise above atmospheric, otherwise the detector window may be damaged. Details of methods for reducing contamination are given by Brandis et al. (1971) and Sutfin (1972).

QUALITATIVE ANALYSIS

Peak Detection

The smallest recognizable peak is statistically defined as one that is larger than three times the standard deviation of the background, although this may well have to be reduced to one standard deviation of the background for biological specimens.

Elements are identified by taking the energy value at the peak centroid and referring to tables such as those of Johnson and White (1970). In many MCAs peak identification is facilitated by means of a K, L, M line marker, which indicates the position of elemental peaks on the spectrum.

Spurious Peaks

Pile-up or Sum Peaks   If pulses appear so close together that the fast inspection channel fails to distinguish them, the energies will sum and a spurious peak, which has a characteristic shape, will appear in the spectrum (Figure 26). This situation is usually found when the spectrum has one or a small number of very large peaks and when the count rate is high. The ratio of the intensity of a sum peak to other peaks in the spectrum will change with changes in count rate.

Escape Peaks   X-rays with energies up to a few keV above 1.74 keV (Si $K\alpha$) can ionize the K shell of a silicon atom close to the front surface of the detector crystal. Consequently, Si $K\alpha$ x-rays may escape from the detector and the energy of the incoming exciting x-rays will be seen as $E - 1.74$ keV. The pulses so formed will appear in the spectrum as a small peak exactly 1.74 keV below a major peak. The ratios of escape peaks to parent peaks are given by Reed and Ware (1972).

Silicon Peaks   Small silicon peaks can be produced as a result of x-rays being absorbed in the silicon dead layer of the detector. When x-rays are absorbed in the dead layer, the electron hole pairs are not collected and the only event the detector sees is a Si $K\alpha$ x-ray that penetrates farther into the detector. Silicon peaks formed in this way are very small and become significant only when there are large numbers of x-rays in the spectrum that have energies just above that of silicon.

Elemental Mapping

The distribution of various elements can be mapped by setting a window in the MCA for a particular peak in the spectrum and using the output to modulate the beam intensity of the CRT of the SEM during the exposure of a photographic film. A number of points concerning this technique should be considered. First, a minimum number of counts is required to produce an image with good spatial discrimination, i.e., one in which the edges are sharp. This will depend on the concentration of the element and the degree of suppression of background due to continuum radiation. Suppression may be accomplished by adjusting the brightness of the CRT screen so that a single dot is not sufficient to fully expose the photographic film. The total number of counts required will be on the order of 50,000.

The spatial resolution depends upon the volume of excitation and the number and size of the picture elements on the screen. With picture elements no larger than 0.1 cm (the diameter of the dots) there will be 80 elements per line on an 8-cm screen. If the excited volume is 2 μm, then the area of the scanned specimen should be at least 160 μm wide to give a sharp picture.

The discrimination of concentration differences will in part be attributable to counting statistics, i.e., SD = (N)$^{\frac{1}{2}}$ (SD = standard deviation, N = total number of counts). If, for example, it is assumed that a screen has 200 lines per frame, the total number of picture elements will be not more than 4 $\times$ 10^4. If the mapping time is 900 s and the maximum counting rate is 5,000 cps, there will be at most 100 counts per element, which gives a standard deviation of 10%.

The continuum can vary with topography and with atomic number of the sample. These effects can easily mask true variations in mapping intensity. Consequently, a separate continuum map should always be prepared for comparison. These points have been considered in detail by Heinrich (1975).

Line Scanning

In line scans, the electron beam scans across the specimen at selected coordinates and the count rate from a particular part of the spectrum, i.e., the output from an MCA window, is converted to a variable voltage signal by the digital to analogue converter in a rate meter. Rate meters have adjustable time constants; therefore, the voltage output from the rate meter represents an average count rate, which will vary according to the selected time constant. It is therefore important to use very slow scanning speeds for line scans when a long time constant is employed on the rate meter, otherwise the electron beam may move to a region of different concentration while the rate-meter response lags behind. This difficulty is overcome if multichannel scaling is used instead of a rate meter. In this mode the MCA places the pulses from a preset energy window into sequential memory channels. The counting time for each channel is selectable. It is thus possible to synchronize the scanning rate of the microscope with sweep rate of multichannel scaling. The analysis is represented as a histogram of frequency of pulses against elapsed time (equivalent to the distance the beam scans across the specimen). The statistical accuracy of multichannel scaling can be improved by repetitive scans. It is possible in some instruments to perform multichannel scaling from more than one window simultaneously, e.g., Edax 9100 (Russ, 1979).

Line scanning is more appropriate for the analysis of sections than for bulk specimens because bulk specimens rarely have a surface smooth enough to avoid signal changes due to variations in topography (take-off angle). If the surface is sufficiently smooth, the method can be used, but it is essential to record the continuum in the same way. The continuum line scan will also reveal changes in specimen density that will alter the concentration of the element.

QUANTITATIVE ANALYSIS

Practical Problems

Mass Loss   Mass loss will invalidate quantitative analyses that are being carried out by references to standards. If mass loss is selective for the specimen matrix it will also invalidate analyses that rely on peak-to-background ratios. This phenomenon has been studied by Hall and Gupta (1974). In sections that are exposed to an electron dose of 10^{-10} C/μm^2, mass loss up to the order of 30% can occur in the first few seconds of analysis. This is reduced by maintaining the specimen at low temperature. It is therefore advisable to restrict the electron dose and to cool the samples if possible.

Contamination   Contamination in the form of polymerized hydrocarbons occurs at the specimen surface in electron microscopes (Brandis et al., 1971). Contaminating hydrocarbons may absorb x-rays of light elements, such as sodium and magnesium, while elements such as sulfur (Adler et al., 1962) and silicon in the contaminants will produce spurious results. Every precaution should be taken to eliminate sources of contamination. Details are given by Brandis et al. (1971) and Sutfin (1972).

Charging   Electron bombardment of a sample may result in the accumulation of negative charge in the sample (Pawley, 1972). Under some circumstances a very high surface potential can develop (Weitzemkamp, 1969; Van Veld and Shaffner, 1971). The effects of charging could be twofold. First, the primary electron beam can be deflected, and, second, ions may migrate within the sample (Scholes and Wilkinson, 1969). It is therefore advisable to avoid or to prevent charging.

To avoid charging, low beam currents can be used or the specimen can be coated with a conducting coat. The latter will also protect the specimen from thermal damage. Evaporated coatings of carbon and aluminum can be used without any marked effect on the analysis (Marshall, 1977), although theoretically it could be expected that the use of carbon, which has a high mass absorption coefficient for sodium and magnesium, would result in significant absorption for these elements. Chromium can also be used with little effect on absorption; there is, however, an increase in background and this would affect quantitative analyses, which are based on peak-to-background ratios. Coating techniques have been discussed by Echlin and Hyde (1972).

Take-off Angle   As take-off angle varies, so will absorption, since the path length of x-rays in the sample will change (Figure 37). As path length increases, absorption increases exponentially, and since a decrease in take-off angle results in an increase in path length, it is advisable to keep the take-off angle as high as possible, certainly above 30°.

Specimen Tilt   Specimen tilt will influence the number of electrons backscattering from the sample (Russ, 1975). As tilt increases, the energy leaving the sample will increase, and consequently the x-ray intensity will decrease. Tilt angle should be kept at zero in order to avoid this effect.

Counting Time   Counting time is important from two points of view. On the one hand, it should be as long as possible, to improve statistical accuracy, and, on the other hand, it should not be so long that errors are introduced due to beam current instability, beam drift, mass loss, or contamination.

The minimum detectable limit of an analysis, i.e., the ability to distinguish trace levels of an element, is proportional to P^2/B (P = peak intensity; B = background intensity) (Ziebold, 1967). This value should be maximized when elements in very low concentrations are being investigated.

The minimum detectable quantity can be estimated for a sample as compared to a standard in the following way:

If the smallest recognizable peak is one that is more than three times the standard deviation (SD) of the background, then the minimum concentration of an element (C_{min}) is proportional to 3 × SD of the background (B), whereas the concentration of an element in a pure standard (C_{st}) is proportional to the peak intensity (P). The minimum concentration

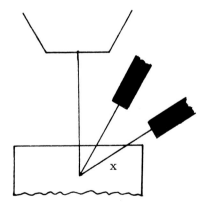

Figure 37.   Increase in take-off angle reduces the path length (x) of x-rays through a sample and therefore reduces absorption.

is:

$$C_{min} = \frac{3\sqrt{B}}{P} \cdot C_{st}$$

It is obvious that C_{min} will be influenced markedly by counting statistics, as is implicit in the Ziebold equation. Thus, if counts are increased tenfold for a particular analysis, C_{min} is decreased:

$$C_{min} = \frac{3\sqrt{200}}{1000} \cdot 100 = 4\%$$

$$C_{min} = \frac{3\sqrt{2000}}{10000} \cdot 100 = 1.3\%$$

If long analysis times are required, then it is important that the beam current should be stable over the counting period and from one analysis to another. If this is not the case, it will be necessary to normalize the data in some way. This can be done by integrating beam current between the start and end of an analysis (Admon, 1977). The initial beam current, I_j^b, and initial specimen current, I_{oj}^s, are measured, and specimen current is recorded throughout an analysis. The latter is integrated to yield the mean specimen current, I_j^m. For each analysis a drift factor is calculated. Thus if the beam current prior to the first analysis is I_1^b, then the drift factor Xj is:

$$Xj1 = (I_j^b/I_1^b)(I_j^m/I_{oj}^s)$$

Each peak intensity for this and subsequent analyses is normalized by multiplying by $1/Xj$. This method compensates for drift during a measurement and from one measurement to another.

Counting may also be normalized by measuring the total intensity and dividing by the accumulated charge. The latter can be derived, if instability is a problem, by measuring initial beam current I_j^b and initial counting rate n_j, and integrating the total counts in the spectrum, N (Mulligan and Lapi, 1976), when the accumulated charge Q is:

$$Q = (N\, I_j^b)/(n_j)$$

Beam current can also be integrated by continuously monitoring aperture current and using it to charge a capacitor to a predetermined level (Vadimsky, 1976).

It is possible to estimate, prior to an analysis, the number of counts, and therefore counting time, required for a given minimum detectability limit and a given sensitivity determination, i.e., the counts required to distinguish between two concentrations (Goldstein, 1976). The minimum detectable concentration (C_{DL}) is estimated from:

$$C_{DL} = \frac{C_{st}}{\overline{P}_{st}} \cdot \frac{\sqrt{2}\, t_{n-1}^{1-\alpha}\, 2\sqrt{\overline{P}_{st}}}{n^{1/2}}$$

and the sensitivity (ΔC) is estimated by:

$$\Delta C = C_1 - C_2 \geq \frac{\sqrt{2}\, t_{n-1}^{1-\alpha}\, 2\sqrt{\overline{P}_2}}{n^{1/2}} \cdot \frac{C_1}{P_1}$$

(where C_{st} = concentration of standard; $\overline{P}_{st}$ = mean peak intensity of standard; C_1 = concentration in region 1; C_2 = concentration in region 2; $\overline{P}_1$ = mean peak intensity of C_1; $\overline{P}_2$ = mean peak intensity of C_2; n = number of determinations; t = Student's t statistic; $1 - \alpha$ = probability level; n − 1 = degrees of freedom). Values of t are obtained from statistical tables.

Relative Error   Error in an analysis may be systematic or statistical. Systematic error is difficult to estimate, but statistical error can be readily determined. Once statistical error is known, it is possible to predict whether an attempt to improve systematic error would or would not be likely to lead to an improvement in the total error. For example, if statistical error is of the order of 20 to 30%, there would be little point in expending effort in trying to reduce systematic error.

The statistical error is given by the standard deviation (SD):

$$SD = \frac{\sqrt{P + 2B}}{P}$$

(P = peak intensity, B = background counts).

MANUAL METHODS OF DATA REDUCTION

Preliminary Treatment of the Spectrum

Background subtraction.   Background can be subtracted manually by interpolation from the low-energy side of the peak. It is best to select for integration an energy band the same width as the peak, on the low-energy side, and then subtract this from the total peak intensity (see Marshall, 1977).

Background can also be subtracted by first acquiring a spectrum from the specimen and then acquiring a spectrum from a blank in the subtractive mode for the same preset live-time. The first spectrum will be reduced until only the elemental peaks remain. The composition of the blank should be as close to that of the specimen as possible but without containing the elements of interest. Both spectra must be obtained under identical operating conditions.

Peak integrals.   If the concentration of a single element is being determined by reference to a standard, then the width of the window that is used to obtain the peak integral is not crucial as long as the same window is used for sample and standard. However, if different elements within a spectrum are being compared, it is necessary to take into account the change in bandwidth that occurs with energy. As discussed previously, peak broadening at FWHM is defined as:

$$FWHM_E = [R^2 + 2.74 (E - 5890)]^{\frac{1}{2}}$$

(R = spectrometer resolution FWHM at Mn Kα; E = peak energy). Ideally the peak integrals should be measured with windows that are some multiple of FWHM (e.g., 1.2 × FWHM) and this should be recalculated for every value of E. The window to be set in channels will be 1.2 FWHM per number of eV per channel.

Since MCAs sort pulses into finite energy bands of 10 or 20 eV, it is usually impossible to set a window that will exactly cover 1.2 FWHM, and interpolation is tedious. Russ (1976a) has pointed out that the total peak area can be calculated from the integral of a whole number of channels in a window. Russ provides a table giving the fraction of peak area in a window of width X times FWHM. This is reproduced in Table 5.

Table 5 can be used to determine total peak area in the following way:

Use a fixed energy window to obtain a peak integral, e.g., 9 channels at 20 eV (bandwidth—180 eV). Say peak integral = 400 counts.

Calculate the number of channels the window should have been at FWHM for energy E (e.g., 3312 eV). Spectrometer resolution is 160 eV.

$$
\begin{aligned}
\text{Bandwidth at FWHM} \quad &= \quad [160^2 + 2.74 (3312 - 5895)]^{\frac{1}{2}} \\
&= \quad 136.11 \text{ eV}
\end{aligned}
$$

The window of 7 channels thus represents (180/136.11) × FWHM, which is 1.32 × FWHM.

Consult the table and determine the fraction of peak area at 1.32 × FWHM, i.e., 0.8799.

$$
\begin{aligned}
\text{Total peak area} \quad &= \quad 400/0.8799 \\
&= \quad \underline{454.6 \text{ counts}}
\end{aligned}
$$

By this means, peak integrals are adjusted for the peak broadening that occurs as a result of the change in resolution of the detector with energy.

Table 5.   Fraction of peak area in window of width X (times FWHM), for values of X from 0.90 to 1.69

X	0.00	0.01	0.02	0.03	0.04	0.05	0.06	0.07	0.08	0.09
0.9	0.7107	0.7160	0.7213	0.7265	0.7316	0.7367	0.7417	0.7466	0.7514	0.7562
1.0	0.7610	0.7656	0.7702	0.7748	0.7792	0.7836	0.7880	0.7923	0.7965	0.8006
1.1	0.8047	0.8088	0.8127	0.8166	0.8205	0.8243	0.8280	0.8317	0.8353	0.8388
1.2	0.8423	0.8457	0.8491	0.8524	0.8557	0.8589	0.8621	0.8652	0.8682	0.8712
1.3	0.8741	0.8770	0.8799	0.8826	0.8854	0.8881	0.8907	0.8933	0.8958	0.8983
1.4	0.9007	0.9031	0.9055	0.9078	0.9100	0.9122	0.9144	0.9165	0.9186	0.9206
1.5	0.9226	0.9246	0.9265	0.9284	0.9302	0.9320	0.9338	0.9355	0.9372	0.9388
1.6	0.9404	0.9420	0.9436	0.9450	0.9465	0.9480	0.9494	0.9507	0.9521	0.9534

Peak deconvolution. In cases of peak overlap, the only procedure that can be easily employed manually to deconvolute the spectrum is to subtract one peak at a time by subtractive counting from reference standards.

Obtaining Elemental Ratios and Concentrations

The intensity of x-ray radiation is directly proportional to the concentration of the assayed element. Thus the basic principle of quantitation is the comparison of the intensity of the unknown with the intensity of a standard of known concentration, using the same operating conditions. This may be formulated as:

$$C_x \simeq \frac{P_x}{P_{kx}} \cdot C_{kx}$$

(where C_x and C_{kx} are the unknown and known concentrations and P_x and P_{kx} are the observed peak intensities of the unknown and known concentrations, respectively).

If a matching standard can be formulated, then the accuracy of the analysis depends upon the intensity measurements and the reliability of the composition of the standard. If a matching standard is not available, then in the case of bulk samples a pure element standard can be used. However, differences in composition of specimen and standard affect the x-ray emission and these "matrix effects" must be corrected by calculation. The matrix corrections are based on complex mathematical models that have been derived partially on a theoretical basis but to a large extent by empirical means. Matrix corrections can be carried out with the aid of a desk calculator, but they are complex and the calculations are lengthy. In the case of biological specimens there are doubts about the validity of the models when they are extrapolated to deal with matrices of low atomic number.

In general it is advisable to select operating conditions that will minimize errors due to matrix effects and to use standards that are as close to the sample in composition as possible. The most important parameters affecting matrix corrections in biological samples are absorption and electron backscatter. These effects can be minimized by using high take-off angles ($> 30°$) and flat, untilted samples. Even if matching standards cannot be devised, the quantitative accuracy can still be 10% relative or better if due attention is paid to measurement (Andersen, 1967a). Details of matrix corrections for biological specimens are given by Andersen (1967b), Hall (1971), and Russ (1974).

Peak intensities and standards. This method of quantitation involves comparing peak intensities from sample and standard to directly yield the concentration of the unknown. The principles have already been discussed with respect to bulk specimens. A more accurate and convenient variation of the method is to prepare a calibration line from standards of different concentrations.

In the case of sections, this method cannot be utilized unless the thickness of the section is known and is the same thickness as the standard. Unfortunately, it is not possible to take the thickness of a section to be the thickness at which the microtome is set to cut. The thickness of a section must be measured by direct means. This can be done in a variety of ways, e.g., resectioning (Yang and Shea, 1975), measuring optical density by photographic means (Casley-Smith and Crocker, 1975), stereoscopy by using latex balls (Heimendahl, 1973), or contamination spots (Lorimer et al., 1976). However, none of these methods, with the possible exception of the last one, lends itself to rapid assessment of section thickness immediately prior to analysis.

Consequently, this author has devised a rapid and simple method for use in STEM or TSEM instruments. All that is required is to make a line scan passing over the edge of a section and to record the transmitted electron signal (Figure 38). The video amplifier controls must be set so that signal clipping does not occur and there is a reasonable signal change between black and white levels. The vertical displacement of the signal is measured and the thickness can then be obtained from a calibration line.

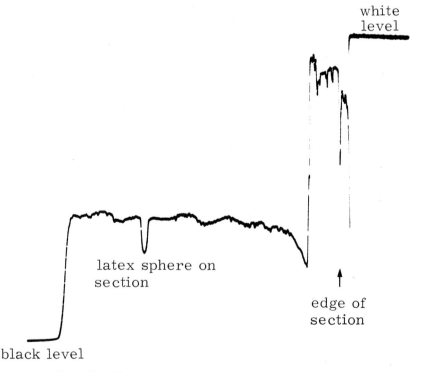

Figure 38.   Line scan profile in STEM across an epoxy resin section.

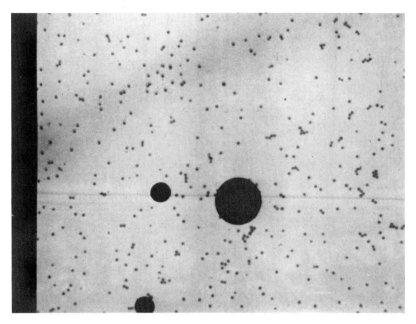

Figure 39.   Line scan across latex spheres (1 μ, 0.5 μ, 0.1 μ) for production of a calibration curve of transmitted electron intensity against thickness.

The latter is obtained from measurements of signal change over a series of latex spheres of known diameter (Figure 39). The signal levels measured for the supporting film (I_o) and the levels measured for the latex spheres (I) (Figure 40) are plotted as log I_o/I against sphere diameter (Figure 41). This method of calibration is valid if the density of the sections and the density of the latex spheres are similar. The density of the latex spheres is 1.05 and the density of Spurr's resin (used in the author's experiments) is 1.06. Provided that all measurements are carried out with the same amplifier control settings, section thickness can be rapidly estimated prior to analysis. Suitable standards can be made for a variety of cations in Epoxy resin by the method of Spurr (1974) and as modified by Chandler (1976). This method of analysis measures concentration as mass per unit volume.

   Peak-to-background ratios and standards.   If peak-to-background (P/B) ratios are used for specimens and standards in sections, concentrations can be expressed in terms of mass per unit mass, and section thickness need not be known. This is essentially similar to Hall's method (see later discussion).

   P/B ratios can also be used to some advantage with bulk samples (Cobet, 1972). In this case the background must be taken from an energy band adjacent to and on the low-energy side of the peak.

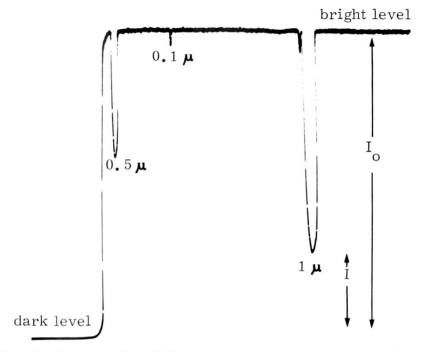

Figure 40. Line scan profile in STEM across latex spheres, showing method of obtaining a calibration curve.

Hall's peak-to-continuum ratio. This method has been developed in two ways (Hall, 1968). It can be used for the determination of absolute concentrations or for relative concentrations. The latter application is the more common one, and indeed it can be made absolute by the use of appropriate standards.

The method depends on the assumption that peak intensity (P) is a measure of elemental mass and that the continuum intensity (W) is a measure of total mass. Thus simultaneous measurement of P and W allows the concentration (C_x) of an element, in terms of mass per unit weight, to be obtained, since:

$$C_x = \frac{\text{elemental mass}}{\text{total mass}} = \frac{P}{W}$$

The position of the energy band in the spectrum for measuring W is not crucial except that it must be entirely free from peaks.

It is necessary in the application of the model to subtract all contributions of extraneous radiation from P and W. Failure to do this can lead to serious errors. The presence of extraneous radiation in microscopes has

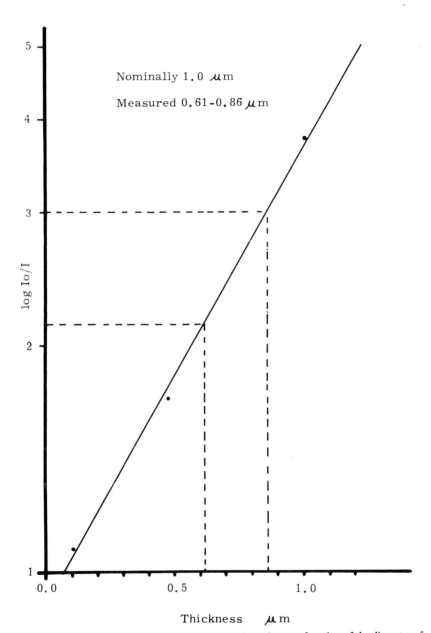

Figure 41. Calibration curve of transmitted electron intensity as a function of the diameter of latex spheres. An example of its use to measure a section nominally 1.0 μ thick is shown. This section varied between 0.61 and 0.86 μ in thickness.

been discussed previously and this can be a severe limitation on the method (Zaluzec and Fraser, 1977; Janossy and Neumann, 1976; Shuman et al., 1976). The method is discussed in detail by Hall et al. (1966, 1972), Marshall and Hall (1968), and Hall (1968, 1971).

Atomic ratio model. Russ (1973, 1974) has developed a method of determining elemental composition in terms of ratios of atomic concentrations. A similar scheme has been proposed by Cliff and Lorimer (1975). Atomic concentrations (C) of elements are proportional to the intensities (P) of the characteristic x-rays of the elements:

$$\frac{C_1}{C_2} = \frac{P_1}{P_2} \cdot \frac{K_2}{K_1}$$

This equation can be applied to thin section analysis and the proportionality constants $(K_{1,2})$ can be measured by using standards (Cliff and Lorimer, 1975) for the particular analysis. However, it is more convenient to calibrate the detector for all elements. Suitable standards for doing this can be prepared as microdroplets of isoatomic solutions (Morgan et al., 1975) or thin layers of salts (Chandler, 1976). The use of standards of this type results in a calibration curve of detector efficiency from which the K values can be obtained to solve the equation:

$$C_1 : C_2 : C_3 = \frac{P_1}{K_1} : \frac{P_2}{K_2} : \frac{P_3}{K_3}$$

COMPUTER METHODS OF DATA REDUCTION

While manual methods are probably adequate for most biological analyses, computer methods are considerably faster. In the analysis of biological samples, computer reduction of data is probably most often of value because it increases the speed of an analysis and not because it increases the accuracy of an analysis. There are of course exceptions to this. For example, accurate background subtraction and peak smoothing will help to identify very small peaks from trace elements, and severe cases of peak overlap can be deconvoluted only by computer. It is not intended here to review in detail the large number of computer methods that now exist, but to indicate the variety and scope of them.

Subtraction of background is carried out in computer programs in two basic ways. These are frequency or digital filtering (Russ, 1977a; Schamber et al., 1977) and fitting calculated backgrounds (Russ, 1977b) based on the equations of Fiori et al. (1976). The frequency-filtering method does not perform well at the low-energy end of the spectrum and does not define absorption edges accurately. Some modification of peak shape may also be produced by the simpler digital-filtering method, but this can be eliminated by using the method in conjunction with least-squares peak fitting (Schamber et al., 1977).

Peak deconvolution can be effectively carried out by computer, and this is very important for some biological situations, e.g., determining low levels of calcium in the presence of potassium. Deconvoluting peak overlaps can be carried out by least-squares fitting of library spectra (Schamber et al., 1977) or by fitting generated peaks (Russ, 1977a).

Programs for the calculation of concentrations by Hall's method and elemental ratios by the atomic ratio model in sections have been developed. The latter requires no user input of data, since the detector efficiency is calculated (Russ, 1974). The atomic ratio model can also be applied to bulk samples. Concentration ratios can also be determined in sections by the program developed by Warner and Coleman (1974, 1975).

Computer programs that employ full matrix correction procedures for bulk specimen analysis, the so-called ZAF procedures, are not readily applicable to biological specimens at the present time, although the programs developed by Warner and Coleman (1974) employ modifications of the conventional ZAF corrections.

LITERATURE CITED

Adler, I., Divornik, E. J., and Rose, H. J. 1962. The detection of sulfur in contamination spots in electron probe x-ray microanalysis. Br. J. Appl. Phys. 13:245.

Admon, U. 1977. The beam stability problem in quantitative analysis. Edax Editor 7 (4):12.

Aitken, D. W., and Woo, E. 1971. The future of silicon x-ray detectors. In: *Energy Dispersion X-ray Analysis*, *X-ray and Electron Probe Analysis*. American Society for Testing and Materials, Philadelphia. (ASTM Special Technical Publication 485, p. 36.)

Andersen, C. A. 1966. Electron probe microanalysis of thin layers and small particles with emphasis on light element determinations. In: *The Electron Microprobe* (McKinley, T. D., Heinrich, K. F. J., and Wittry, D. B., eds.), p. 58. John Wiley and Sons, New York.

Andersen, C. A. 1967a. The quality of x-ray microanalysis in the ultra-soft x-ray region. Br. J. Appl. Phys. 18:1033.

Andersen, C. A. 1967b. An introduction to the electron probe microanalyzer and its application to biochemistry. Meth. Biochem. Anal. 15:147.

Anderson, K., Brookes, K. A., and Finbow, D. C. 1976. The Corinth analytical electron microscope. In: *Developments in Electron Microscopy and Analysis* (Venables, J. A., ed.), p. 69. Academic Press, London.

Beaman, D. R., and Isasi, J. A. 1972. *Electron Beam Microanalysis*. American Society for Testing and Materials, Philadelphia. (ASTM Special Technical Publication 506.)

Beaman, D. R., and Solosky, L. F. 1972. Accuracy of quantitative electron probe microanalysis with energy dispersive spectrometers. Analyt. Chem. 44:1598.

Birks, L. S. 1963. *Electron Probe Microanalysis*. John Wiley & Sons, New York.

Birks, L. S. 1969. *X-ray Spectro-Chemical Analysis.* John Wiley & Sons, New York.

Bolon, R. B., and McConnell, M. D. 1976. Evaluation of electron beam tails and x-ray spatial resolution in the SEM. Scanning Electron Microscopy 1976:163.

Bovey, P., Wardell, I., and Williams, P. M. 1977. X-ray microanalysis in the STEM. *Eighth Intern. Conf. X-ray Optics Microanalysis*, p. 117A.

Brandis, E. K., Anderson, F. W., and Hoover, R. 1971. Reduction of carbon contamination in the SEM. Scanning Electron Microscopy 1971:505.

Broers, A. N. 1973. High resolution scanning electron microscopy of surfaces. In: *Microprobe Analysis* (Andersen, C. A., ed.). John Wiley & Sons, New York.

Casley-Smith, J. R., and Crocker, K. W. J. 1975. Estimation of section thickness, etc. by quantitative electron microscopy. J. Microscopy 103:351.

Chandler, J. A. 1976a. X-ray microanalysis in the electron microscope. In: *Practical Methods in Electron Microscopy* (Glauert, A. M., ed.), p. 319. North Holland Publishing Company, Amsterdam.

Chandler, J. A. 1976b. A method for preparing absolute standards for quantitative calibration and measurement of section thickness with x-ray microanalysis of biological ultrathin specimens in EMMA. J. Microscopy 106:291.

Cliff, G., and Lorimer, G. W. 1975. The quantitative analysis of thin specimens. J. Microscopy 103:203.

Cobet, U. 1972. Quantitative electron beam analysis of thick biological tissue samples. Phys. Med. Biol. 17:736.

Echlin, P., and Hyde, P. J. W. 1972. The rationale and mode of application of thin films to non-conducting materials. Scanning Electron Microscopy 1972:137.

Elad, E. 1972. Drain feedback—a novel feedback technique for low-noise cryogenic preamplifiers. IEEE Trans. Nucl. Sci. NS 19:403.

Everhart, T. E., Herzog, R. F., Chung, M. S., and Devore, W. J. 1972. Electron energy dissipation measurements in solids. In: *Proc. Sixth Intern. Conf. X-ray Optics and Microanalysis* (Schinoda, G., Kohra, K., and Ichinokawa, T., eds.), p. 81. University of Tokyo Press, Japan.

Fano, U. 1947. Ionization yield of radiations. II. The fluctuation of the number of ions. Phys. Rev. 72:26.

Fiori, C. E., Myklebust, R. L., Heinrich, K. F. J., and Yakowitz, H. 1976. Prediction of continuum intensity in energy dispersive x-ray microanalysis. Analyt. Chem. 48:172.

Fitzgerald, R. 1964. Beam stability in the electron probe microanalyzer. Advances in X-ray Analysis 7:369.

Fitzgerald, R., Klaus, K., and Heinrich, K. F. J. 1968. Solid-state energy-dispersion spectrometer for electron-microprobe x-ray analysis. Science 159:528.

Fowler, B. A., and Goyer, R. A. 1975. Bismuth localisation within nuclear inclusions by x-ray microanalysis—effect of accelerating voltage. J. Histochem. Cytochem. 23:722.

Geiss, R. H., and Huang, T. C. 1975. Quantitative x-ray energy dispersive analysis with the transmission electron microscope. X-ray Spectrometry 4:196.

Goldstein, J. I. 1976. Statistics of x-ray analysis. *Proc. Eleventh Ann. Conf. Microbeam Analysis Society*, p. T1A.

Goldstein, J. I., and Williams, D. B. 1978. Spurious x-rays produced in the scanning transmission electron microscope. Scanning Electron Microscopy 1978:(1) 427.

Green, M. 1963a. The efficiency of production of characteristic x-radiation. In: *X-ray Optics and X-ray Microanalysis* (Pattee, H. H., Cosslett, V. E., and Engström, A., eds), p. 185. Academic Press, New York.

Green, M. 1963b. A Monte Carlo calculation of the spatial distribution of characteristic X-ray production in a solid target. Proc. Phys. Soc. 82:204.

Green, M., and Cosslett, V. E. 1961. The efficiency of production of characteristic x-radiation in thick targets of a pure element. Proc. Phys. Soc. 78:1206.

Haine, M. E., and Einstein, P. A. 1952. Characteristics of the hot cathode electron microscope gun. Br. J. Appl. Phys. 3:40.

Hall, T. A. 1968. Some aspects of the microprobe analysis of biological specimens. In: *Quantitative Electron Probe Microanalysis* (Heinrich, K. F. J., ed.), p. 269. National Bureau of Standards, Special Publication 298, Department of Commerce, Washington.

Hall, T. A. 1971. The microprobe assay of chemical elements. In: *Physical Techniques in Biological Research*, 2nd ed. (Oster, G., ed.), Vol. 1 (c), p. 157. Academic Press, New York.

Hall, T. A. 1977. Reduction of background due to backscattered electrons in energy-dispersive x-ray microanalysis. J. Microscopy 110:103.

Hall, T. A., and Gupta, B. L. 1974. Beam-induced loss of organic mass under electron microprobe conditions. J. Microscopy 100:177.

Hall, T. A., Hale, A. J., and Switsur, V. R. 1966. Some applications of microprobe analysis in biology and medicine. In: *The Electron Microprobe* (McKinley, T. D., Heinrich, K. F. J., and Wittry, D. B., eds.), p. 805. John Wiley and Sons, New York.

Hall, T. A., Rockert, H. O. E., and Saunders, R. L. de C. 1972. *X-ray Microscopy in Clinical and Experimental Medicine*. Charles C Thomas, Springfield, Illinois.

Harada, Y., Tamura, N., and Goto, T. 1976. Application of field emission electron gun to STEM & SEM. In: *Developments in Electron Microscopy and Analysis* (Venables, J. A., ed.). Academic Press, London.

Heimendahl, M. von 1973. Specimen thickness determination in transmission electron microscopy in the general case. Micron 4:111.

Heinrich, K. F. J. 1975. Scanning electron probe microanalysis. Advances in Optical and Electron Microscopy 6:275.

Janossy, A. G. S., and Neumann, D. 1976. Quantitative x-ray microanalysis: microcrystal standards and excessive background. Micron 7:225.

Johansen, B. V. 1973. A high performance Wehnelt grid for transmission electron microscopes. Micron 4:121.

Johnson, G. G., and White, E. W. 1970. X-ray emission wavelengths and keV tables for nondiffractive analysis. Am. Soc. Test. Mater. Data Ser., DS546.

Joy, D. C. 1974. SEM parameters and their measurement. Scanning Electron Microscopy 1974:327.

Kandiah, K. 1966. *Radiation Measurements in Nuclear Power*. Institute of Physics and the Physical Society, London.

Kenik, E. A., and Bentley, J. 1977. Influence of x-ray induced fluorescence on energy dispersive x-ray analysis of thin foils. *Eighth Intern. Conf. X-ray Optics Microanalysis*, p. 114A.

Kramers, H. A. 1923. Theory of x-ray absorption and of the continuous x-ray spectrum. Phil. Mag. 46:836.

Kyser, D. F., and Geiss, R. H. 1977. Spatial resolution of x-ray microanalysis in STEM. *Eighth Intern. Conf. X-ray Optics Microanalysis*, p. 110A.

Lorimer, G. W., Cliff, G., and Clark, J. N. 1976. Determination of the thickness and spatial resolution for the quantitative analysis of thin foils. In: *Developments in Electron Microscopy and Analysis* (Venables, J. A., ed.), p. 153. Academic Press, New York.

Marshall, A. T. 1974. X-ray microanalysis of frozen dried and frozen-hydrated biological specimens. *Eighth Intern. Cong. Electron Microscopy*, Canberra. 2:74.

Marshall, A. T. 1975a. Electron probe x-ray microanalysis. In: *Principles and Techniques of Scanning Electron Microscopy*, Vol. 4 (Hayat, M. A., ed.), p. 103. Van Nostrand Reinhold, New York.

Marshall, A. T. 1975b. Background reduction and spectrometer alignment in the JEM 100B electron microscope. Edax Editor 5 (4):9.

Marshall, A. T. 1977. Iso-atomic droplets as models for the investigation of parameters affecting X-ray microanalysis of biological specimens. Micron 8:193.

Marshall, A. T., and Forrest, Q. G. 1977. X-ray microanalysis in the transmission electron microscope at high accelerating voltages. Micron 8:135.

Marshall, D. J., and Hall, T. A. 1968. Electron probe x-ray microanalysis of thin films. Br. J. Appl. Phys. (J. Phys. D.) Ser. 2, 1:1651.

Mizuhira, V. 1976. Elemental analysis of biological specimens by electron probe x-ray microanalysis. Acta Histochem. Cytochem. 9:69.

Morgan, A. J., Davies, T. W., and Erasmus, D. A. 1975. Analysis of droplets from isoatomic solutions as a means of calibrating a transmission electron analytical microscope (TEAM). J. Microscopy 104:271.

Moseley, H. 1913. The high frequency spectra of the elements. Phil. Mag. 26:1024.

Mulligan, T. J., and Lapi, L. A. 1976. Normalization procedure for x-ray microanalysis which accounts for fluctuations in beam current. Scanning Electron Microscopy 1976:195.

Murray, R. T. (1973). An energy dispersing x-ray detector coupled to a Philips EM 300 electron detector. J. Phys. E. (Scient. Instrum.) 6:19.

Nakagawa, S., and Yanaka, T. 1975. A low heating power LaB_6 cathode gun with control system for scanning electron microscope. J. Electron Microscopy 24:275.

Nicholson, W. A. P., Robertson, B. W., and Chapman, J. N. 1977. The characterisation of x-ray spectra from thin specimens in the transmission electron microscope. Proc. EMAG 1977 (in press).

Pawley, J. B. 1972. Charging artifacts in the scanning electron microscope. Scanning Electron Microscopy 1972:153.

Reed, S. J. B. 1972. Pulse pile-up rejection in Si (Li) X-ray detection systems. J. Phys. E. 5:997.

Reed, S. J. B. 1975. *Electron Microprobe Analysis*. Cambridge University Press, Cambridge.

Reed, S. J. B., and Ware, N. G. 1972. Escape peaks and internal fluorescence in x-ray spectra recorded with lithium-drifted silicon detectors. J. Phys. E. 5:582.

Revell, R. S. M. 1970. Recommended gun conditions for the EM 300. Phillips Bulletin April 1970.

Russ, J. C. 1972. Resolution and sensitivity of x-ray microanalysis in biological sections by scanning and conventional transmission electron microscopy. Scanning Electron Microscopy 1972:73.

Russ, J. C. 1973. X-ray microanalysis in the biological sciences. J. Submicrosc. Cytol. 6:55.

Russ, J. C. 1974. The direct element ratio model for quantitative analysis of thin sections. In: *Microprobe Analysis as Applied to Cells and Tissues* (Hall, T., Echlin, P., and Kaufmann, R., eds.), p. 269. Academic Press, London.

Russ, J. C. 1975. A simple correction for backscattering from inclined samples. *Proc. Tenth Ann. Conf. Microbeam Analysis Soc.*, p. 7A.

Russ, J. C. 1976a. Integration of a peak with an energy window. Edax Editor 6 (4):46.

Russ, J. C. 1976b. X-ray microanalysis at high beam voltages. Scanning Electron Microscopy 1976:143.

Russ, J. C. 1977a. Processing of energy-dispersive x-ray spectra. X-ray Spectrometry 6:37.

Russ, J. C. 1977b. Automatic fitting of calculated background in energy dispersive x-ray spectra. *Eighth Intern. Conf. X-ray Optics Microanalysis*, p. 102A.

Russ, J. C. 1977c. Selecting optimum KV for STEM microanalysis. Scanning Electron Microscopy 1977:335.

Russ, J. C. 1979. New methods to obtain and present SEM x-ray line scans. *Proc. Fourteenth Ann. Conf. Microbeam Analysis Society*, p. 292.

Russ, J. C., Baerwaldt, G. C., and McMillan, W. R. 1976. Routine use of a second generation windowless detector for energy-dispersive ultra-light element x-ray analysis. X-ray Spectrometry 5:212.

Sandborg, A., and Lichtinger, R. 1977. An energy dispersive spectrometer for elements from 6 to 92. *Eighth Intern. Conf. X-ray Optics and Microanalysis*, p. 107A.

Schamber, F. W., Wodke, N. F., and McCarthy, J. J. 1977. Least-squares fit with digital filter: the method and its application to EDS spectra. *Eighth Intern. Conf. X-ray Optics Microanalysis*, p. 98A.

Scholes, S., and Wilkinson, F. C. F. 1969. Specimen damage during microprobe analysis of silicate glasses. *Fifth Intern. Conf. X-ray Optics Microanalysis*, p. 438.

Shuman, H., Somlyo, A. V., and Somlyo, A. P. 1976. Quantitative electron probe microanalysis of biological thin sections: methods and validity. Ultramicroscopy 1:317.

Spurr, A. R. 1974. Macrocyclic polyether complexes with alkali elements in epoxy resin as standards for x-ray analysis of biological tissue. In: *Microprobe Analysis as Applied to Cells and Tissues* (Hall, T. A., Echlin, P., and Kaufmann, R., eds.), p. 213. Academic Press, London.

Statham, P. J. 1976a. Beam switching: a technique to improve the performance of pulse-processing electronics for E. D. S. *Proc. Eleventh Ann. Conf. Microbeam Analysis Society*, p. 11A.

Statham, P. J. 1976b. The generation, absorption and anisotropy of thick-target bremsstrahlung and implications for quantitative energy dispersive analysis. X-ray Spectrometry 5:154.

Statham, P. J. 1977. Pile-up rejection: limitations and corrections for residual errors in energy-dispersive spectrometers. X-ray Spectrometry 6:94.

Sutfin, L. V. 1972. High resolution X-ray microanalysis of thin specimens in the scanning electron microscope. Scanning Electron Microscopy 1972:65.

Vadimsky, R. G. 1976. Modification of a field emission source SEM for quantitative x-ray microanalysis. Analyt. Letters 9:521.

Van Veld, R. D., and Shaffner, T. J. 1971. Charging effects in scanning electron microscopy. Scanning Electron Microscopy 1971:19.

Warner, R. R., and Coleman, J. R. 1974. Quantitative analysis of biological material using computer correction of x-ray intensities. In: *Microprobe Analysis as Applied to Cells and Tissues* (Hall, T., Echlin, P., and Kaufmann, R., eds.), p. 249. Academic Press, London.

Warner, R. R., and Coleman, J. R. 1975. A biological thin specimen microprobe quantitation procedure that calculates composition and px. Micron 6:79.

Weitzenkamp, L. A. 1969. Measurement of fibre potentials in a scanning electron microscope. J. Phys. E. (Scient. Instrum.) 2:561.

Wiesner, J. C. 1976. Measurement of electron optical parameters for the scanning electron microscope. Scanning Electron Microscopy 1976:675.

Williams, D. B., and Goldstein, J. I. 1977. A study of spurious x-ray production in a Philips EM 300 TEM/STEM. *Eighth Intern. Conf. X-ray Optics Microanalysis*, p. 113A.

Yang, G. C. H., and Shea, S. M. 1975. The precise measurement of the thickness of ultrathin sections by a "re-sectioned section" technique. J. Microscopy 103:385.

Zaluzec, N. J., and Fraser, H. L. 1977. Contamination and absorption effects in x-ray microchemical analysis of thin metal films. *Eighth Intern. Conf. X-ray Optics Microanalysis*, p. 112A.

Ziebold, T. O. 1967. Precision and sensitivity in electron microprobe analysis. Analyt. Chem. 39:858.

Chapter 2

PREPARATION OF SPECIMENS Changes in Chemical Integrity

A. John Morgan

Zoology Department,
University College,
Cardiff, Wales

INTRODUCTION

Modern microprobe analyzers can, under favorable conditions, detect elements in quantities of $< 10^{-18}$ g and with a spatial resolution of 20–30 nm

(Hall, 1977a). The first instrument was developed and introduced by Castaing (1951), and the technology was quickly and advantageously adopted by metallurgists and materials scientists. However, it has only been within the last decade that microprobes have captured the increasingly widespread attention of biologists studying biochemical and physiological processes in soft tissues (Chandler, 1978). The great attraction of microprobe analysis for the biologist is that it provides a facility for extending his knowledge of tissues, cells, organelles, and even molecules from the structural to a chemical or functional level. It is interesting, therefore, to reflect briefly on the relative slowness of biologists to exploit this undoubtedly powerful and exciting technology.

In simple terms, x-ray microprobe analysis capitalizes on the fact that x-rays are generated when electrons invested with sufficient kinetic energy collide with solid specimens. The generated x-rays characterize the atomic constituents of the specimen. Microprobes are often conventional electron microscopes furnished with a detector and analyzer system for collecting and integrating the x-rays emanating from the specimen. Thus, it is possible to determine the structural and chemical (element) features of a given specimen within a single, albeit composite, instrument.

This dual facility, besides creating a situation whereby a dialogue between morphologists and biochemists/physiologists becomes increasingly probable and with mutual benefits, also discloses new and otherwise inaccessible biological data by virtue of the uniqueness of the technology (Coleman and Terepka, 1974). This has inevitably resulted in interpretative problems. For example, it is not always possible to explain microprobe data derived from the analysis of tissue sections on the basis of established physiological concepts.

Optical and electron microscopists have traditionally been impressed by the uniformity of appearance of a given organelle or cell type. In other words, the vast majority of cells are seldom considered to differ qualitatively from others of similar type except when they exhibit extremes of functional activity, e.g., during division, senescence, or when exposed to pathogenic agencies. Clearly, however, cells and organelles are continuously variable, and quantitative differences may be imposed by a variety of subtle factors, such as positional effects within a tissue or organ, developmental age, fluxes in functional state, and interactions with neighboring cells, etc. This concept of "cellular ecology" envisages cells and organelles as individual three-dimensional entities that are dynamically creating and reacting harmoniously with their permanently fluctuating local environment. The techniques of stereology (Weibel and Bolender, 1973) attempt to quantify the structural differences within and between cellular populations. Microprobe analysis, on the other hand, can be used to measure the chemical heterogeneities within and between populations of cells and organelles.

Therefore, it is not always necessary, or even desirable, to explain micro-probe data according to established physiological concepts. Provided that the biologist has confidence in his microprobe technique (including prepara-tive and instrumental parameters), he must be prepared to adopt novel hypotheses to accommodate his data.

Apart from interpretative difficulties, the most important single reason for biologists' inertia toward microprobe techniques is related to the nature of biological specimens, and the resultant technical problems of preparing such specimens for microprobe analysis without disturbing their chemical integrity.

Organic specimens are subject to three kinds of changes when exposed to electron bombardment in a vacuum: radiation damage, contamination, and mass and element loss (Chandler, 1975a). These are less serious impedi-ments in materials science but all three artifacts may introduce serious errors in quantitative x-ray microanalysis of soft biological specimens. A consideration of these artifacts, a discussion of methods for either minimiz-ing or monitoring their influences, and a survey of their significance in quantitative microprobe analysis are beyond the scope of the present chapter; the reader may obtain detailed information elsewhere (Morgan et al., 1978).

The physical and chemical integrity of biological specimens can be grossly and irrevocably disturbed during processing for the electron micro-scope. Unlike many metallurgical specimens, for example, biological specimens often have a high water content and cannot therefore be introduced into an electron microscope without preliminary "processing." At the very least, water may have to be withdrawn from the specimen by one of several techniques; alternatively, the water can be immobilized by freezing and the specimen maintained in the frozen-hydrated state while it is in the microscope.

The functional duality of the microprobe as a high-resolution electron microscope and an analytical tool presents the biologist with an uncom-fortable paradox. Optimal structural representation and viable chemical integrity are mutually exclusive goals. Structure is not "preserved" by classical processing procedures involving fixatives, etc.; indeed, structural and chemical integrity are necessarily modified by such intervention (Baker, 1960; Hayat, 1981). In contrast, the microprobe analyst often needs to preserve the chemical integrity of tissues in a state as similar as possible to that in vivo. All preparative techniques, therefore, involve some form of com-promise.

General Microprobe Approach

The fact that energy-dispersive x-ray detectors, in particular, are readily available as straightforward attachments for most electron microscopes encourages the assumption that microprobe analysis is a convenient supplementary tool in routine morphological surveys. However, as already mentioned, this simple approach is seldom tenable because of the incompatability of the various morphological and chemical requirements. The morphologist wishing to perform microprobe analysis must therefore readjust his previously held conceptions of cell structure and, consequently, of preparative procedures. The cell is no longer to be considered a series of discrete, relatively empty, membrane-bound compartments, but rather a fragile unit comprising an infinite array of water-filled loci, each with its own characteristic complement of highly diffusible organic and inorganic solutes plus less exchangeable organic molecules, separated tenuously by dynamic boundaries—the membranes. It is useful, therefore, to outline some of the readjustments that may be necessary in the approach of the biological microprobe analyst to the task of preparing his specimens for analysis.

1.   The potential user should be heavily committed to the technique. The problems and rigors of preparing specimens for analysis should not be underestimated, and the many parameters involved should be thoroughly appreciated if viable biological data are to be obtained. As we shall see later, it is not always necessary in specific instances to maintain, or even attempt to maintain, the complete chemical integrity of a specimen. Indeed, in some cases it may be advantageous to either extract certain defined element phases; for example, the microprobe provides a complete elemental analysis and does not distinguish between exchangeable and bound compartments, but they might be distinguished by an appropriate preparative technique. In other cases it will be desirable to correlate the chemical information with precise ultrastructural information. Nevertheless, one should always be aware that by exposing tissues to extraneous fluid phases, or even by *physically* dehydrating the specimen, we are immediately imposing chemical limitations on the significance of the microprobe data that may be retrieved from the specimen.

2.   Very specific questions must be asked at the outset of a given program so that the aims are clearly defined. The nature of these questions will largely determine the type of preparative regime best suited for the specific application. Often, however, it may be advisable to adopt several complementary procedures (Morgan et al., 1978).

3.   When the user elects to use a technique that obviously involves biochemical compromise, it is important to establish the magnitude of the imposed artifact. If, for example, a particular stage in a wet chemical regime is extracting an element of interest to a much greater extent than other stages, then it may be possible to circumvent this weakness by employing alternative regimes. Checks of this type are usually undertaken by flame spectrophotometry, the scintillation counting of radioactive tracers, or, on much smaller samples ($< 10^{-1}$ mg), by bulk analysis in the microprobe (Davies and Morgan, 1976).

It must be recognized that even the smallest percentage loss of an element from specimens betrays the existence of the other equally serious artifact—the translocation of elements from their natural loci to sites they do not normally occupy. Unfortunately, the translocation of elements is not easily monitored. A practical solution to this problem is, whenever possible, to check data against data obtained from specimens prepared by cryo-procedures, where the degree of chemical compromise has been minimized.

   4.   Ideally, microprobe data should be checked against data obtained by entirely different techniques. For example, Gupta et al. (1978a) have elegantly proven the utility of their frozen-hydrated sectioning and quantitative microprobe analysis procedures (discussed in detail later) by comparison with data obtained by microelectrode measurements (see Table 17). Such correlative data are seldom available and are not easily obtainable, but they are extremely important since they serve to check the viability of the preparative procedure and the instrumental parameters, including the method of quantitation. In addition, they facilitate the interpretation of microprobe data.

   5.   It is useful to remind ourselves as biologists that we are often more concerned with differences (the difference between controls and experimentally, or pathologically, induced states, for example) than with the definition of absolute physiological states. Consequently, it is advisable to establish 1) whether differences do occur, by using gross techniques (see Foster and Erasmus, 1979) before undertaking relatively high-resolution microprobe studies, and 2) to rigorously normalize all technical aspects of the microprobe analysis so that any variability monitored is directly attributable to the biological tissues analyzed.

Aims

The primary purpose of this chapter is to define some of the chemical changes that occur in biological specimens during their exposure to various processing procedures, so that the potential user may obtain a better perspective for making the essentially personal decision about what represents an acceptable compromise that will meet his own specific requirements. A major part of this chapter deals with wet chemical (or conventional) preparative techniques since they represent the extreme in chemical compromise. Many of the principles raised will obviously apply to the other techniques that rely on the intervention of fluid phases. In addition, some selected applications of each preparative technique will be briefly described to illustrate: 1) some of the specific modifications that have been introduced to improve the suitability of the given technique for microprobe purposes, and 2) the biological applicability of each technique.

   Before proceeding to discuss each preparative procedure individually, I would like to consider briefly data derived from a systematic comparative

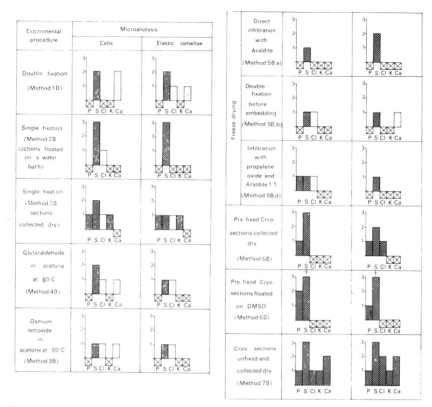

Figure 1. Summary of microprobe data obtained from the analysis of smooth muscle cells and elastic lamellae in rat aortas subjected to various preparative regimes for electron microscopy (including wet chemical, freeze-drying, freeze-substitution, and cryo-microtomy methods). The vertical scale of the histograms represents (P-b)/b values adjusted to the nearest whole number from 1 to 3. ■ = elements considered to be normal constituents of the aortic compartment analyzed. □ = elements considered to have been introduced via one of the preparative fluids. ⊠ = elements not detected. (Reproduced from Morgan et al., 1975.)

study of the influence of various preparative procedures on the chemical composition of certain structural components within the rat aorta (Morgan et al., 1975). This consideration serves to emphasize some of the points already made and on which attention will be focused during the course of this chapter (Figure 1). First, elements are extracted from and introduced into biological specimens during exposure to preparative fluids. Second, the chemical composition of individual tissue components varies according to the precise procedures to which they were exposed. Third, some elements are more readily extracted and displaced than others. Finally, we can surmise that different tissues exposed to similar preparative procedures will suffer qualitatively similar changes.

CONVENTIONAL (WET CHEMISTRY) PREPARATORY TECHNIQUES

Wet chemistry techniques are standard procedures in all biological EM (electron microscope) laboratories. They usually involve the sequential exposure of tissues to several aqueous media, including fixatives (Hayat, 1980), buffers, dehydration agents, resins (Hayat, 1980), and stains (Hayat, 1975, 1980). Unfortunately, the chemical constitution and integrity of biological specimens are seriously affected by the intervention of these fluid phases.

Loss of Elements During Fixation, Dehydration, and Embedding

The total loss of one or more elements (including especially the biologically important electrolyte series, but also structural elements, such as sulfur and phosphorus and heavy metals) from one or more plant or animal tissue has been quantitatively measured by several authors (Table 1). Several general conclusions can be drawn from the results of these important studies:

1.   The degree of loss of most elements is determined by the type of fixative/buffer combination to which the tissue was exposed. For example, Vassar et al. (1972) described significant differences in the effect of glutaraldehyde, acetaldehyde, and formaldehyde on the leakage of potassium from human erythrocytes. Similarly, glutaraldehyde and osmium produce qualitatively similar but kinetically different changes in the magnesium and potassium permeability of Ehrlich ascites tumor cells (Penttila et al., 1974). In general, these differences between fixatives are of little more than theoretical interest. A more realistic practical perspective is that all losses are highly restrictive and potentially serious.

2.   Losses occur at each step in the preparative regime, with the major losses usually recorded during the initial "fixation" stage (Figures 2 and 3). The large initial loss of elements is probably due to the leaching of unbound ionic phases, followed later by the relatively slow release of less readily exchangeable compartments. This point is well illustrated by the experiments reported by DeFilippis and Pallaghy (1975), who showed that the loss from plant tissues during processing of ^{65}Zn and ^{203}Hg labels was significantly reduced when the radioisotopes residing in the free spaces of leaf tissue were previously exchanged with unlabeled ZnCl and $HgCl_2$ solutions (Figure 3). Thus, care must be exercised in the interpretation of sequential loss data, because a negligible recorded loss during a late stage of processing does not unequivocally exonerate that stage of processing. It may simply mean that compartments soluble in that particular medium have already been extracted.

The chemical state of binding of an element therefore affects its solubility. In general, elements like sulfur and phosphorus, which contribute to macromolecular moieties, are less readily lost than the electrolytes.

Table 1. Summary of element losses from plant and animal tissues during various conventional (wet chemical) procedures

Element	Tissue	Preparative technique	Percent loss	Analytical technique	Reference
K	Uterus	1% OsO_4	75	Flame photometry	Garfield et al., 1972
	Erythrocytes	3% gluteraldehyde in water	95	Flame photometry	Vassar et al., 1972
		2% formaldehyde in PO_4 buffer, pH 7.2	70*		
		4% formaldehyde in PO_4 buffer, pH 7.2	70		
		1.65% glutaraldehyde in PO_4 buffer, pH 7.2	70		
		3.3% glutaraldehyde in PO_4 buffer, pH 7.2	70		
		2% acetaldehyde in PO_4 buffer, pH 7.2	70		
Rb	Leaf Tissue	1% OsO_4 in cacodylate-acetate buffer	90		Hall et al., 1974
Na	Smooth muscle	5% glutaraldehyde in physiological saline, pH 7.2, 36°C or 4°C, 4 h	increased by 200% over controls	Flame photometry	Schoenberg et al., 1973
Ca	Aorta	3% glutaraldehyde in PO_4 buffer, pH 7.4, 2 h	47	Flame photometry	Morgan et al., 1975
		3% glutaraldehyde in PO_4 buffer, 1% OsO_4 PO_4 buffer, alcohols, propylene oxide			
	Tendon	3% glutaraldehyde in PO_4 buffer, pH 7.4, 2 h	51		
		3% glutaraldehyde in PO_4 buffer, 1% OsO_4, PO_4 buffer, alcohols, propylene oxide	64		
	Pancreas	0.5%, 1%, or 3% glutaraldehyde	57	Liquid scintillation technique	Howell and Tyhurst, 1976
			42–58 (depending on fix. conc.)		
	Smooth muscle	5% glutaraldehyde in physiological saline, pH 7.2, 36°C or 4°C, 4 h	increased by 35% over controls	Flame photometry	Schoenberg et al., 1973
[86]Sr	Isolated mito-chondria	6.25% glutaraldehyde in 0.1 M PO_4 buffer, pH 7.4	25–35	Liquid scintillation counting	Greenawalt and Carafoli, 1966
		12.5% glutaraldehyde in 0.1 M PO_4 buffer, pH 7.4	25		
		12.5% glutaraldehyde in 1 mM succinate	13		
		12.5% glutaraldehyde + 3 mM ATP	26		
		12.5% glutaraldehyde + 10 mM succinate + 3 mM ATP	28		
		12.5% glutaraldehyde + 1% OsO_4 in 0.1 M PO_4 buffer, pH 7.4	63		
		1% OsO_4 in veronal-acetate buffer, pH 7.4	50		
		10% formaldehyde ± succinate and ATP	40		

Mg	Ehrlich ascites tumor cells	5% glutaraldehyde + 1% OsO_4 in Krebs-Ringer PO_4 buffer, pH 7.45	$\simeq 55$	Flame photometry	Penttila et al., 1974
	Smooth muscle	5% glutaraldehyde in physiological saline, pH 7.2, 36°C or 4°C, 4 h	0	Flame photometry	Schoenberg et al., 1973
^{31}Si	Diatoms	2% glutaraldehyde, wash, 1% OsO_4, alcohols	25	Liquid scintillation counting	Mehard and Volcani, 1975
	Spleen	2% glutaraldehyde, wash, 1% OsO_4, alcohols	35		
	Lung	2% glutaraldehyde, wash, 1% OsO_4, alcohols	55		
	Liver	2% glutaraldehyde, wash, 1% OsO_4, alcohols	70		
	Kidney	2% glutaraldehyde, wash, 1% OsO_4, alcohols	90		
	Mitochondria (liver and diatom)	2% glutaraldehyde, wash, 1% OsO_4, alcohols	40–60		
^{68}Ge	Diatoms	2% glutaraldehyde, wash, alcohols	60	Liquid scintillation counting	Mehard and Volcani, 1975
	Spleen	2% glutaraldehyde, wash, alcohols	35		
	Liver mitochondria	2% glutaraldehyde, wash, alcohols	60		
	Diatom nuclei	2% glutaraldehyde, wash, alcohols	80		
	Diatom vesicles	2% glutaraldehyde, wash, alcohols	65		
	Diatom microsomes	2% glutaraldehyde, wash, alcohols	60		

* The erythrocyte potassium content equilibrated at approximately similar levels during prolonged fixation in the three aldehydes, but the kinetics of potassium loss displayed considerable differences.

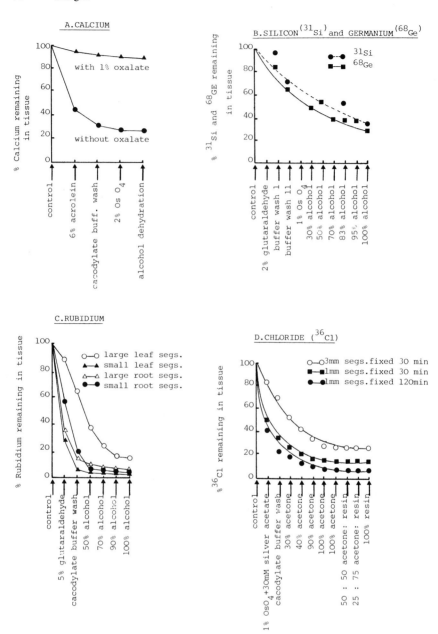

Figure 2. The loss of selected ions from plant and animal tissues during various electron microscopic preparative sequences. **A:** ^{45}Ca loss from chorioallantoic membranes. **B:** ^{31}Si and ^{68}Ge loss from double-labeled rat liver. **C:** Rubidium loss from plant leaf segments. **D:** ^{36}Cl loss from leaf segments. In **A** and **D**, precipitation agents were also included in the preparative fluids. (**A** redrawn from Coleman and Terepka, 1972. **B** redrawn from Mehard and Volcani, 1975. **C** redrawn from Hall et al., 1974. **D** redrawn from Harvey et al., 1976a.)

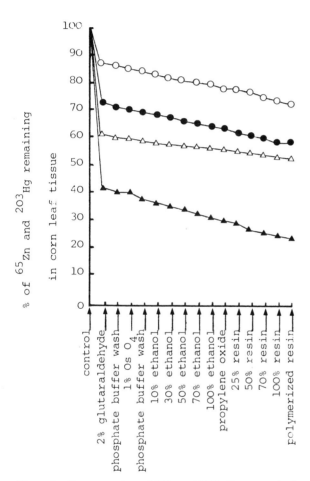

● ● ^{203}Hg loss from fully labelled tissue

▲ ▲ ^{65}Zn loss from fully labelled tissue

○ ○ ^{203}Hg loss from tissue after prior removal of free ^{203}Hg

△ △ ^{65}Zn loss from tissue after prior removal of free ^{65}Zn

Figure 3. Percentage loss of ^{65}Zn and ^{203}Hg from corn leaf segments (before and after the exchange of labels occupying the tissue free space with unlabeled Zn^{2+} and Hg^{2+} ions) during the course of a wet chemical preparative procedure. (Redrawn from DeFilippis and Pallaghy, 1975.)

Macromolecules can themselves be extracted (see Hayat, 1980, 1981, for extensive reviews of the extraction of proteins, lipids, and polysaccharides during EM processing). For example, available microprobe data (Andersen, 1967; Morgan et al., 1975) suggest that nuclear phosphorus (residing primarily in nucleoproteins) is not as easily extracted as cytoplasmic phosphorus (residing primarily in membrane lipid and relatively low molecular weight compounds).

    3. The rate of loss of an element depends on a number of factors, each of which is amenable to experimental manipulation, including: specimen size (Hall et al., 1974; Harvey et al., 1976a) (see Figures 2C and 2D); the type of fixative used (Vassar et al., 1972; Penttila et al., 1974); the processing temperature (Penttila et al., 1974); and the concentration of the fixative (Vassar et al., 1972; Penttila et al., 1974) (see Figure 4).

    Rates of element loss during processing have seldom been accurately measured, but it is quite clear that loss rates can be exceptionally high. For example, the rate of loss of potassium from Ehrlich ascites tumor cells during the first 2 min of fixation in 3% glutaraldehyde is $\approx 15\%$ of the original K^+ content per minute at $0°C$, and $\approx 40\%$ of the original K^+ content per minute at $37°C$ (calculated by the author from data presented by Penttila et al., 1974). Major chemical losses occur, therefore, well within the time needed to adequately fix tissues for ultrastructural studies. Thus, even drastic reduction of fixation time is not a practical proposition for the retention of labile tissue components.

Loss of Elements During Flotation and Staining of Sections

Embedded specimens are usually sectioned and subsequently stained in aqueous solutions of heavy metal salts for morphological observations. The sections are cut onto and collected from the surface of water in microtome troughs. Thus, both sectioning and staining involve the exposure of thin (i.e., considerably less than one cell diameter in thickness) specimens to relatively large volumes of aqueous media. The fact that the chemical integrity of most specimens has already been irrevocably altered (see above) should not prevent discussion of some of the implications of staining and flotation, because many workers have been tempted to use these routine procedures, having otherwise prepared their specimens by procedures that afford fewer chemical compromises, e.g., freeze-drying, freeze-substitution, and "wet-cryo" techniques.

    The loss of bone mineral during water flotation was qualitatively demonstrated by Boothroyd (1964). In our laboratory we have also recorded substantial losses of extracellular calcitic mineral from earthworm calciferous glands during section flotation (Figure 5). These losses, which may be attributable in part to an incomplete infiltration with resin, can be

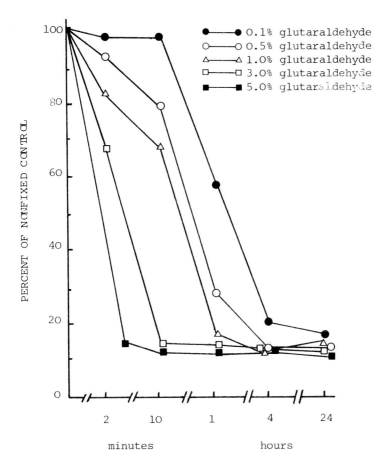

Figure 4.   The effect of fixation in different concentrations of glutaraldehyde at 0°C on the rate of loss of potassium from Ehrlich ascites tumor cells. (From Penttila et al., 1974. Used by permission.)

avoided by cutting onto dry knives and by avoiding staining. A second problem with flotation and staining, and which is especially serious in the case of heavily mineralized tissues, is the fact that sections are exposed to media that contain increasingly high concentrations of the extracted elements. The non-specific precipitation of mineral (see Coleman and Terepka, 1972, for a discussion) and secondary binding of ions to high-affinity sites within the cellular and extracellular matrix are strong possibilities under such circumstances.

A systematic quantitative study on the loss of electrolytes during the flotation of sections was reported by Harvey et al. (1976b). The authors

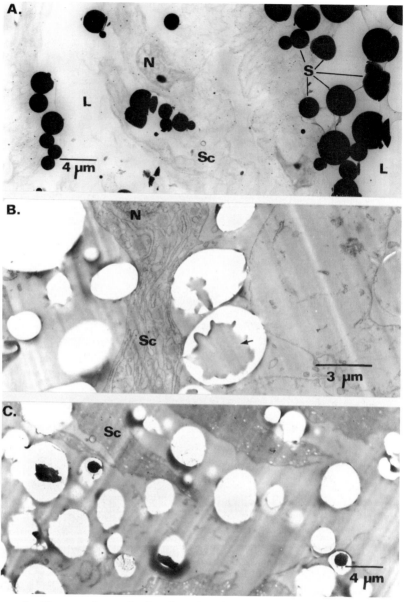

Figure 5. Electron micrograph of sections through the calciferous gland of the earthworm *Lumbricus terrestris* to show the leaching of mineral during water flotation and staining. All specimens were double fixed in 3% glutaraldehyde and 1% osmium and embedded in Spurr resin. **A:** Section cut onto a dry glass knife and unstained. Note the intact calcitic spherules (S) in the lumen (L) of the gland, which have survived exposure to fixatives, alcohols, and embedding media. Sc = secretory cells; N = nucleus. **B:** Section cut onto a dry knife and stained in uranyl acetate and lead citrate. Mineral has been leached from the spherules, leaving holes, some of which have a resin "core" (*arrow*), suggesting that some spherules possess hollow centers. **C:** Section cut onto a water bath, but unstained. Mineral has again been almost completely removed from the spherules. Compare with **B**. (Courtesy of Mrs. C. Winters.)

Table 2. Proportion of radioisotope lost into aqueous trough media during thin sectioning of freeze-substituted leaf segments (*Suaeda maritima* L.)

Trough liquid	^{22}Na (percent label lost)*	^{36}Cl (percent label lost)*
Distilled H_2O	87.30 ± 7.94	58.47 ± 14.07
Saturated $MgSO_4$	86.32 ± 9.76	61.09 ± 3.11
Saturated $Ca(NO_3)_2$	40.33 ± 16.2	75.46 ± 16.6

Table drawn from data of Harvey et al., 1976b. Used by permission.

* The figures represent activity (c/min) in trough medium as percent of total activity in trough medium and sections. The sections (100 nm thick) were in the trough for approximately 2 min.

measured the loss of ^{22}Na and ^{36}Cl from plant tissues embedded in Spurr's resin and prepared initially by freeze-substitution (Table 2). Losses between 40 and 90% were recorded during a standardized 2-min flotation period. Flotation on saturated $MgSO_4$ and $Ca(NO_3)_2$ solutions did not significantly improve the retention. Furthermore, the authors failed to produce supportive evidence for Spurr's claim (Spurr, 1972a) that the water permeability of Spurr's resin is low, and that the leaching of ions from tissues embedded in this resin is consequently minimized. Similar findings were made by Agostini and Hasselbach (1971) on the loss of ^{45}Ca from isolated sarcoplasmic reticulum vesicles during on-grid fixation and staining (Table 3). Losses during various staining procedures are equally serious. Staining in ammonium molybdate, phosphotungstic acid, and uranyl acetate solutions removed 37–65% of the ^{45}Ca label from untreated, whole-mounted sarcoplasmic reticulum samples (Agostini and Hasselbach, 1971, Table 3). In addition, the stains were able to remove significant proportions of the ^{45}Ca label residual after glutaraldehyde fixation. Similarly, Yarom et al. (1974b), in a microprobe study of osmium + potassium pyroantimonate-fixed myocardial tissues, found that no calcium was detectable in mitochondria, triads, nuclei, or nucleoli after 10-min staining in uranyl acetate, while calcium was detectable only in the nucleus after 3 min of lead citrate staining.

Phase Transformations in Mineralized Tissues

Apart from the dissolution of the mineral phase from bone during conventional preparative procedures involving fixation, dehydration, flotation, and staining in aqueous media (Renaud, 1959; Boothroyd, 1964, 1968; Dudley and Spiro, 1961; Furseth, 1969; Glimcher and Krane, 1968; Posner, 1972; Termine, 1972), these procedures cause mineral phase transformations detectable by x-ray diffraction. For example, Termine (1972) has shown that while the hydroxyapatite crystalline phase in bone is relatively insoluble, the amorphous calcium phosphate phase is more easily solubilized (from 5 to 100% loss being recorded, depending on the degree of hydration

Table 3. ^{45}Ca lost from ^{45}Ca-loaded sarcoplasmic reticulum vesicles mounted on EM grids during various staining procedures

A. ^{45}Ca loss from unfixed specimens

Controls*	Water-treated grids	2% Ammonium molybdate, pH 7.0	1% Phosphotungstic acid (PTA), pH 7.0	1% Ammonium molybdate & 0.5% PTA, pH 7.0	1% Uranyl acetate, pH 4.5
0% (n = 32)	41% (n = 32)	37% (n = 32)	62% (n = 32)	65% (n = 32)	45% (n = 32)

B. ^{45}Ca loss from sarcoplasmic specimens during fixation on grids, and from the fixed samples during various staining procedures

Control*	2.5% Glutaraldehyde + 10 mM potassium oxalate + 30 mM KCl + 20 mM imidazole, pH 7.0	2% Ammonium molybdate, pH 7.0	1% PTA, pH 7.0	1% Uranyl acetate, pH 4.5
0% (n = 15)	53% (n = 15)	67% (n = 15)	76% (n = 15)	59% (n = 15)

From Agostini and Hasselbach, 1971.

* Controls represent untreated, ^{45}Ca-loaded samples mounted on grids. Droplets of the reagents were placed directly onto individual grids, and removed after standardized periods (0.5–3 min for stains, 10–30 min for fixative) with blotting paper.

of the sample). Furthermore, the amorphous phase is readily transformed to apatite during preparation, e.g., in a case where 25% loss of amorphous mineral was observed, 40% of that remaining was transformed to apatite crystals (Termine, 1972). These rather cryptic artifacts are of considerable significance in mineralization studies. They are reduced by the adoption of freeze-substitution (Landis et al., 1977b), and cryotechniques (Landis et al., 1977a), and can only be eliminated by procedures that maintain the in vivo state of hydration of the tissue (Pautard, 1972).

Redistribution of Elements

The movement of elements from their normal loci to regions they do not occupy in vivo is a major problem that must be vigilantly controlled or at least monitored by all microprobe analysts, whatever their specific biological goals. The artifact is more likely to occur in tissues that are exposed to fluid phases, but cannot be overlooked as a minor impediment even in freezing techniques (see later). There are two aspects of the artifact that are worthy of brief discussion.

First, there is the extent of redistribution. Quite simply, the movement of elements to new loci can be tolerated if the movement occurs across distances that are either too small to be resolved by the analytical capabilities of the microprobe being used, or (and of more practical significance) if the movement is confined within the volume of even relatively major structures, e.g., whole organelles or cells, that are being analyzed with large probe diameters.

Second, the pervasive influences of preparative fluids on tissue chemistry are extremely difficult to expose. Redistribution cannot definitely be pinpointed and measured unless the true distribution pattern in vivo has been previously defined. In general, the recording of element loss from a specimen is indicative of translocation within the specimen. In theory at least, two important exceptions exist. Element losses during processing may represent the complete and selective elution of one or more of the relatively exchangeable tissue fractions, leaving the less exchangeable fractions at their in vivo sites. On the other hand, in cases where no element loss is recorded it may not be valid to assume even a negligible endogenous redistribution.

The seriousness of the artifact is amplified, and not diminished, by the fact that few workers have been able to qualitatively, even less quantitatively, demonstrate its occurrence. Coleman and Terepka (1974) proposed at least four criteria that may prove useful in identifying redistribution artifacts.

1. *The occurrence of unusual intracellular crystals, especially of calcium phosphate salts, may indicate translocation.*

Cell cytoplasm is usually rich in phosphate groups, while cellular free calcium levels are normally very low, $< 10^{-7}$ M. Aggregations of calcium phosphate salts intracellularly may, therefore, suggest pathological changes in cellular function (Trump and Mergner, 1974) or the spurious accumulation of precipitates in normal cells; both events presumably caused by the failure of cell membranes to exclude Ca^{2+}. For example, Maunder et al. (1977) suggested that the elevated Ca:P ratios recorded in myonuclei of patients suffering from Duchenne muscular dystrophy (Table 4) may not represent the true chemistry of the diseased cells, but that the variance from normal is secondarily induced, i.e., by the entry of calcium (presumably) into the cells during processing as a result of the predisposing influences of the disease.

The difficulty with this criterion is establishing what represents "unusual crystals"; one can establish the origin of the precipitates only by the systematic exposure of tissue to several different preparative procedures (Coleman and Terepka, 1972).

2. *Distinct chemical boundaries within specimens may indicate the absence of major shifts in distribution.*

This criterion serves as a good rule-of-thumb guideline, and has been used by many authors to validate the use of various non-freezing and cryo-

Table 4.   Examples of some of the elements that may enter tissues during various wet chemical processing stages

Processing stage		Introduced element	Implications and comment
FIXATIVES	Glutaraldehyde	Ca (contaminant)	Some commercial grades of glutaraldehyde contain various amounts of calcium; the presence of electron-opaque deposits in a variety of tissues can be correlated with the presence of calcium in the fixative (Oschman and Wall, 1972). The implications are that while calcium uptake pathways may be localized, the distribution of calcium in the living state cannot be. This has led some workers to use re-distilled glutaraldehyde (Oschman and Wall, 1972; Oschman et al., 1974; Skaer et al., 1974).
	Osmium tetroxide	Os	The Mα emission line of osmium at 1.914 keV overlaps the Kα line of potassium at 2.013 keV. Can be resolved with crystal spectrometer (Chandler and Battersby, 1976a).
	Potassium pyroantimonate	Sb	Ca Kα and Kβ peaks at 3.69 and 4.02 keV are obscured by the major Sb Lα and Lβ peaks (3.60, 3.84, and 4.10 keV). These peaks can be resolved with crystal (wavelength) spectrometers (Yarom et al., 1974b). Complex deconvolution or peak stripping procedures are required to determine the presence of Ca by EDS (Simson et al., 1979).
BUFFERS	K-phosphate ⎫ S-collidine ⎬ Na-cacodylate ⎭	Contaminants and normal constituents	All three buffers were shown to contain significant quantities of contaminants; calcium and potassium or sodium (Oschman and Wall, 1972). S-collidine buffer contained the least amount of calcium. Cacodylate buffer also contains arsenic as a normal constituent—the As Lα emission at 1.282 keV overlaps the Mg Kα emission at 1.253 keV in energy-dispersive spectra and may create sufficient background signal to swamp weak Na Kα emissions at 1.041 keV.

RESINS	Araldite Epon 812 Spurr's medium	Cl	Workers studying endogenous chlorine distribution have developed or exploited various low-chlorine embedding media, e.g., 50% glutaraldehyde-urea mixture (Pease and Peterson, 1972; Yarom et al., 1974a; Van Steveninck et al., 1974), which does not contain chlorine and has the advantage that no dehydration and long infiltration periods are necessary; 50% glutaraldehyde-carbohydrazide mixture (Heckman and Barrnett, 1973; Van Steveninck et al., 1974), which contains no chlorine; Epon 826 (Ingram and Hogben, 1968), which contains 30 mM/kg organically bound chlorine compared with 1400 mM/kg in Epon 812. Pallaghy's (1973) modification (see Table 12 for recipe) of Spurr's resin (Spurr, 1969) contains no chlorine and retains the low viscosity feature of Spurr's.
CHLOROFORM VAPOR		Cl	Chloroform vapor is routinely used for stretching resin-infiltrated sections floating on water troughs. The vapor is readily absorbed by sections, thus producing significant chlorine signals within the specimen (Pallaghy, 1973). Pallaghy (1973) suggests that methyl cyclohexane can be a useful alternative; although flotation on aqueous media is generally not to be recommended for sections destined for microprobe analysis (see above).
STAINS	Uranyl acetate Lead citrate Potassium permanganate Ammonium molybdate	U Pb Mn Mo	The major emission lines of the heavy metals commonly used for staining EM sections arise within the energy range of many of the biological elements, e.g., Pb (Mα -2.342 keV and Mβ -2.442 keV) emissions are adjacent to S (Kα -2.307 keV) and Cl (Kα -2.621 keV) lines; U (Mα -3.165 keV and Mβ -3.336 keV) emissions are adjacent to the K (Kα -3.312 keV) line. In general, sections to be analyzed should not be stained with heavy metal salts.

preparative procedures. Gupta et al. (1978b), in a study of electrolyte distribution in epithelial cells and intercellular spaces of rabbit ileum, stated that "... diffusion (artifacts during specimen preparation) ... can only affect the (microprobe) results by 'smoothing-out' the gradients that exist in the control tissue *in vitro*. These artifacts could not have generated gradients ... although they may reduce the magnitude of such gradients." Transcellular ionic gradients can be disturbed during fixation with aldehydes and osmium. For example, fixed smooth muscle cells passively accumulate sodium and calcium from extracellular fluid, with a concomitant loss of intracellular potassium (Schoenberg et al., 1973) (Table 1). Similar fluxes may occur in fixed cells during routine washing in buffers, but whether these introduced ions survive subsequent preparative stages to interfere with the analysis has not been established.

Distinct chemical boundaries can exist in cells whose chemical integrity has been severely disturbed. Heterogeneous distribution patterns per se can be misleading and are not sufficient validation of the efficacy of a preparative procedure. For example, we have found that osmium-fixed nuclei invariably contain more calcium than nuclei prepared by dry cryomicrotomy, even though the wet chemical technique produced an overall loss of $> 50\%$ calcium from the tissue. This recruitment of calcium to specific loci may be explained by the passive influx of calcium (Schoenberg et al., 1973), followed by an interaction between Ca^{2+}, osmic acid, and osmic acid–reactive groups within the cells (Krames and Page, 1968). Significantly, perhaps, cytoplasmic calcium was detected in myocardial tissue only after osmium tetroxide/pyroantimonate fixation, but not in cells fixed by glutaraldehyde or pyroantimonate alone (Yarom et al., 1974a).

3. *The use of gelatine models loaded with elements to test whether the elements are displaced during processing.*

The extreme simplicity of such models, with, for example, the absence of physical boundaries (membranes), makes their usefulness even for the comparison of the efficacy of different preparative procedures highly questionable.

4. *Translocation obviously occurs if different preparative procedures yield different element distribution patterns.*

This represents the best practical criterion. Erasmus and Davies (1979) have, for example, demonstrated that the calcium distribution pattern within the major structural components of the vitelline cells of *Schistosoma mansoni* differs according to the preparative method used (Figure 6). Fortunately, the vitelline cells contain structures with significant inherent electron opacity, which facilitates their identification in unfixed, freeze-dried cryosections. However, the vast majority of cells do not contain such easily distinguished components; in this context the preparation and analysis of

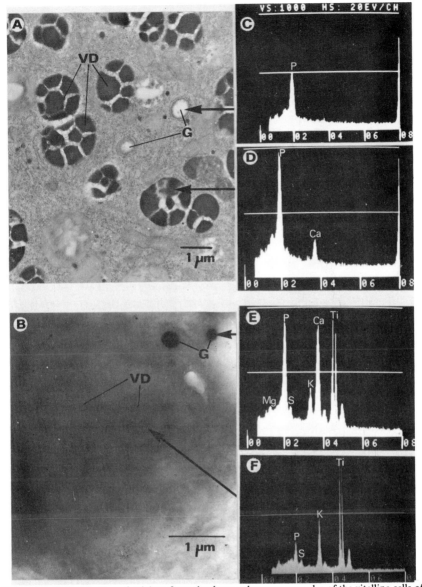

Figure 6. Redistribution of calcium from the dense calcareous granules of the vitelline cells of *Schistosoma mansoni* to the vitelline droplets during osmication (either after glutaraldehyde fixation or during osmium tetroxide single-fixation). **A:** Electron micrograph of a section through a glutaraldehyde and osmium-fixed vitelline cell, stained in uranyl acetate and lead citrate. Note the vitelline droplets (VD) and the calcareous granules (G) in the distended cisternae of the granular endoplasmic reticulum. **B:** Electron micrograph of an unfixed, unstained, freeze-dried cryosection through the vitelline cell. The calcareous granules (G) and vitelline droplets (VD) are easily identifiable. **C:** Spectrum obtained from the analysis of a "calcareous spherule" prepared by double-fixation, but unstained. Note especially the absence of detectable calcium. **D:** Spectrum obtained from the analysis of a vitelline droplet prepared as in C. Note especially the presence of a very significant calcium signal (in contrast with **F**). **E:** Spectrum obtained from the analysis of a calcareous granule in a cryosection. Note the approximately equal phosphorus and calcium signals. **F:** Spectrum obtained from the analysis of a single vitelline droplet prepared as in **E**. Note the absence of a significant calcium signal. (Courtesy of D. A. Erasmus and T. W. Davies.)

frozen sections may be viewed as a means of checking data obtained from similar samples prepared by less rigorous techniques. Obviously, cryo-procedures should be used wherever prevailing circumstances are favorable, since many of the chemical artifacts encountered with techniques involving fluid phases are bypassed (Morgan and Bellamy, 1973; Morgan et al., 1975).

Introduction of Exogenous Elements

Processing fluids can contribute detectable elements to the specimen. These elements are vexatious 1) because they invalidate the microprobe observations if the element concerned is also naturally occurring and the subject of study—chlorine in resin being a good example; 2) because if the contributed element is not a "biologically interesting" element and is not normally present in the tissue, its presence may still present difficulties, since its characteristic x-ray emission lines may be adjacent to or overlap the emission lines of certain biological elements [this is a problem often encountered with energy-dispersive spectrometers, whose resolving power is significantly inferior to that of wavelength-dispersive spectrometers (≈ 150 eV at 5.9 keV and ≈ 10 eV, respectively) (Chandler, 1975a)]; or 3) because the presence of the element may contribute significantly to the background signal if it emits high-energy x-rays; thus it may preclude the detection of low concentrations of certain (especially low atomic number) naturally occurring elements.

It is possible to distinguish two categories of introduced elements:

1.   Those that occur in the extracellular fluid of the tissue itself. These elements are translocated into the tissue during fixation, etc., by virtue of either a change in membrane permeability and/or an increased affinity of cellular constituents as a direct consequence of chemical fixation. Examples are furnished by the work of Krames and Page (1968) and Schoenberg et al. (1973). Moreover, Elbers (1966) showed by a microconductiometric technique that the cell membrane of *Limnaea* eggs became permeable to ions within a few seconds of fixation in osmium tetroxide.

2.   Those elements that are truly exogenous to the specimen and are either chemical constituents or contaminants of a given processing fluid. Some indication of the elements that can be introduced from given processing fluids and some of the analytical implications of their presence are outlined in Table 4.

Modified Procedures

Clearly, the basic EM wet-chemical preparative procedure can be subject to an infinite number of relatively minor modifications, many of which do not greatly improve the efficacy of the basic method (Table 1). However, modifications have been introduced that were designed to maintain the chemical integrity of biological specimens or to meet specific requirements related to the problems of analyzing biological specimens in the microprobe. Some of these modifications are discussed briefly.

Simplified Embedding Procedures The direct embedding of fixed specimens in a polymerizable mixture of 50% glutaraldehyde and urea was first suggested by Pease and Peterson (1972), and subsequently used by Yarom et al. (1974a, 1975a) for the localization of calcium in myocardial tissue. The advantage of this embedding procedure over more conventional regimes is that it avoids dehydration. In addition, Yarom et al. (1975a) showed that up to 10 times more calcium is retained in the sarcoma of myocardial tissue embedded in glutaraldehyde-urea than in samples dehydrated and embedded in Epon 812 (Table 5).

The direct embedding of glutaraldehyde-fixed tissues in Epon 812, having previously increased the glutaraldehyde concentration by stages from 2.5% to 50%, was introduced by Yarom et al. (1975b). The technique was shown to reduce the leaching of calcium and zinc from rat tongue muscle that occurs during dehydration in alcohol.

The data presented by Yarom et al. (1975a,b) in support of their respective preparative procedures serve only to emphasize the fact that dehydration and embedding agents can markedly deplete tissue chemistry; the data should not disguise the fact that many of the major losses and redistribution artifacts have been previously and irrevocably introduced during primary fixation.

A realistic perspective for these techniques, and perhaps for many other preparative procedures in the context of microprobe analysis, was drawn by

Table 5. Relative calcium content [expressed as relative mass fractions, $R = (P - b)/(W - W_b)$] in myocardium fixed in a mixture of 1% osmium and 2% potassium pyroantimonate*

Specimen	Embedding	R of specimen normalized to R of a standard	
		Nucleus	Sarcomere
Rat heart, control	Dehydration + Epon 812	0.0148	0.0012
Rat heart, control	Glutaraldehyde/ urea	0.0269	0.0115
Rat heart, 6 h after isoproteranol	Glutaraldehyde + Epon 812	0.0141	0.0045
Rat heart, 6 h after isoproterenol	Glutaraldehyde/	0.0160	0.0450

Table drawn from data of Yarom et al., 1975a.

* Note that these microprobe readings were obtained by wavelength-dispersive analysis in EMMA 4. The energy-dispersive spectrometer cannot resolve Sb $L\alpha$ and Ca $K\alpha$ radiations. W = Continuum measured on the specimen. W_b = Continuum measured on the support film.

Yarom et al. (1975b): "Elemental shifts undoubtedly still occur, but with strict adherence to identical methodology, comparative studies should give valid results of a semi-quantitative and perhaps diagnostic nature in clinical and experimental studies. Moreover . . . it remains to compare differently prepared samples in order to establish some basic values."

Precipitation Procedures Attempts have been made to counter the leaching influences of processing fluids by employing various precipitation reactions. These reactions commonly involve the sequestration in situ of an endogenous tissue ion with an introduced heavy metal ion, thus producing an electron-dense deposit that, ideally, is highly insoluble. For example, Cl^- ions have been precipitated with gold (Läuchli et al., 1974; Läuchli, 1975) and silver ions (Van Steveninck et al., 1973; Harvey et al., 1976a), Na^+ with pyroantimonate (Tandler et al., 1970), PO_4^{3-} with lead ions (Tandler and Solari, 1969), and Ca^{2+} with oxalate ions (Diculescu et al., 1971; Coleman and Terepka, 1972). The precipitating agent is usually included in the prefixative and also in some or all of the subsequent preparative fluids. Ideally, both the fixative molecules and the precipitating agent should have similar penetration kinetics, so that the capture of endogenous ions before they drift from their in vivo sites and the fixation of their microenvironments are simultaneous events. How closely this optimal arrangement is approached in any of the available reactions is not known.

A discussion of the mechanisms involved in these various histochemical reactions and of the controversy surrounding their specificity and general applicability is beyond the scope of this chapter. Detailed reviews on these topics are available elsewhere (Läuchli, 1975; Chandler, 1975a, 1977a, 1978; Bowen and Ryder, 1978; Simson et al., 1979). Nevertheless, four general comments pertinent to our present discussion should be made:

1. The effectiveness of many of the precipitation reactions in reducing elemental loss during the various stages of fixation, etc., is minimal (Table 6).

2. Elemental loss during the subsequent stages of dehydration, embedding, flotation, and staining have been recorded after a given element has been immobilized by precipitation (Harvey et al., 1976a—Figure 2d; Coleman and Terepka, 1972—Figure 2a; Yarom et al., 1974b, 1975a). Precipitate drift and therefore element redistribution artifacts are persistent and major restraints.

3. The formation of discrete precipitates implies that elements are drawn to focal points from their in vivo locations in the surrounding tissue; but whether this ion migration occurs across distances within the resolving power of existing microprobe instruments is unestablished, although this need not detract from the usefulness of the techniques in low-resolution microprobe investigations.

4. Ironically, the popularity of the pyroantimonate technique bears testimony to its lack of specificity. It was a technique originally introduced to specifically localize sodium (Komnick, 1962), but with the advent of the microprobe the precipitated deposits were found to contain several cations, including sodium, potassium, calcium, and magnesium (Tisher et al., 1972). Although the reaction has been used with apparent success in several clinical studies, notably of calcium and zinc distribution in sperm (Chandler and Battersby, 1976a) and muscular tissues (Yarom et al., 1974b, 1975a), the technique is generally considered unsuitable for the localization of monovalent electrolytes (Chandler and Battersby, 1976a) and does not prevent gross calcium fluxes during the various stages of processing (Yarom et al., 1974b, 1975a).

The microprobe does not distinguish between bound and unbound element fractions. Consequently, although the data obtained are obviously not worthless, the microprobe nonetheless fails to explore ionic compartments and thus establish the functional potential of a tissue or cellular component. Precipitation reactions may in the future be employed in the identification and measurement of ionic species. For example, Chandler and Battersby (1976a) hint that by a careful scrutiny of Ca:Sb ratios in pyroantimonate-treated sperm cells they were able to qualitatively distinguish differences in the degree of calcium binding to macromolecules in various subcellular regions.

In pursuit of this extremely important goal of measuring subcellular ionic compartments, we need not be limited by the vagaries of the precipitation reactions currently used in electron microscopy. The microprobe facilitates the identification of non–electron-dense histochemical reaction products (Ryder and Bowen, 1974; Chandler, 1975a). Therefore, techniques such as Kashiwa's (1966, 1968) glyoxal-bis-hydroxyanil technique for localizing ionic calcium at light microscope level could in theory be appropriated for electron microprobe usage. Ultimately, these ion precipitation reactions may be advantageously combined with cryoprocedures, so that the loss and migration of the sequestered product would be minimized or effectively eliminated.

Applications

Contrary, perhaps, to the impression given in the above discussion, wet chemical procedures need not be banished from microprobe laboratories; not least because of the severe morphological limitations of those techniques that more faithfully maintain the in vivo chemical integrity of biological specimens. The present author subscribes to the view that wet chemical techniques can be used, often with considerable advantage, especially if

Table 6. Summary of quantitative measurements of the loss of given elements from biological specimens during exposure to certain precipitation reactions

Precipitation reaction	Endogenous ion/radioactive tracer	Tissue	Processing regime	Percent loss		Reference
				With precipitant	Without precipitant	
Pyroantimonate for monovalent and divalent cations	Na^+	Uterus	3% glutaraldehyde + 2% pyroantimonate	90	95	Garfield et al., 1972
			1% OsO_4 + 2% pyroantimonate	75	75	
	Ca^{2+}	Uterus	3% glutaraldehyde + 2% pyroantimonate	55	30	
			1% OsO_4 + 2% pyroantimonate	85	0	
	Mg^{2+}	Uterus	3% glutaraldehyde + 2% pyroantimonate	30	20	
			1% OsO_4 + 2% pyroantimonate	35	10	
Oxalate for calcium	Ca^{2+}	Chorioallantoic membranes	6% acrolein in 0.1 M cacodylate – HCl + 1% sodium oxalate; buffer + oxalate wash; 2% OsO_4 + 1% oxalate; graded alcohols	12	75	Coleman and Terepka, 1972
Sulfide for heavy metals	^{65}Zn	Corn leaf, Barley root, Chlorella	0.3% or 1% glutaraldehyde in PO_4 buffer + 30% Na_2S	7.6–9.5	62–70	DeFilippis and Pallaghy, 1975
	^{203}Hg	Corn leaf, Barley root	0.3 or 1% glutaraldehyde in PO_4 buffer + 3% Na_2S	72–81	34–41	

Analyte	Element	Tissue	Processing regime			Reference
Sulfide for heavy metals	Cd*	Bivalve tissues	glutaraldehyde + H_2S	14	28	George et al., 1976
			glutaraldehyde + H_2S; embedded in glutaraldehyde/urea	20	(—)**	
			glutaraldehyde + H_2S; alcohols	33	(—)	
			glutaraldehyde + H_2S; alcohols; embedded in Spurr's resin	41	(—)	
Silver ions for chloride	[36]Cl	Root tissue	1% OsO_4 in 0.1 M cacodylate-acetate buffer, pH 6.5 + either 0.5% Ag acetate or 1% Ag lactate; buffer wash; acetone (+ brief wash in 0.05 M HNO_3); propylene oxide; Spurr's resin	<4	(—)**	Läuchli et al., 1974
		Leaf tissue	1% OsO_4 in cacodylate-acetate + 30 mM Ag acetate; buffer wash; acetone; Pallaghy modification of Spurr's resin	70–95 (depending on fixative time and sample size)	(—)**	Harvey et al., 1976

* Details of processing regime not provided in this communication.

** Not measured.

attempts are made to optimize the conditions of preparation, to monitor the chemical fluxes that inevitably tend to occur, and to verify the microprobe data wherever possible with microprobe data obtained from specimens prepared by different procedures, and by comparison with data obtained by other analytical techniques, e.g., flame spectrophotometry, scintillation counting of radioactive tracers, and specific ion microelectrodes. To illustrate this important contention and perhaps to redress the perspective of our discussion of wet chemical procedures, I will outline very briefly here a series of specific examples of applications where the microprobe analysis of conventionally prepared specimens has enhanced our understanding of biological processes. Many similar examples are available, most of which are cited by Coleman and Terepka (1974), Chandler (1975a,b; 1977a,b; 1978), and Ghadially (1979).

Analysis of Natural Inclusion Bodies

Calcium-rich spherules in the earthworm (*Morgan, 1980*).   Most invertebrates possess mineral forms in various tissues, which are often concentrically layered and spherical; they are usually very rich in calcium plus ancillary cations, and the dominant anionic group is either carbonate or phosphate (Simkiss, 1976). We were able to identify examples of the calcium carbonate and of the calcium phosphate spherules in the calciferous glands and chloragogenous tissue, respectively, of the earthworm *Lumbricus terrestris* (Figure 7). In addition, we found that Sr^{2+} injected directly into the worms was localized preferentially in the "carbonate spherules" during a period up to 24 h post-injection. In contrast, Zn^{2+} and Pb^{3+} were accumulated preferentially by the "phosphate spherules." These major and fundamental chemical differences were unlikely to be introduced during the wet chemical preparation. An advantage of conventional processing over cryoprocedures in this particular instance is that the superior morphological detail it affords facilitates a simultaneous study of the cytoplasmic relationships of the spherules and the distribution of elements within the substructure of individual spherules.

Distribution of Deliberately Introduced Compounds

Distribution of the antimonial drug Astiban within the vitelline cells of the parasite Schistosoma mansoni (*Erasmus, 1974, 1975*).   Astiban (sodium antimony dimercaptosuccinate) distribution was monitored by locating the x-ray emission from the antimony atoms within the molecule. The drug was found to be specifically localized within the vitelline droplets of the vitelline cells; no antimony signals emanated from mitochondria or ribosomal complexes. The advantage of wet chemical preparation was the ability to correlate antimony distribution patterns with both the developmental state of the rapidly dividing vitelline cells and the degree of cellular damage imposed

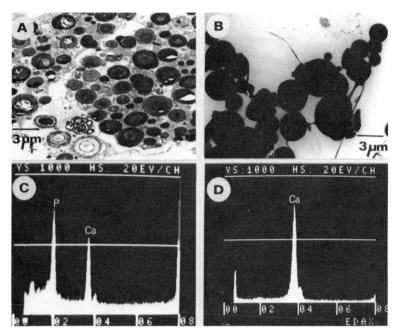

Figure 7.   The intracellular chloragosome granules (**A**) and extracellular calciferous gland spherules (**B**) of *Lumbricus terrestris* can be distinguished structurally and chemically in tissues prepared by wet chemical procedures (compare with Figure 13). The chloragosome granules consist primarily of calcium and phosphate, and also zinc. (**C**) Gland spherules yield only a calcium signal, and are known to consist of a form of calcium carbonate. The advantage of wet chemical techniques in mineralization studies is that the relationship of the mineral to the secretory cells is more easily clarified. In addition, the distribution of individual elements within mineralized structures can occasionally be studied. For example, the denser regions of the chloragosome granule possess significantly higher Ca:P ratios than the less dense regions.

by drug treatment. Dry cryomicrotomy was used to confirm the general antimony localization pattern and to establish changes in the element composition of the cells induced by drug sequestration.

Pathology

Calcium accumulation in the liver of rats experimentally infected with Fasciola (Foster and Erasmus, 1980).   The total element contents of selected areas of livers of infected and non-infected rats were compared by a bulk analysis procedure in the microprobe (Davies and Morgan, 1976). One of the most significant changes induced by the presence of the parasite was a rise in calcium concentration in the tissue immediately adjacent to the locus of infection (Figure 8). This gross calcium elevation was subsequently attributed to an accumulation of calcified necrotic debris within the cytoplasm of eosinophils assembled within the region in response to parasite activity.

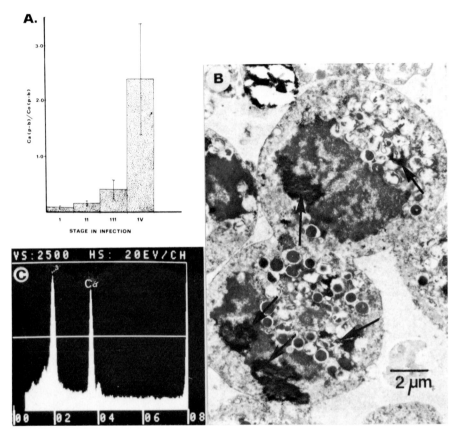

Figure 8.   Calcification of rat liver infected with *Fasciola hepatica*. **A:** Bulk analysis in the transmission electron analytical microscope of liver samples immediately adjacent to the site of infection, and taken at intervals after parasitic invasion. The analysis was performed on tissue ashed in a low-temperature oxygen plasma; the ash was solubilized in 0.5 N HNO_3 containing a known quantity of cobalt as an internal reference, and droplets of the solution were analyzed (Davies and Morgan, 1976). "Stage in infection" refers to histological categories defining the extent of the cellular response and damage caused by the parasite. Note that the calcium concentration increases rapidly from normal (stage 1) as the host tissue becomes more involved in inflammatory and fibrotic reactions. **B:** Electron micrograph of a section through stage IV tissue. The tissue contains large numbers of round polymorphic leukocytes exhibiting amorphous eosinophilia. The gross calcium increase measured by the bulk technique is attributable to the precipitation of calcium phosphate crystals (arrows in **B** and spectrum **C**) in the necrotic debris within these cells. **C:** Spectrum obtained from the analysis of a crystalline deposit in the polymorphic leukocytes. (Courtesy of J. R. Foster.)

Changes in the Ca:P ratio of normal and dystrophic human muscle (Maunder et al., 1977).   An elevated Ca:P ratio was measured in the myonuclei (and interstitial cell nuclei) of patients suffering from Duchenne muscular dystrophy compared with non-diseased controls (Table 7). The authors considered that the difference was secondary; (presumably) the calcium elevation occurred during tissue processing as a consequence of changes in membrane permeability *caused* by the disease state. Thus, the wet chemical procedure produces a "specific chemical artifact" that can be used as a diagnostic criterion for *comparing* diseased and non-diseased muscle fibers, and especially for differentiating diseased fibers that are ultrastructurally normal in subclinical states.

Identification of Histochemical Reaction Products

Differential staining of adrenaline and noradrenaline secretory cells (Lever et al., 1977). Rat adrenal medulla and sympathetic ganglia samples were fixed by glutaraldehyde/formaldehyde → potassium dichromate → osmium sequence and plastic embedded. Chromium signals were detected by microprobe analysis in the secretory granules in both adrenomedullary noradrenaline cells and type II sympathetic small granulated cells; chromium was not detected in adrenomedullary adrenaline cells nor in two other sympathetic granulated cell types (Figure 9). Since chromium is known to bind to the stable Schiff monobase formed by the interaction of glutaraldehyde and noradrenaline, the authors suggested that noradrenaline is present in the granules of the type II sympathetic small granulated cells, as well as in the adrenomedullary noradrenaline cells. Adrenalin, on the other hand, does not form a stable residue and is eluted during glutaraldehyde

Table 7.   Ca:P ratios (derived from the relative mass fractions of calcium and phosphorus × 10³) in myonuclei of structurally undamaged fibers in male patients suffering from Duchenne muscular dystrophy (DMD) and in controls

	Case no.*	Age	Ca:P × 10
	1	17 days	4.7
	2	6 years	3.6
Non-diseased	3	7 years	4.1
controls	4	16 years	2.5
	5	32 years	2.6
	6	34 years	4.3
			$\bar{x} = 3.6$
			$SD = \pm 0.8$
	1	3 months	5.0
	2	4 years	8.8
Patients with	3	5 years	11.5
diagnosed	4	6 years	10.4
DMD	5	8 years	5.6
	6	9 years	6.5
			$\bar{x} = 8.0$
			$SD = \pm 2.4$

Table drawn from data of Maunder et al., 1977.

* Average number of fibers analyzed per patient = 7.

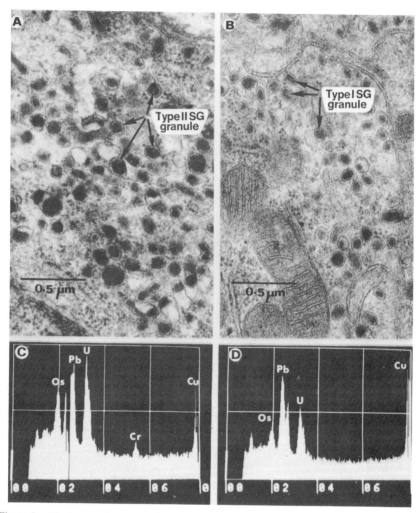

Figure 9.   Electron probe x-ray microanalysis of chromium binding in the inclusion granules of sympathetic small granulated cells (SG), after fixation in glutaraldehyde-paraformaldehyde mixture, then incubated in 0.2 M cacodylate buffered 3% $K_2Cr_2O_7$ before osmication in 1% OsO_4, etc. Chromium binds specifically to the Schiff monobase formed by glutaraldehyde and noradrenaline during fixation. The combination of histochemistry and microprobe analysis showed that there are two types of SG cells having two different secretory products. Analysis of cytoplasmic fields containing membrane-bound inclusion granules shows positive chromium binding in type II SG cells, i.e., noradrenaline secreting cells (**A** and **C**), and no chromium binding in type I SG cells (**B** and **D**). (From Lever et al., 1977. Used by permission.)

fixation. The reliability and utility of the preparative and analytical methods were confirmed by the recording of a significant reduction in the chromium content of adrenomedullary granules after reserpine treatment (Santer et al., 1978).

Fine structural localization of Alcian Blue-positive mucopolysaccharides (Bowen et al., 1975). Alcian Blue staining is a specific histochemical test for certain acid mucopolysaccharides at light microscope level. The technique is readily modified for higher-resolution localization studies in electron microscopes equipped with x-ray detectors because of the copper atoms in the stain molecule.

Copper was detected in the mucus on the surface of the epithelium of planarian worms stained in Alcian Blue at pH 2.5 (Figure 10) and in the mucus gland cells, but not in the cyanophil gland cells.

This example demonstrates the usefulness of microprobe analysis for the localization of specific histochemical stains, and especially where the reaction product is relatively non-electron dense. This principle was also extended to demonstrate enzymatically released products (Ryder and Bowen, 1974; Bowen et al., 1976).

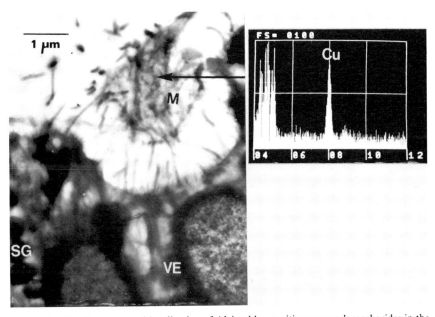

Figure 10.   The fine structural localization of Alcian blue-positive mucopolysaccharides in the ventral epithelium of the planarian, *Polycelis tenuis*, using EM and x-ray microanalysis. Analysis of the mucus (M) at the surface of the epithelium yielded a significant copper signal (derived from localized Alcian blue molecules). SG = secretory granules; VE = ventral epithelial cells. The titanium signal in the x-ray spectrum was derived from the grid material. (Courtesy of I. D. Bowen.)

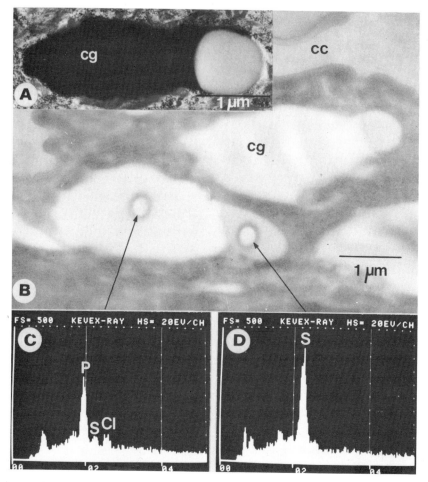

Figure 11. Sulfur in epidermal keratohyalin granules. **A**: Composite keratohyalin granule (cg) from newborn rat epidermis. The granule contains an electron-lucent "single granule component" and a dense component. Stained section. **B**: Micrograph showing composite keratohyalin granules (cg) of newborn rat epidermis in glutaraldehyde-fixed, unstained section. Contamination rings (deposited during analysis) are seen over the two components of the composite granule. cc = Cornified cell. **C**: Spectrum from the analysis of the dense component of a composite granule. **D**: Spectrum obtained from the analysis of the lucent "single granule component" of a composite granule. Note especially the difference in the sulfur composition of the two components. (From Jessen et al., 1976. Used by permission.)

Bound Elements

Sulfur in epidermal keratohyaline granules. Jessen et al. 1976) were able to correlate differences in the morphology of keratohyalin granules in rat epidermis with differences in chemical composition (Figure 11). One type of granule (termed "single granules") present in the nuclei and cytoplasm of epidermal cells in both newborn and adult rats is rich in sulfur (S content 2.5%–3.6%). Other types, which differ morphologically in newborn and adult rats, contain a sulfur-poor (S content 0.6%–0.9%) and a sulfur-rich component. These results were qualitatively similar to earlier findings in keratohyalin granules in lingual and esophageal tissues (Jessen et al., 1974).

WHOLE MOUNTING AND AIR DRYING

Whole mounting essentially involves the direct mounting of fresh whole cells onto appropriate EM specimen supports, either by smearing, centrifugation, or spraying from fluid suspension. The extreme simplicity of the technique is a great advantage where routine sampling of cellular populations is required, since preparation time is minimal.

The basic technique has been used fairly extensively for the preparation of two types of samples for microprobe analysis: 1) cells suspended in voluminous extracellular fluids, and 2) more compact tissues whose cells often contain relatively dense cytoplasmic structures (see Table 8 for examples).

Chemical Restraints

Not surprisingly, at least two major changes in the chemical integrity of whole-mounted, air-dried specimens can occur: contamination and chemical redistribution.

Contamination   In the specific case of fluid-suspended cells, it is highly likely that the cell surfaces will be contaminated by elements from the surrounding fluid environment that are also significant intracellular constituents (Figure 12). Early workers using this preparatory procedure disregarded the problem of chemical contamination (Duprez and Vignes, 1967; Carroll and Tullis, 1968; Beaman et al., 1969). More recently, workers studying human red blood cells have attempted to remove the surface contaminants by washing in various isotonic media (Kimzey and Burns, 1973; Kirk et al., 1974; Roinel et al., 1974; Roinel and Passow, 1975; Lechene et al., 1976). Kimzey and Burns (1973) found that elements within the washing medium also adhere to the cell surface during air drying. The microprobe failed to detect magnesium, calcium, and rubidium in unwashed red blood cells, but after washing in a series of isotonic media that contained only the chlorine salt of these cations as the osmotically active solute (i.e., 2.20% $MgCl$; 1.30% $CaCl_2$; 0.90% $NaCl$; 1.19% KCl, 1.86% $RbCl_2$), substantial quantities of the cations were found associated with individual cells. Therefore, Kimzey and Burns (1973), when analyzing the intracellular potassium concentration, washed the cells free of plasma in a

Table 8.   Some examples of the biological applications of the whole mounting and air-drying preparative procedure

A. FLUID-SUSPENDED	
CELLS	
Erythrocytes	Duprez and Vignes (1967), Carroll and Tullis (1968), Beaman et al. (1969), Kimzey and Burns (1973), Gullasch and Kaufmann (1974), Kirk et al. (1974), Roinel and Morel (1973), Roinel et al. (1974), Roinel and Passow (1975), Colvin et al. (1975), Lechene et al. (1976), Kirk and Lee (1978)
Mast cells in peritoneal fluid	Padawer (1974)
Spermatozoa	Chandler (1973, 1975a), Chandler and Battersby (1976b), Werner and Gullasch (1974), Maynard et al. (1975)
Protozoan—*Tetrahymena pyriformis*	Gullasch and Kaufmann (1974)
Vorticellid spasmonemes	Routledge et al. (1975)
Platelets	Costa et al. (1974, 1977a,b), Tanaka et al. (1975), Takaya (1975b), Skaer et al. (1976)
B. FRESH-TISSUE	
SPREADS	
Zymogen granules	Takaya (1975a)
Neurosecretory granules	Takaya (1975c)
Nuclei of connective tissue cells	Takaya (1975d)
Human melanosomes	Takaya (1977)
Nuclei of Ehrlich tumor cells	Mizuhira (1976)
Earthworm chlorogosome granules	Morgan (1980)

K-free medium containing rubidium as a potassium replacement (10 mM $RbCl_2$, 141 mM NaCl, 1.0 mM $MgCl_2$, 1.3 mM $CaCl_2$, 0.8 mM $NaHPO_4$, 5 mM $NaHPO_4$, 11.1 mM glucose, pH 7.4). More recently, workers interested in the distribution of all the analyzable lighter elements within red cells have employed simpler washing media, e.g., Lechene's group washes with several changes of acid sucrose (0.285 mM sucrose + 0.005 M $MgSO_4$, pH 6.0), and Roinel's group washes with several changes of lithium chloride/sucrose solutions (0.020 M sucrose + 0.150 M LiCl). The basis of all these methods is that while the washing medium removes the contaminative elements from the cell surface, the medium does not contribute elements that are major intracellular constituents to the adherent layer (Figure 12). Conveniently, also, the red cell membranes resist the free exchange of intracellular cations for cations in the washing medium (Whittam and Wheeler, 1970; Hoffman, 1966; Kimzey and Burns, 1973). Thus, in the special case of mammalian erythrocytes, the precise choice of washing medium is not critical. Whether washing media can be used to remove the surface precipitations from other fluid suspended cells, e.g., sperm cells (Chandler and Battersby, 1976a), has not been investigated.

Because of the simplicity and validation of the preparative technique [confirmed, for example, by the microprobe analysis of erythrocyte "standard" populations artificially loaded with known quantities of electrolytes after increasing membrane permeability with nystatin (Roinel et al., 1974; Lechene et al., 1976)], the red blood cell represents one of the few examples

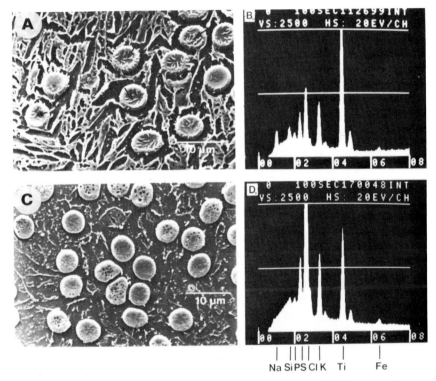

Figure 12.   Element composition of washed (serum-free) and unwashed human erythrocytes. The cells were smeared onto EM grids and (for present purposes) were freeze-dried in order to demonstrate visually the presence of serum on the unwashed cells. Similar chemical data were obtained from air-dried cells. A: Scanning image of unwashed erythrocyte smear; note the serum deposited on the individual cells (arrows). B: Spectrum from the analysis of a single unwashed cell; note the high sodium signal, derived primarily from the contaminating serum. C: Scanning image of an erythrocyte population that was previously washed in two changes of 155 mM choline chloride/Tris-HCl solution. D: Spectrum from the analysis of a single washed cell; note the presence of high chlorine signal derived primarily from adherent washing medium, and the absence of a detectable sodium signal. (Courtesy of H. B. Jones.)

where the microprobe has been used to study the fundamental heterogeneity of a population of cells and, furthermore, to monitor changes in the population distribution of chemicals during such biological events as development (Kirk and Lee, 1978) and aging (Lechene et al., 1976).

Contamination is an equally serious problem in the analysis of intracellular structures in air-dried tissue smears or in thick air-dried frozen sections where section thickness exceeds the diameter of the structure of interest. Here, the contaminants are elements suspended in the surrounding cytoplasm that deposit on the structure of interest during the slow withdrawal of water. For example, potassium is a relatively major constituent of chloragosome granules in air-dried 10-μm-thick cryostat sections (Figure 10), but a comparison of granules in ultrathin freeze-dried cryosections shows that much of this potassium is derived from cytoplasmic fluid. However, the Ca:Zn:P ratio of the granules is similar in both preparations. Therefore, the population distribution of these elements can be derived from air-dried preparations (Morgan, 1979). In general, the viability of air drying for a specific application should be determined by comparison with samples prepared by more rigorous procedures, e.g., freeze-drying (Skaer et al., 1976) and cryo-ultramicrotomy (Chandler and Battersby, 1976b—Table 9). It is pertinent to add that, from the strict morphological viewpoint, air-drying results in gross shrinkage and deformation artifacts in soft tissues (Boyde and Wood, 1969; Nermut 1977). These structural artifacts can be a blessing for the microprobe analyst, since the less hydrated components of the tissue are left standing proud of the collapsed adjacent cytoplasm. Mineralized structures, such as the granules in earthworm chloragocytes, are therefore easily localized and analyzed in the SEM (Figure 13). In addition, the shrinkage may reveal cell boundaries (Boyde and Wood, 1969), subsurface bodies, such as nuclei or mitochondria (Boyde et al., 1968, 1969), and the A, I, and F bands in striated muscle (Boyde and Williams, 1968).

Chemical Redistribution   In the case of fluid-suspended cells that are to be analyzed whole with a relatively dispersed electron beam, the problem to be avoided is that of solute movement across the cell membranes into the drying and, therefore, increasingly more concentrated external medium—the washing medium in the case of red blood cell populations. Lechene et al. (1976) found that the method of mounting red blood cells onto specimen supports greatly affects the quality of the x-ray information derived from samples. These authors either assessed or commented on four different mounting procedures.

Air-dried smears.   Errors arise due to the unevenness of the suspension medium layer across a given sample. Lechene et al. (1976) found that the mean x-ray signals from individual cells varied in different regions of the same support. They also located evidence of elemental loss from cells

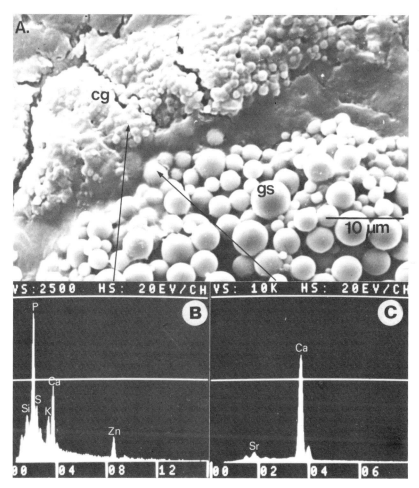

Figure 13.   Composition of chloragosome granules (cg) and calciferous gland spherules (gs) in *Lumbricus terrestris* 2 h after intra-coelomic injection of 0.1 ml 5% Sr Cl₂ solution. **A:** Scanning electron micrograph of an air-dried 10-μm thick cryosection through the calciferous gland. Note the extracellular spherules (gs) in the gland lumen, and the intracellular chloragosome granules (cg) in the chlorogogenous tissue enveloping the gland. **B:** Spectrum derived from the analysis of a single chlorogosome granule. Note the calcium, phosphorus, and zinc peaks; potassium and chlorine are probably cytoplasmic "contaminants." No strontium was detected. Sections used for microprobe analysis were mounted on 1-mm thick carbon discs and coated with carbon. Analysis was performed with a stationary beam (0.2 μm diameter) in a SEM. **C:** Spectrum derived from the analysis of a single gland spherule (cg) approximately 5 μm away from the granule represented in **B**. Note the fundamental difference in the mineral, and the presence of strontium.

in the form of small salt crystals containing sodium and potassium outside the cells. Not surprisingly, they noted that the lower the x-ray intensities recorded within cells, the higher they were in the adjacent external medium.

Sandwiching suspended cells between a beryllium support and a celloidin film (Roinel et al., 1974).   Lechene et al. (1976) did not assess this method, but suggested that its disadvantage is its relative complexity and the fact that the mounted cells are likely to be surrounded by large volumes of trapped suspension medium.

Coating of smeared cells with dibutyl phthalate to trap cellular ions (Kirk et al., 1974).   Lechene et al. (1976) noted low and variable mean x-ray intensities in these samples, probably attributable to x-ray absorption in the unevenly thick dibutyl phthalate coating.

Spraying suspension of cells in isotonic media onto (pyrolytic graphite) supports, followed by air drying.   This was the technique of choice, since it yielded the highest and most reproducible x-ray intensities. No sodium, potassium, sulfur, or iron leakage from the cells was detected. The success of this method in limiting ionic redistribution during drying rests simply on the dispersal of cells, so that each cell is often deposited on the support suspended in its own *small* fluid droplet.

In the case of cells (either fluid-suspended or in compact tissues) whose internal structures are to be analyzed, the problem of element redistribution is potentially more serious. If the elements are tightly bound, e.g., calcium and phosphorus in platelets (Skaer et al., 1976; Costa et al., 1977b) and earthworm chloragosome granules (Morgan, 1980), then the problem obviously does not arise. However, Costa et al. (1977b) warned that significant local compositional changes may occur within individual dense granules during drying, thus emphasizing once again that the resolution of analysis (not the resolution capabilities of the analytical instrument) determines the permitted preparative procedure.

Intracellular redistribution of exchangeable electrolytes has been systematically studied in various fluid-suspended cells by only two groups of workers. On the one hand, Gullasch and Kaufmann (1974) and Werner and Gullasch (1974) concluded from a fairly limited number of analyses that the electrolyte distribution patterns in air-dried erythrocytes, *T. pyriformis*, and rat spermatozoa failed to correspond with the patterns in identical cells prepared by cryoprocedures. In contrast, Chandler and Battersby (1976b), in a more extensive quantitative study of human spermatozoa, concluded that subcellular element compartments were identifiable in air-dried cells, and the element distribution patterns were not significantly different from those in ultra-thin freeze-dried frozen sections (Table 9). Chandler and Battersby (1976a) also claimed a good agreement between the subcellular calcium and zinc distribution patterns in air-dried and pyroantimonate treated spermatozoa, respectively.

Table 9. Comparison of relative element concentrations* in subcellular regions of human spermatozoa prepared by whole mounting/air drying and by cryo-ultramicrotomy**

	Head Air-dried	Head Cryo	Midpiece Air-dried	Midpiece Cryo	Acrosome Air-dried	Acrosome Cryo	Seminal Fluid Air-dried	Seminal Fluid Cryo
Na	35.7 (5.4)	28.6 (12.2)	47.2 (12.3)	39.2 (6.8)	38.0 (10.5)	43.5 (6.1)	0.33 (0.22)	8.5 (3.8)
P	66.2 (20.8)	86.0 (27.6)	58.4 (33.3)	71.1 (19.5)	47.8 (10.3)	73.8 (34.0)	0.42 (0.29)	27.2 (7.6)
S	10.6 (2.7)	12.5 (3.1)	10.2 (3.4)	11.3 (4.1)	9.9 (2.3)	12.4 (4.7)	0.69 (0.18)	7.9 (1.5)
Cl	17.0 (3.6)	18.1 (7.0)	13.7 (3.7)	12.1 (6.3)	14.6 (2.0)	17.9 (7.2)	0.53 (0.42)	10.0 (3.2)
K	28.6 (9.5)	36.8 (20.7)	25.3 (14.3)	36.8 (10.2)	22.7 (4.4)	36.4 (6.5)	0.78 (0.25)	22.1 (10.0)
Ca	8.8 (2.3)	5.6 (2.0)	18.5 (14.3)	11.6 (11.2)	9.9 (3.4)	7.5 (2.5)	0.32 (0.42)	10.0 (2.9)
Zn	5.6 (0.9)	5.6 (2.0)	7.4 (1.9)	4.7 (3.1)	5.7 (1.4)	8.7 (5.2)	0.13 (0.15)	3.8 (0.9)
Na/K	1.24 (0.26)	0.9 (0.49)	1.86 (1.14)	1.08 (0.22)	1.67 (0.10)	1.19 (0.15)	0.42 (0.28)	0.36 (0.10)

Table drawn from data of Chandler and Battersby, 1976b. Used by permission.

* The figures ($\pm$ SD) represent characteristic emission $\div$ white radiation ($\times 10^3$) and corrected for relative detection efficiency.

** The frozen sections were freeze-dried prior to analysis. NB: No statistical differences were noted ($p = 0.01$) for any of the element concentrations within a given subcellular region between samples prepared by the two methods.

We (Morgan et al., 1978) have previously concluded that the technique of tissue whole mounting followed by air drying is useful for: 1) maintaining the chemistry of whole cells, while being less suitable for the preservation of local intracellular ionic compartments, and 2) for preserving the gross chemical integrity of certain discrete, often highly mineralized intracellular structures. This practical perspective has not changed since then, and probably also applies to the simple modification of whole mounting/air-drying–"heat fixation" (Coleman et al., 1972, 1973a,b, 1974)—where fresh specimens are dehydrated rapidly by passage through a propane flame.

CRYOPROCEDURES

". . . the water in my flesh elongated to crystal slivers that would pierce and shatter the walls of my cells."

Annie Dillard, 1964, Pilgrim at Tinker Creek

Water forms an important and major component of the structural and physiological matrix of living cells. The water contains, in addition to molecular and macromolecular moieties, a complement of highly diffusible ions and electrolytes in a permanent state of dynamic flux. To maintain the chemical discontinuities of this fragile three-dimensional structure (which is synonymous with cytoplasm) through the trauma of removal from the living organism, subsequent processing and analytical examination are the tasks confronting the microprobe analyst studying physiological processes. The various techniques generically designated "cryoprocedures" offer the best possibility of achieving this ultimate aim. The problem that this target presents is highlighted by Marshall's (1972) estimate that 50% of the K^+ would diffuse out of a 1-nm-thick fully hydrated section in 10^{-4} s. This phenomenally high mobility of cellular ions can be impeded in one of two general ways (Figure 14).

1.   By freezing the tissue as rapidly as possible (at ~5–10 K/s—Gersh and Stephenson, 1954; Riehle, 1968; Bullivant, 1970; Christensen, 1971; Moor, 1971; Bank and Mazur, 1973; Appleton, 1974a; Costello and Corless, 1978; Nei, 1978), i.e., at a rate that exceeds the self-diffusion rate for water, so that there is minimal ionic movement (Fletcher, 1970; Moor, 1973) and ice-crystal formation. Water is then removed from the *frozen* specimen at temperatures below the re-crystallization temperature of ice either by substitution with organic fluids (*freeze-substitution*) or by exposure to a vacuum (*freeze-drying*).

2.   Alternatively, the frozen specimen is maintained in the fully *frozen-hydrated state* throughout all stages of pre-preparation, including sectioning, transfer, and examination in the microprobe analyzer.

Rapid freezing is, therefore, the extremely critical starting point of all cryoprocedures. Many techniques have been described for achieving the necessary fast cooling rates in fresh biological tissues that confine the growth of ice crystals so that they are too small to cause gross morphological and chemical damage. For example, small pieces of tissue have been quenched in various boiling fluids, including nitrogen, fluorocarbons, hydrocarbons (Rebhun, 1972; Somlyo et al., 1977), and helium (Sleytr and Umrath, 1974); by direct contact with polished silver or copper surfaces cooled by a boiling fluid (Eränkö, 1954; Van Harreveld and Crowell, 1964; Van Harreveld et al., 1965, 1974; Christensen, 1971; Dempsey and Bullivant, 1976a,b); and in liquid nitrogen slush (Umrath, 1974; Hodson and Williams, 1976; Appleton, 1978a; Barnard and Sevéus, 1978; Sevéus, 1978). It is important to stress that all of these techniques yield specimens in which there is a fairly discrete layer approximately 10–12 μm wide, near the tissue/coolant boundary, which is ostensibly free of ice-crystal damage (i.e., vitrified zones). A vitrified zone is the maximum depth of the well-preserved band obtainable by any quenching technique in those instances where the

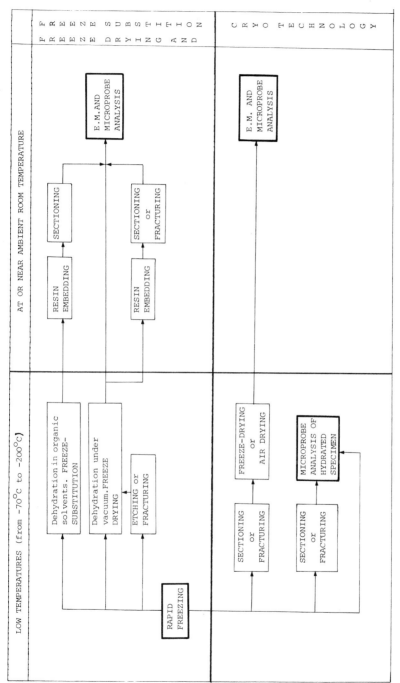

Figure 14.   A summary of the principal low-temperature specimen preparation techniques for biological specimens.

fresh tissue is not protected by cryoprotectants or frozen under high pressure, and is a consequence of the low thermal conductivity of ice (Dempsey and Bullivant, 1976a—see below). Analytical studies on samples subsequently processed by freeze-substitution, freeze-drying, or cryosectioning should be confined to this layer. A cautionary note should be added: while this layer is morphologically well preserved, the superficial tissue constituting it is the most likely to have suffered physical and dehydration damage during dissection and delays prior to freezing.

Novel freezing arrangements have been introduced by various groups of workers to satisfy personal requirements. Thin layers of cells in tissue culture (Asahina et al., 1970) or small droplets of yeast cells (Moor, 1964) can be readily frozen without apparent damage. Similarly, Bachmann and Schmitt (1971) and Plattner et al. (1973) achieved ice-crystal–free preparations of unprotected cells by reducing specimen size to approximately 10 μm diameter via spray freezing into liquid propane. Freezing between two liquid N_2-cooled metal hammers, one of which forms parts of a microtome chuck (Sjöström et al., 1974) and ballistic cryofixation (Monroe et al., 1968; Bullivant, 1970) permitted the arrest and study of muscle in predetermined physiological states. Such techniques are of only limited applicability. The various rapid specimen-immersion devices described by Luyet and Gonzales (1951), Rebhun (1965, 1972), Somlyo (1976), Somlyo et al. (1977), and Costello and Corless (1978) can significantly increase the freezing rate (Costello and Corless, 1978), but the advantages gained by reducing freezing damage may be lost by delays in loading the fresh samples into the immersion device (but see Somlyo et al., 1977). The exciting prospect of increasing freezing rates by high-pressure techniques (Riehle, 1968; Riehle and Hoechli, 1973; Bald and Robards, 1978), which allow the vitrification of small specimens due to a lowering of the freezing point of water and of the rates of nucleation and ice-crystal growth (Dempsey and Bullivant, 1976a), awaits further technological refinement. It should be stressed that in the context of maintaining the distribution of physiologically active components in biological tissues, the *simplicity* of the freezing method is almost as crucial as the rate of cooling the agent affords.

Preparations free of ice crystals can be produced either by partially removing tissue water before freezing or by freezing in the presence of cryoprotectants (Rebhun, 1972; Franks, 1977). It is unclear how cryoprotectant agents, such as glycerol, ethylene glycol, and DMSO, function; Rebhun (1972) and Nei (1978) suggest that the effectiveness of glycerol solutions stems from a combination of a reduced ability of water to form ice and an increased concentration of cell cytoplasm due to partial water removal by osmotic forces. Tissues frozen in the presence of high glycerol concentrations display considerable shrinkage (Bullivant, 1965; Pease, 1966a,b, 1967; Nei, 1978). Water movements are inevitably accompanied by ion and electrolyte drift. In addition to these little-documented influences on membrane permeability and on the distribution of freely diffusible substance (Echlin, 1975), evidence has recently emerged showing that these simple cryoprotectants introduce subtle but significant changes in micro-tubular (Rebhun, 1972) and intramembranous structure (Plattner et al., 1973; McIntyre et al., 1974).

Recently, dilute solutions of high molecular weight polymers, polyvinyl pyrrolidone (PVP), hydroxyethyl starch (HES), and dextran have been employed as cryoprotective agents (Echlin et al., 1977, 1978; Franks et al., 1977; Skaer et al., 1977) and were found to have little adverse physiological effects on a wide range of plant and animal tissues, and also to have good embedding and sectioning properties at low temperatures. Furthermore, the authors claimed that these polymers do not penetrate cells, do minimize ice-crystal damage, and, most importantly, do not cause ions to leak out of cells and tissues under conditions where they retain their effective cryo-protectant properties. Thus, while the use of low molecular weight cryopro-tectants cannot be widely recommended for analytical studies, the future usage of polymeric cryoprotectants warrants further serious consideration and practical assessment. It is noteworthy that Wilson et al. (1979) found, in a freeze-etch study of barley roots incubated in a 25% solution of PVP, that ice crystal formation was *not* significantly suppressed compared with untreated fresh roots. Furthermore, a rapid withdrawal of cellular water, with accompanying morphological changes were observed even during short incubation periods of 4–10 min. in the polymer solution.

Freeze-Substitution

In freeze-substitution, the sample is rapidly frozen and the ice is slowly replaced by organic solvent molecules at low temperature (Fernández-Morán, 1960, 1961; Rebhun, 1961, 1972; Bullivant, 1965; Pearse, 1968; Woolley, 1974). The dehydrated specimen is usually infiltrated with embed-ding medium at or near room temperature and sectioned for observation in the microscope. In theory, freeze-substitution offers a good compromise between good morphological detail and a minimal loss and translocation of endogenous tissue elements during processing.

A number of organic solvents have been used to dissolve ice in frozen biological specimens (Table 10). Generally, small pieces of tissue have to be substituted for periods of several days at temperatures of about $-70°C$ to $-80°C$ for complete dehydration (Rebhun, 1965; Van Harreveld et al., 1965). The substitution period depends critically on sample size, on the organic solvent employed, and on the temperature of substitution. For example, extraction studies involving Sudan Black B-loaded agar blocks (Fisher and Housley, 1972; Harvey et al., 1976b) as crude tissue models (c.f.

Table 10.   Some commonly used freeze-substitution organic media

Fluid	Tissue origin	Experimental application		Reference
		Morphology	Microprobe analysis	
Acetone	A	✓		Malhotra and Van Harreveld (1965)
	A	✓		Rebhun (1965)
	A	✓		Rebhun and Sander (1971)
	B	✓		Fisher (1972)
	B	✓**		Fisher and Houseley (1972)
	B	✓		Hereward and Northcote (1972)
	A	✓		Kuhn (1972)
	B	✓	✓	DeFilippis and Pallaghy (1973)
	B	✓	✓	Pallaghy (1973)
	B	✓**		Steinbiss and Schmitz (1973)
Diethyl ether	B	✓**		Luttge and Weigl (1965)
	B	✓	✓	Läuchli et al. (1970)
	B	✓	✓	Läuchli et al. (1971)
	B	✓	✓	Spurr (1972a,b)
	B	✓**		Steinbiss and Schmitz (1973)
	A&B	✓	✓	Mehard and Volcani (1975)
	A	✓	✓	Mehard and Volcani (1976)
Ethanol	A	✓		Bullivant (1965)
	B	✓**		Neeracher (1966)
	A	✓		Van Harreveld and Steiner (1970)
	A	✓		Nath (1972)
Ethylene glycol	A	✓		Pease (1967a)
Propyelene glycol	A	✓		Woolley (1974)
Methanol	B	✓**		Fisher and Houseley (1972)
Propylene oxide	B	✓**		Fisher and Houseley (1972)
n-Hexane	B	✓		Neumann (1973)
	B		✓	Neumann et al. (1974)
Tetrahydrofuran	A	✓		Rebhun and Sander (1971)
Pure acrolein	A	✓		Afzelius (1962)
	A&B	✓	✓	Van Zyl et al. (1976)
Methanol/acrolein mixtures	A	✓		Zalokar (1966)
Acetone/acrolein mixtures	B	✓**		Steinbiss and Schmitz (1973)
	B	✓	✓	DeFilippis and Pallaghy (1975)
	A&B	✓	✓	Van Zyl et al. (1976)
Glycerol/water mixtures	A	✓		Fernández-Morán (1957)

A = animal tissue. B = botanical tissue.
** Including autoradiographic localization of water-soluble ions or molecules.

Zalokar, 1966; Woolley, 1974) have indicated that acetone probably sub-
stitutes ice at a much faster rate than either propylene oxide or diethyl
ether.

Element Loss During Substitution From the viewpoint of x-ray microprobe analysis, the choice of substitution medium rests solely on its
properties of element retention; morphological considerations are obviously
important, but they must remain secondary to the maintenance of chemical
integrity. A few workers have attempted to qualitatively or semi-qualita-
tively assess the loss of selected elements from botanical tissues by microprobe analysis or by autoradiography. Gielink et al. (1966), in an autoradiographic study, claimed that the loss of water-soluble and exchangeable calcium could be prevented by substitution in acetone at $-80°C$. Läuchli et al.
(1970) reported minimal losses of introduced ^{86}Rb from corn roots during
anhydrous-ether substitution followed by infiltration with Spurr's resin. In
contrast, Spurr (1972a) found that most of the sodium and calcium content
of the cell walls and cytoplasm of tomato leaves (measured in EMMA 4)
was lost during diethyl ether substitution at $-80°C$ for 8 days, followed by
Spurr's resin infiltration. However, these losses were averted by the addition
of benzamine to the substitution medium; other additives (picrolonic acid,
2,6-dinitrophenol, alizarin, oxalic acid, cyanuric acid) were considerably less
effective.

Four groups of authors have quantitatively and systematically
measured the loss of endogenous elements and/or introduced radioisotopes
from both animal (Ostrowski et al., 1962a,b; Mehard and Volcani, 1975;
Van Zyl et al., 1976) and plant tissues (DeFilippis and Pallaghy, 1975;
Harvey et al., 1976b; Van Zyl et al., 1976). In addition, Van Zyl et al.
(1976) measured the loss of [22]Na radioisotope from agar tissue models. The
results of these various investigations are summarized in Table 11, and are
discussed in detail elsewhere (Morgan et al., 1978). However, two significant conclusions may be drawn from the results of these important studies.

1. The loss of even the labile electrolytes can be limited to $<5\%$ by
the rigorous maintenance of anhydrous conditions during all stages of substitution and embedding (see also Pallaghy, 1973). Anhydrous solvents can
be produced by the addition of aluminum oxide (Läuchli et al., 1970) or
activated molecular sieve (Pallaghy, 1973; Fisher, 1972) well before and
while the solvent is used in the preparative regime. Moreover, shrinkage
artifacts that may arise during substitution are almost completely
eliminated in plant tissues under anhydrous conditions (Fisher, 1972).

2. Unquestionably, the best substitution fluid for the retention of the
more labile elements is diethyl ether, with acetone representing a good
second choice. A minor disadvantage of diethyl ether compared with
acetone is the considerably prolonged substitution period required (Harvey
et al., 1976b).

Table 11. Summary of published data* on the effect of various freeze-substitution solvents on the total retention of elements in plant and animal tissues, determined by either flame photometry or scintillation counting

Endogenous element	Tissue	Freeze-substitution medium	Percent retention	Reference
P	Rat liver	methanol, −79°C, 6–9 days	61.3	Ostrowski et al. (1962a)
		ethanol, −79°C, 6–9 days	79.7	
		butanol, −79°C, 6–9 days	83.0	
		propanol, −79°C, 6–9 days	83.7	
		acetone, −79°C, 6–9 days	98.4	
		acetone, −79°C, 3 days	97.5	Ostrowski et al. (1962b)
		acetone, −20°C, 3 days	98.5	
		acetone, +4°C, 3 days	94.1	
		methanol, −79°C, 3 days	60.6	
		methanol, −20°C, 3 days	46.1	
		methanol, +4°C, 3 days	44.8	
N	Rat liver	methanol, −79°C, 6–9 days	95.5	Ostrowski et al. (1962a)
		ethanol, −79°C, 6–9 days	97.2	
		butanol, −79°C, 6–9 days	97.6	
		propanol, −79°C, 6–9 days	99.4	
		acetone, −79°C, 6–9 days	100.0	
		acetone, −79°C, 3 days	100.9	Ostrowski et al. (1962b)
		acetone, −20°C, 3 days	101.2	
		acetone, +4°C, 3 days	100.3	
		methanol, −79°C, 3 days	96.7	
		methanol, −20°C, 3 days	96.5	
		methanol, +4°C, 3 days	95.7	

Isotope	Specimen	Treatment	%	Reference
68Ge**	Diatom	dry ether, −80°C, 3 days	100	Mehard and Volcani (1975)
	† Liver, spleen, kidney, lung	dry ether, −80°C, 3 days	100	
	† Isolated liver, nuclei, microsomes	dry ether, −80°C, 3 days	100	
65Zn	Agar block	dry ether, −78°C, 3–5 days	100	DeFilippis and Pallaghy (1975)
	Carrot root	dry ether, −78°C, 3–5 days	100	
	Carrot root	dry 20% acrolein in ether, −78°C, 3–5 days	100	
203Hg	Agar block	dry ether, −78°C, 3–5 days	98.4	
	Carrot root	dry ether, −78°C, 3–5 days	96.1	
	Carrot root	dry 20% acrolein in ether, −78°C, 3–5 days	95.1	
	Barley root	dry 20% acrolein in ether, −78°C, 3–5 days	95.9	Van Zyl et al. (1976)
22Na	Barley root	dry methanol, −78°C, 6 days	1	
	Cockroach nerve	dry methanol, −78°C, 6 days	7	
	Agar block	dry methanol, −78°C, 6 days	0	
	Barley root	dry acetone, −78°C, 6 days	75	
	Cockroach nerve	dry acetone, −78°C, 6 days	61	
	Agar block	dry acetone, −78°C, 6 days	12	
	Barley root	dry ethanol, −78°C, 6 days	10	
	Cockroach nerve	dry ethanol, −78°C, 6 days	23	
	Agar block	dry ethanol, −78°C, 6 days	2	
	Barley root	dry acrolein, −78°C, 6 days	70	Van Zyl et al. (1976)
	Cockroach nerve	dry acrolein, −78°C, 6 days	36	
	Agar block	dry acrolein, −78°C, 6 days	2	
	Barley root	dry ether, −78°C, 6 days	99	
	Cockroach nerve	dry ether, −78°C, 6 days	99	
	Agar block	dry ether, −78°C, 6 days	100	
	Barley root	dry 20% acrolein, −78°C, 6 days	99.5	
	Cockroach nerve	dry 20% acrolein, −78°C, 6 days	97.5	

Table 11.   Continued

Endogenous element	Tissue	Freeze-substitution medium	Percent retention	Reference
^{36}Cl	Barley root	dry 20% acrolein in ether, $-78°C$, 6 days	96	Harvey et al. (1976b)
	Cockroach nerve	dry 20% acrolein in ether, $-78°C$, 6 days	94	
^{86}Rb	Barley root	dry 20% acrolein in ether, $-78°C$, 6 days	101	
	Cockroach nerve	dry 20% acrolein in ether, $-78°C$, 6 days	104	
Na	Leaf tissue	dry acetone, $-72°C$ (1 day), $-40°C$ (2 days)	98	
	Leaf tissue	dry ether, $-72°C$ (1 day), $-40°C$ (20 days)	>99	
^{22}Na	Leaf tissue	dry acetone + Spurr resin embedding dry ether + Spurr resin embedding	96.85 99.43	
K + Rb	Leaf tissue	dry acetone, $-72°C$ (1 day), $-40°C$ (2 days) dry ether, $-72°C$ (1 day), $-40°C$ (20 days)	<98.5 <98.5	
$^{36}Cl**$	Leaf tissue	dry acetone + Spurr resin embedding dry ether + Spurr resin embedding	95.18 99.25	

* Based on data from Morgan et al. (1978)

** Radioisotope loss was monitored during the various stages of freezing, substitution, and resin infiltration, and, in the case of ^{22}Na and ^{76}Cl loss, during section flotation (Harvey et al., 1976b)—see table.

† These are mean values derived from the data on individual tissues and organelle fractions (Mehard and Volcani, 1975).

Effects of Adding Fixatives to Substitution Media   The inclusion of osmium tetroxide to either acetone or ethanol does not significantly improve the ultrastructural appearance of sea urchin eggs (Rebhun, 1972) or rat spermatozoa (Woolley, 1974). In fact, osmium does not interact with tissues during substitution; osmium oxidation occurs post-substitution at temperatures above about $-25°C$ (Van Harreveld et al., 1965; Hereward and Northcote, 1972; Woolley, 1974). Furthermore, the addition of 1% OsO_4 did not improve the ion retention properties of methanol, acetone, ethanol, acrolein, or diethyl ether; while recorded losses (of ^{86}Rb) during n-hexane substitution were exacerbated by the inclusion of 1% OsO_4 (Van Zyl et al., 1976).

Woolley (1974) demonstrated that chemical fixation by 5% glutaraldehyde in ethylene glycol is perceptible at temperatures at least as low as $-20°C$ and that fixation probably occurs, albeit very slowly, at $-50°C$. Contrary to the suggestion of Pease (1967a), Woolley (1974) found no evidence of fixation by 5% formaldehyde in ethylene glycol even at $-20°C$. The influence of aldehyde inclusion on the retention of tissue elements during freeze-substitution has not been investigated.

A significant improvement in the ultrastructural appearance of ether-substituted tissues, paralleled by a minimal loss of chemical advantages, is provided by the inclusion of 20% or $<20\%$ acrolein in dry ether (DeFilippis and Pallaghy, 1975; Van Zyl et al., 1976—see Table 11). The cellular detail in acrolein/ether substituted specimens is comparable with that seen in conventionally fixed, embedded, etc., material (Figure 15), while the recorded losses of ^{86}Rb, [22]Na, and [31]Cl are often less than 6% (Van Zyl et al., 1976).

Redistribution of Elements   The possibility of translocation of mobile tissue components is a frequent criticism of all preparative techniques involving exposure to fluids, even where minimal percentage losses of these components have been recorded (Echlin and Moreton, 1974; Panessa and Russ, 1975). Unfortunately, this criticism cannot be disregarded even in the case of specimens processed by the optimalized freeze-substitution procedures mentioned above. Nevertheless, there exists a small body of evidence, much of it based on fairly preliminary observations, to suggest that wholesale ion redistribution does not occur during these preparative regimes.

1.   As mentioned earlier, ion-loaded agar blocks are poor tissue models; by their very nature, these models would be expected to be considerably more susceptible to pervasive leaching than compact biological tissues. However, evidence gained from ion-retention studies in agar blocks suggests that ion fluxes are not extensive during freeze-substitution in either ether or 20% acrolein/ether under anhydrous conditions (DeFilippis and Pallaghy, 1976; Van Zyl et al., 1976).

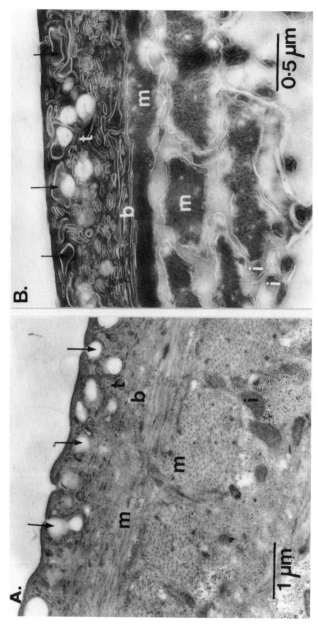

Figure 15. A comparison of ultrastructural preservation of the tegument and underlying musculature of the blood parasite, *Schistosoma mansoni*, in freeze-substituted and conventionally fixed specimens. **A:** Electron micrograph of a glutaraldehyde-fixed and osmium post-fixed specimen. **B:** Electron micrograph of a specimen prepared by freeze-substitution at −80°C in a 20% acrolein-ether mixture. Both sections were stained in uranyl acetate and lead citrate. t = tegument; b = basement membrane; m = muscle; i = mitochondria. The same gross structural features are to be seen in both micrographs. Note that the tegumental surface folds (arrows) are full of trapped host-serum material in the freeze-substituted specimen. Material has been extracted from the folds during the conventional preparative regime. (Courtesy of Tudor W. Davies.)

2. Microprobe studies of freeze-substituted plant and animal tissues (Läuchli et al., 1970; Spurr, 1972a,b; Pallaghy, 1973; DeFilippis and Pallaghy, 1975; Forrest and Marshall, 1976; Mehard and Volcani, 1975; Van Zyl et al., 1976) indicate that ions are heterogeneously distributed at a cellular and even at an intracellular level.

3. The overall loss of soluble ions from *intact* cells during freeze-substitution may be even less than published data suggest (Table 11). A significant proportion of the small recorded loss may conceivably arise from damaged cells at the cut edges of the specimen (Harvey et al., 1976b). An outstanding advantage of the good morphological qualities of freeze-substituted tissues (compared with cryosectioned tissues, for example) is that it becomes possible to avoid performing probe analyses on these grossly damaged superficial cells.

4. Forrest and Marshall (1976), in a study of the sodium-transporting cells of brine-fly larvae, showed that energy-dispersive x-ray spectra obtained from freeze-substituted and frozen-hydrated sections were generally similar, thus indicating that gross element redistribution does not occur, at least over distances exceeding cellular dimensions. In a less satisfactory comparative study, these same authors inferred that the intra-axoplasmic element ratios (K:P:S:Cl) in freeze-substituted and frozen-hydrated cockroach nerve cord displayed similar quantitative variations.

Conclusions   In the particular case of freeze-substitution, it is possible to outline certain firm conditions under which the chemical integrity of biological specimens is best approximated:

1. Rapid freezing (see above) is preferable to immersion in the organic medium at the substitution temperature of $-70°C$ to $-80°C$.
2. Diethyl ether is the best-known substitution medium. The addition of 1–20% acrolein in ether improves the ultrastructural appearance of the tissue without seriously affecting its chemical composition.
3. *All* processing fluids must be kept completely anhydrous.
4. Substituted tissues are best infiltrated with low-viscosity resin. Infiltration and embedding theoretically represent the weakest stages of the regime, since the substituted tissues usually need to be brought up to room temperature while still in contact with the organic substituting solvent. Infiltration at lower temperatures (Harvey et al., 1976b; Edelmann, 1978a) or sectioning without embedding in a cryo-ultramicrotome (suggested by Morgan et al., 1978) may represent useful alternatives.

Consequently, it is possible to present a fairly specific freeze-substitution regime for microprobe analysis applications that incorporates all of these features (Table 12).

Table 12.   A comprehensive freeze-substitution regime* that (theoretically) affords optimal ion retention and minimal redistribution

Process	Methodology
1. Rapid freezing	Various available methods (see text), e.g., N_2 slush, pre-cooled metal surfaces, liquid propane cooled with liquid N_2, Freon-22.
2. Freeze substitution	Rapid transfer of small specimens to *diethyl ether*** in stoppered tubes. The tubes contain 1 part freshly activated molecular sieve to 2 parts anhydrous solvent: Substitution over 1 day at −70 to −80°C ↓ Substitution over 20 days between −80°C and −40°C
3. Resin infiltration	Infiltration in a dry atmosphere with low viscosity resin [Spurr (1969), or Pallaghy's (1973) modification of Spurr's resin]: 50% diethyl ether:50% ERL 4206 (vinylcyclohexene dioxide); −40°C, 4 h ↓ 100% ERL 4206; −40°C, 16 h ↓ 100% ERL 4206; −40°C to −15°C, 3 h ↓ 50% ERL 4206:50% complete medium; −15°C, 6 h ↓ Complete medium ⎰ ERL 4206—10 g / Dibutyl phthalate—12 g / Nonenyl succinate anhydride—26 g / DY 064—1.6 g
4. Embedding and storage	Samples embedded in anhydrous resin (kept in sealed capsules at 60°C for 18 h). Tissue blocks stored over silica gel in a dessicator.
5. Sectioning	Sectioning on a dry knife, i.e., no flotation on a water bath. Staining also to be avoided in the case of tissues to be analyzed.

* This regime based on those presented by Pallaghy (1973), Harvey et al. (1976b), and Van Zyl et al. (1976). Resin after Pallaghy (1973).

** Alternatively, 1–20% acrolein in ether. All solvents, including resin mixtures, dried for approximately 48 h with activated molecular sieve (activated by heating under a vacuum) prior to usage using a 1:5 molecular sieve:solvent ratio.

Freeze-Drying

The attraction of the "physical" dehydration of cryo-fixed tissues in the context of microprobe analysis of biological specimens is that aqueous and/or inorganic solvents are not necessarily involved, and hence the possibility of closely approximating tissue chemical integrity is a realistic possibility. Furthermore, the technology of freeze-drying is uninvolved and well established, but most importantly can conveniently be applied to prepare either (1) resin-embedded or unembedded bulk specimens, (2) thick or thin sections of embedded specimens, or (3) thick or thin cryosections for visualization and analysis in TEM, SEM, or STEM modes, where appropriate.

Excellent reviews of the use of freeze-drying for the preparation of biological tissues specifically for morphological examination have been presented by Sjöstrand (1967), Pearse (1968), Boyde and Wood (1969), Burstone (1969), Rebhun (1972), Nermut (1977), and Umrath (1977).

Theory and Instrumentation   The rate of sublimation of ice is a function of the vapor pressure of ice at its drying temperature, and of the partial pressure of water vapor above the ice (Boyde and Wood, 1969). The combination of specimen temperature and the necessary drying time is, therefore, a very important practical parameter.

In theory, drying temperatures of around $-120°C$ should minimize tissue damage, but in practice drying temperatures between $-60°C$ and $-100°C$ (i.e., temperatures below the recrystallization point of ice, or of the lowest eutectic point of mineral salts in the tissue) are more usually used (Ingram et al., 1972, 1974; Ingram and Ingram, 1975; Skaer et al., 1974; Sjöstrand and Kretzer, 1975; Höhling et al., 1976; Zs.-Nagy et al., 1977; Pieri et al., 1977; Frederik and Klepper, 1976; Barkhaus and Höhling, 1978; Edelmann, 1978b). The importance of drying at temperatures below about $-60°C$ was emphasized by Ingram et al. (1972), who claimed that the morphological damage incurred at these low temperatures was only discernible at EM level and, most importantly, that intracellular-extracellular ion boundaries were maintained.

The period required to completely remove free tissue water is not easily established, and consequently the user has a fairly wide choice (Rebhun, 1972). Obviously, the dehydration period depends on the chosen temperature and on the size and nature of the specimen. It is useful, therefore, to quote some of the pertinent drying parameters from selected recent publications (see Table 13). By comparison, the freeze-drying regimes adopted by earlier workers (cited by Rebhun, 1972) generally included considerably longer drying periods, often of several days.

Most tissue freeze-drying units have three characteristic features: 1) the ability to produce and maintain a vacuum of 10^{-3} torr or better; 2) a specimen-support platform whose temperature can be lowered to the

Table 13.   Examples of the conditions used by some recent authors for the freeze-drying of bulk specimens

Tissue	Vacuum pressure	Specimen temperature	Drying period	Authors
Rat testis ("small pieces")	10^{-5} torr	$-85°C$	15 h at $-85°C$ + unspecified period over which specimen temperature was raised to $+30°C$	Frederik and Klepper (1974)
Epiphyseal plate ("small pieces")	*	$-80°C$	3 days	Barkhaus and Höhling (1977)
Rat liver and brain cortex	$1-3 \times 10^{-5}$ torr	$-100°C$ to $-120°C$	4 h at $-100/-120°C$ + 90–100 min during which specimen temperature was raised to room temperature	Zs.-Nagy et al. (1977)
Fungal hyphae (7–10 μm diameter)	10^{-4} torr	$-100°C$	8–16 h	Roomans and Boekestein (1978)

* Not quoted by the authors.

required drying temperature, and also raised to chosen temperatures between the drying temperature and about $+60°C$ to complete the drying process and/or to facilitate resin infiltration and polymerization; (3) a system whereby vapor fixatives and/or resins can be introduced into the main drying chamber without incurring vacuum loss. Freeze-driers possessing these features are commercially available; others possessing all or some of these features and designed to meet personal requirements have also been described (e.g., Frederik and Klepper, 1974; Sjöstrand and Kretzer, 1975; Zs.-Nagy et al., 1977; Geymayer et al., 1978; Edelmann, 1978b). The device described by Edelmann (1978b) is particularly deserving of comment because, unlike the others, it relies exclusively on cryosorption pumping produced by cooled molecular sieve (final pressure about 10^{-3} torr) and not on a vacuum generated via a diffusion-pumped system. The compactness this affords makes it possible to permanently incorporate this device inside the freezing compartment of certain cryomicrotomes, thus eliminating the need for special additional devices for transferring frozen sections from the cryomicrotome to a separate pre-cooled freeze-drying chamber.

Edelmann's morphological and histochemical findings (Edelmann, 1978a,b) confirm that there is no advantage in using vacuum pressures lower than about 10^{-3} torr, since the rate of water loss from the specimen is more dependent on the rate of water transport across the already dried surface layers of the specimen than on the speed of water transport from the specimen surface to the pump or cold trap (Sjöstrand, 1967; Malhotra, 1968; Boyde and Wood, 1969; Ingram and Ingram, 1975; Rebhun, 1972).

Vapor Fixation of Freeze-Dried Specimens   Although the process of freeze-drying is often considered to be non-destructive (Meryman, 1966), several authors have distinguished fairly gross morphological damage in various freeze-dried tissues (Bell, 1956; Mackenzie, 1965; Pearse, 1968; Burstone, 1969). Volume shrinkage, for example, is a frequently encountered artifact in embedded specimens (Miyamoto and Moll, 1971; Hanzon and Hermodsson, 1960; Ingram et al., 1974). Since Hanzon and Hermodsson (1960) demonstrated that volume shrinkage could be reduced by fixing the dried tissues in osmium before embedding, several authors have fixed their dried specimens at atmospheric or reduced pressures with osmium vapor (Ingram et al., 1974; Frederik and Klepper, 1974; Malhotra, 1968; Bondareff, 1967) or formaldehyde vapor (Bondareff, 1967), with resultant improvement in morphological detail.

However, osmium interferes with analyses and considerably increases the background signal in energy-dispersive spectra, so that in general it is advisable to avoid osmication except for morphological surveys. Moreover, Somlyo et al. (1977) found that calcium was not detectable in cardiac mitochondria after osmium vapor fixation of freeze-dried cryosections, thus indicating that osmium can displace ions under certain conditions.

Applications   As mentioned above, several different types of specimen (i.e., combinations of thick or thin, embedded or unembedded) may be prepared for EM visualization and microprobe analyses by cryofixation, followed by freeze-drying. Each type of specimen presents its own preparative and analytical problems.

Bulk specimens.   Bulk specimens are, by definition, too thick to allow the transmission of electrons. Specimens are usually freeze-dried after freeze fracturing and subsequently analyzed (Ryder and Bowen, 1977; Zs.-Nagy et al., 1977; Pieri et al., 1977), or the dried specimens are infiltrated with resin and the polymerized block is fractured prior to analysis (Ingram et al., 1974; Ingram and Ingram 1975). Alternatively certain amenable specimens may be freeze-dried and analyzed without recourse to either freeze-fracture or embedding, e.g., insect salivary gland chromosomes (Trösch and Lindemann, 1975) and fungal hyphae (Roomans and Boekestein, 1978).

The preparation of bulk samples is simple, and gross ion displacement is not a major limiting factor, especially in the case of unembedded specimens, e.g., the average concentration of Na^+ and K^+ in freeze-dried fungal hyphae, measured by quantitative microprobe analysis, compared favorably with values determined by flame photometry (Roomans and Boekestein, 1978). However, the microprobe analysis of bulk specimens is fraught with major difficulties:

1. Since the specimens are infinitely thick, the electron beam will penetrate to a considerable depth and will diffuse in pear-drop fashion, so that x-rays are generated from a volume considerably larger than the surface feature upon which the electron beam was focused. At an accelerating voltage of only 10 kV, the beam penetrates freeze-dried liver sections to a depth of at least 6 μm, with most of the x-ray information originating from the upper 4 μm; higher accelerating voltages will obviously increase the depth of beam penetration (Zs.-Nagy et al., 1977). Thus, even at 10 kV, only the large cellular compartments, such as nucleus and cytoplasm, can safely be analyzed–even when the surface morphology permits the identification of finer cellular detail (Zs.-Nagy et al., 1977).

2. The weak x-rays emitted by the lighter elements will tend to be absorbed by the tissue volume through which they have to travel to reach the x-ray detector. In addition to internal x-ray absorption, atomic number effects and secondary fluorescence are parameters that have to be considered in the case of electron-opaque specimens, but whose significance is negligible in thin specimens (for definitions and detailed discussion, see Russ, 1978). Various authors have performed quantitative analyses on freeze-dried bulk specimens and have compensated for these three physical parameters in one of three ways: (1) by the so-called ZAF correction expression in conjunction with organic standards (Roomans and Boekestein 1978), a method commonly used with metallurgical and mineralogical thick specimens; (2) by the conversion of measured count rates in the unknown sample to concentration units, through direct comparison with the slope of curves derived from characterized organic standards, which in turn were prepared and analyzed in a fashion similar to that used for the tissue sample (Ingram et al., 1972, 1974, 1975); (3) by the mass-fraction method of Hall in conjunction with crystalline standards (Hall et al., 1973), a method originally introduced for the quantification of thin specimens where atomic number, absorption, and fluorescence corrections (ZAF) can be neglected (Zs.-Nagy et al., 1977; Pieri et al., 1977). Not surprisingly, the protagonists claim good accuracy for their own methods of quantification. For example, Ingram et al. (1974) claimed an accuracy of 90–95% in their estimates of electrolyte concentrations in various vertebrate soft tissues. The method adopted by Zs.-Nagy's group (Zs.-Nagy et al., 1977; Pieri et al., 1977) was only validated by the analysis of simple crystalline standards, and to this extent its efficacy for bulk biological specimens must be considered equivocal (cf. Roomans and Boekestein, 1978). Nevertheless, these authors were able to describe profound age-dependent changes in the electrolyte content of the nuclei and cytoplasm of rat liver and brain cells (Figure 16), which proved amenable to biochemical interpretation and have subsequently gained support from independent experimental approaches (Pieri et al., 1977).

Thin sections of plastic-embedded specimens. The analysis of bulk samples is primarily undertaken so that statistical fluctuations resulting from low count rates emanating from soft biological tissues are reduced, but this inevitably results in a sacrifice of structural and analytical resolution, with attendant quantitative problems (see above). Considerably higher resolution analysis (approximately equal to section thickness) and an immediate simplification of quantitative expressions can be achieved by preparing thin specimens (Russ, 1978). Thin sections of soft materials can be prepared only by investing the specimens with sufficient physical rigidity to withstand the mechanical stresses of sectioning. In the context of freeze-drying technology, this is achieved in one of two fundamentally different ways: (1) by quench-freezing and cryo-sectioning of the frozen block (see below); (2) by infiltrating the previously dried specimen with polymerizable resin. The latter technique has been used by a fairly limited number of authors, who subsequently localized various water-soluble tissue components by microprobe or histochemical or autoradiographic techniques (Skaer et al., 1974; Höhling and Nicholson, 1975; Morgan et al., 1975; Frederik and Klepper, 1976; Edelmann, 1978a,b; Barkhaus and Höhling, 1978).

Apart from facilitating the preparation of thin sections, the plastic matrix is useful because (1) the sections can be stained and morphologically studied by conventional microscopy in correlative studies; and (2) the embedded specimen, which is about 80% plastic, is more homogeneous in composition and probably more stable under the electron beam (cf. Hall and Gupta, 1974). However, over all, the chemical integrity of this type of preparation is potentially at risk at three stages:

1. First, and in common with all cryoprocedures, there are those changes that may occur before and during freezing (see above).
2. Second, there are changes that undoubtedly occur during drying that at least approach a level of significance in high-resolution analytical studies on thin specimens. The withdrawal of water from biological tissues inevitably leads to a shift in the distribution of chemicals suspended in that water. These chemicals probably do not move beyond the nearest membrane or similar structural boundary. Consequently, the analysis of ions normally suspended in extracellular fluid is precluded after freeze-drying, since these components will have presumably deposited on the outer surface of adjacent cells.
3. Finally, resin infiltration and sectioning are potential sources of error. Elements may be lost and redistributed either by solubilization in the

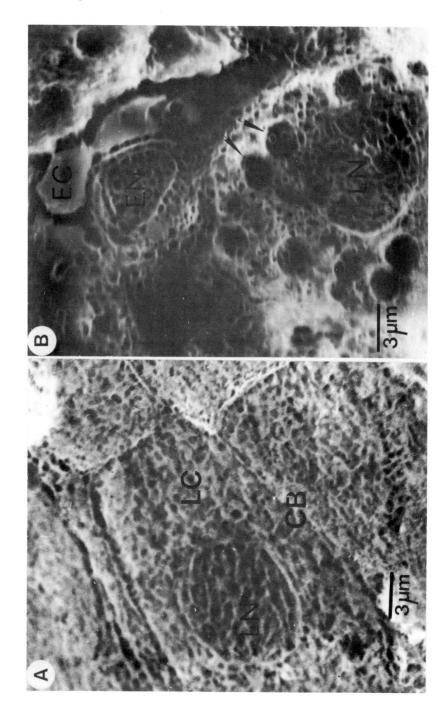

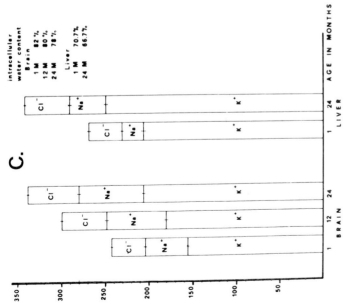

Figure 16. X-ray microanalysis of electrolytes in freeze-dried/freeze-fractured bulk specimens. **A:** Low-power SEM micrograph of the fractured surface of old rat liver. LN = liver cell nucleus; LC = liver cell cytoplasm; CB = cell border. **B:** Low-power SEM micrograph of the fractured surface of a young rat liver. LN = liver cell nucleus; EN = endothelial cell nucleus; EC = erythrocyte; arrows indicate holes, probably corresponding to the sites of lipid droplets. **C:** Histogram of total monovalent electrolyte contents (measured by microprobe analysis of freeze-dried bulk specimens) in the nuclei of large brain cortical cells and hepatocytes at different ages. Water content data are estimated, assuming a water loss of 4% between 1 and 24 months of age. (**A** and **B** from Zs.-Nagy et al., 1977. **C** from Pieri et al., 1977. Used by permission.)

resin or by the surface tension forces at the advancing resin front. However, Ingram et al. (1974, 1975) found no evidence of electrolyte leaching or redistribution in mouse kidney cells, frog red blood cells, frog skin epithelial cells, adenopodia, and mud-puppy retinal cells, all of which were infiltrated with Epon 826 after freeze-drying. In general, the chemical artifacts introduced by floating sections on a water bath are sufficiently deleterious (see above—Tables 2 and 3) to warrant the routine cutting of sections on dry knives, together with its attendant manipulative problems. As an exception, in the case of some highly mineralized structures, brief water flotation may be permissible (Skaer et al., 1974; Barkhaus and Höhling, 1978), but exploratory comparative studies of "floated" and "dry" sections are important prerequisites (cf. Skaer et al., 1974).

Thick or thin cryosections. Since the preparation of these specimens involves a very specialized and widely used technology, a consideration of this methodology is reserved for a separate section (see below).

Conclusion Although, historically, freeze-drying as a technology dates back to the turn of the century (Altmann, 1889 and 1890, q.v. Pearse, 1968—see Pearse, 1968, for a good outline of the historical development of freeze-drying technology), and is in wide contemporary usage for the preparation of specimens for structural examination in SEM, it is not widely used (except as an adjunct to cryosectioning) for the preparation of embedded and unembedded biological tissues for microprobe analysis. Theoretically, freeze-drying could be expected to maintain the chemical integrity of tissues at least as well as freeze-substitution, and yet there is a current upsurge of interest in the latter technique. Perhaps the lack of interest is born of the comparatively poor structural qualities of freeze-dried tissues (Figure 16). Interestingly, therefore, Edelmann (1978b) listed the following steps, which seem to be important for good morphological preservation (in sectioned, embedded muscle); the same parameters may reasonably be expected to optimalize the preservation of tissue chemical integrity by the freeze-drying method:

1. Very fast freezing.
2. Freeze-drying exclusively below a critical but indeterminate temperature (probably at least as low as −60°C) and with a relatively weak vacuum (cf. Appleton, 1974a).
3. Embedding (in Spurr's resin) at a low temperature (see also Müller, 1957; Sjöstrand and Kretzer, 1975), e.g., embedding at −15°C, yielded better morphological preservation in muscle than in specimens embedded at +20°C.
4. Polymerization at temperatures lower than those commonly employed, e.g., morphological preservation, was better when the resin was polymerized at +40°C compared with +70°C. Polymerization at

considerably lower temperatures [e.g., of methacrylate at -27°C (Sjöstrand and Kretzer, 1975)] may confer further morphological *and* chemical advantages.

CRYO-ULTRAMICROTOMY

It is not the purpose of this discussion to review the many detailed technical aspects of the available methods for cutting thin sections of frozen biological materials. Specific preparative sequences have been presented by many previous authors for the sectioning of both 1) *pre-fixed specimens* (see Bernhard and Leduc, 1967; Bernhard and Viron, 1971; Christensen, 1969; Iglesias et al., 1971; Tokuyasu, 1973; Doty et al., 1974; Howell and Tyhurst, 1976; Simard, 1976) and 2) *unfixed, fresh-tissue specimens* (see Appleton, 1968, 1969, 1972a,b, 1977, 1978a; Hodson and Marshall, 1969, 1970; Davies and Erasmus, 1973; Sjöström et al., 1973; Sjöström and Thornell, 1975; Hodson and Williams, 1976; Somlyo et al., 1977, 1978; Ali et al., 1977, Sevéus, 1978; Barnard and Sevéus, 1978). Discussion is therefore confined to the more general manipulative aspects of the two distinct methodologies, giving priority to those stages at which the chemical integrity of the specimen is most vulnerable to disruptive effects.

Pre-fixed Specimens

This version of the cryopreparative sequence generally involves an immediate and relatively brief chemical fixation of fresh tissue, followed by cryoprotection, quench freezing, and finally cryosectioning onto a dry knife or, alternatively, onto a trough of fluid, e.g., DMSO (Bernhard and Viron, 1971; Sjöström et al., 1973), glycerol (Sjöström, 1975), cyclohexene (Hodson and Marshall, 1970), or sucrose (Tokuyasu, 1973). The basic "wet cryo procedure," as introduced by Bernhard and Leduc (1967), and later modified by Bernhard and Viron (1971)—see Table 14—has been adopted in several minor variant forms by numerous workers (e.g., Tokuyasu, 1973; Doty et al., 1974; Howell and Tyhurst, 1974).

Fine structural components are fairly readily identifiable in tissue sections prepared by wet cryo procedures (Figure 17; see also Doty et al., 1974; Sjöström, 1975; Sjöström and Thornell, 1975), but the exposure to aqueous media even for short periods seriously reduces the content of water-soluble tissue components (Figure 1). However, the technique can and does contribute significantly to two areas of microscopic investigation.

First, the good structural preservation characteristics and reproducibility of the technique mean that it can be regarded as an abbreviated version of conventional wet chemical EM preparative technique, and thus can be exploited to increase the productivity of EM structural studies (Doty et al., 1974). One presumes that such a reduction in processing time represents a considerable advantage in morphological surveys of surgical biopsy specimens, for example.

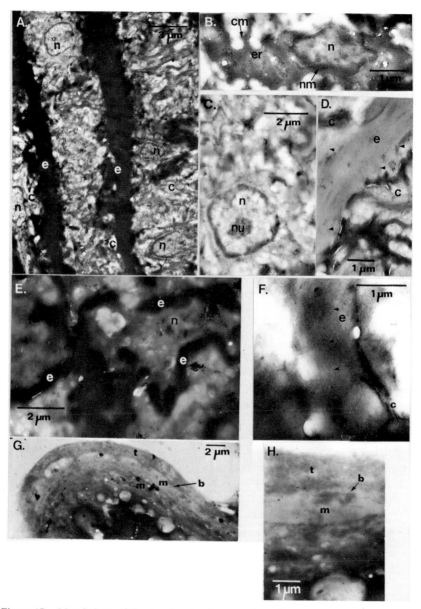

Figure 17.   Morphology of fixed and unfixed cryosections. The amount of structural detail to be seen in unstained cryosections is considerably less than in conventionally prepared specimens. However, this relative lack of structural information is not always restrictive in analytical studies—the extent of the imposed limitation depends on the tissue [simple tubular structures (cf. Figures 20 and 21) or cells with distinctive organelles (cf. Figure 6) present few problems; similarly, the structure of cryosectioned striated muscle is fairly easily interpreted], and on the spatial resolution of the analyses to be undertaken. **A, B, C,** and **D** are micrographs

Table 14.   Outline of a basic wet cryo procedure [after Bernhard and Leduc (1967) and Bernhard and Viron (1971)]

1. Fixation with 1–4% glutaraldehyde in 0.1–0.2 M cacodylate buffer, pH 7.2, for 15 min to 4 h at 3°C or room temperature.
2. Overnight wash in buffer solution.
3. Encapsulation in pure gelatin or 20% thiolated gelatine.
4. Cryo-protection (or "partial dehydration") in 30–50% glycerol for 15 min at room temperature.
5. Sectioning with glass knives at −35°C. Sections collected on a trough containing 40–50% DMSO solution.
6. Removal of sections from the trough with plastic Marinozzi (1964) rings, floated on distilled water for a few seconds to remove DMSO, and finally picked up on Formvar-coated grids and air dried.
7. Contrasting with either positive or negative stains. Best staining results achieved with negative stains.
8. Air drying for examination in the electron microscope.

Second, and of considerable consequence, is that the technique has been successfully exploited for the cytochemical and electron microprobe localization of various enzymatic and non-enzymatic histochemical reaction-products [see Bowen and Ryder (1977, 1978) for references]. For example, Bowen et al. (1976), pursuing a principle established earlier by Ryder and Bowen (1974) and Bauer and Sigarlakie (1973), used a wet-cryo technique combined with microprobe analysis in TEM to identify a relatively non–electron-dense azo-dye reaction product, and thus to reveal fine structural loci of acid hydrolase activity (Figure 18). The fixed tissue was incubated after fixation and buffer washing in a medium consisting of the substrate and naphthyl AS BI phosphoric acid + 2,5-dibromoaniline as the coupler (i.e., a simultaneous coupling procedure), and was subsequently processed according to a wet-cryo schedule similar to that outlined in Table

←———:

of unstained glutaraldehyde-fixed cryosections of the rat aorta. Most of the major tissue components are readily seen (A); at higher magnifications some of the cellular constituents and substructure of the elastic lamellae can also be seen. c = collagen; cm = cell membrane; e = elastic lamellae; er = endoplasmic reticulum; n = nucleus; nu = nucleolus; nm = nuclear membrane; elastic microfibrils are indicated by arrows. E and F are micrographs of unstained, unfixed cryosections of rat aorta. In comparison to A–D, considerably less structural detail is seen in E and F. Because the arterial wall is predominantly biphasic in structure, low-resolution analyses of cells can be obtained by positioning the probe between elastic lamellae and relying on the analytical spectra for positional confirmation. In very many complex tissues composed of several different cell types, such interpretive problems are magnified. G and F are micrographs of an unstained, unfixed cryosection through the parasitic worm *Schistosoma mansoni*. Sufficient detail of the gross components of this structurally heterogeneous specimen can be seen to enable analyses to be performed of the tegumental syncytium (t), basement membrane (b), and underlying musculature (m). Indeed, many of the features seen in conventionally prepared and freeze-substituted specimens are discernible (cf. Figure 15). Note that deeper structures are not so well preserved, due to poorer freezing conditions. (Courtesy of Tudor W. Davies.)

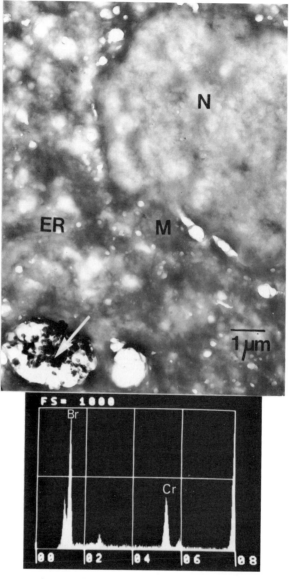

Figure 18.   Electron microprobe x-ray analysis of acid phosphatase. The micrograph represents part of an ultrathin cryosection from pre-fixed rat liver tissue exposed to a simultaneous-coupling test for acid phosphatase activity (i.e., Naphthol AS BI, enzymatically released from naphthyl AS BI phosphoric acid, is coupled with diazotized 2,5-dibromoaniline to produce a fine insoluble red azo dye). Granules of azo-dye reaction product can be seen (arrow) in the pericanalicular region. The endoplasmic reticulum (ER), mitochondria (M), and nucleus (N) are negative. Microprobe analysis of the pericanalicular granules confirms the occurrence of brominated azo dye. Chromium signal arises from instrumental sources. (Courtesy of I. D. Bowen.)

14. The advantage of wet-cryo procedures in this particular instance is that the use of organic solvents is limited, and therefore the possible solubilization and drift of azo dye during dehydration and resin embedding (Livingston et al., 1969) are averted, especially in the simultaneous coupling as opposed to the post-coupling variant (Ryder and Bowen, 1974; Bowen et al., 1976). Bowen and Ryder (1978) further advised that sections be cut and collected on dry knives.

In general, therefore, although pre-fixed cryosections cannot be recommended for the localization of ions and electrolytes, the technique removes some of the chemical constraints associated with conventional EM preparative techniques and can be used, in conjunction with electron microprobe analysis, for the appropriation of established light microscope histochemical tests to EM cytochemistry without the further constraint of needing an electron opaque final reaction product. This principle can be extended to include the location of introduced molecules possessing an appropriate non-biological "marker" element. The combination of wet cryomicrotomy and microprobe analysis can thus represent a viable alternative to autoradiography, especially where the molecule concerned is either insoluble in aqueous media or is actively incorporated into structural macromolecules, such as proteins or nucleic acids (Bowen, personal communication).

Fresh, Unfixed Specimens

It is extremely difficult to critically evaluate the success of a given cryomicrotomy procedure, such as that utilized by Appleton (see Table 15),

Table 15. A summary of the sequence used by Appleton (see text for references) for the preparation of "dry" ultrathin frozen tissue sections

1. Rapid dissection and tissue sampling.
2. Freezing of 1 mm^3 tissue specimens in liquid nitrogen slush (prepared by placing liquid nitrogen, $-196°C$, under a rough vacuum in an evaporating unit until it freezes solid; breaking the vacuum yields melting nitrogen slush, $-210°C$, which is stable for about 6–7 min).
3. Storage under liquid nitrogen.
4. Mounting in a cryostat and trimming to a pyramid (temperature range $-70°C$ to $-90°C$).
5. Ribbons of sections are drawn using a special weak-vacuum suction device. Sections are cut in the temperature range of about $-70°C$ to $-80°C$, i.e., the lowest temperature at which true sections are obtained from a given tissue.
6. Sections mounted onto Formvar-coated grids, and flattened using light pressure from a cold polished copper rod.
7. Freeze-dry at atmospheric pressure in the cryostat chamber (3 h at $-80°C$).
8. Either: 1) freeze-drying completed by gradual rise of specimen temperature from $-80°C$ to ambient room temperature in a dry nitrogen atmosphere, or 2) direct transfer of sections to a microscope cold stage.
9. Carbon coating of freeze-dried sections under vacuum.

in maintaining the distribution of the more mobile tissue components at their in vivo locations. Direct corroboration of microprobe data by independent methods is seldom feasible. All "dry" cryomicrotomy procedures can be subdivided into at least four distinct manipulative phases (dissection and pre-preparation, freezing and storage, sectioning, and drying—c.f. Table 15), each tending to violate the chemical integrity of the specimen. While the distribution of diffusible elements cannot be guaranteed by any practically viable cryoprocedure currently available, we can increase the likelihood of obtaining authentic biological data by recognizing and understanding the separate points of weakness in a given protocol, and by vigorously striving to limit their disruptive influences.

Dissection and Pre-Preparation   Tissue chemistry presumably begins to deteriorate immediately after an animal or plant is anesthetized or killed. These changes proceed during dissection and tissue sampling. Some delay is obviously inevitable and accompanies all preparative techniques, but these delays must especially be minimized and controlled before dry cryomicrotomy, since this is the recognized preparative procedure for microprobe studies of dynamic physiological processes.

The effect of pre-freezing delays may be reduced by immersing the expeditiously excised tissue in a suitable physiological medium. Small specimens from selected areas may be cut, mounted, and oriented on ultramicrotome chucks, while submerged under the physiological medium. For example, epithelia of frog skin and toad urinary bladder were immersed for short periods (30 s) immediately before shock freezing in a Ringer's solution containing 20% albumin (Dörge et al., 1975; Rick et al., 1978). Gupta and co-workers (Gupta, 1976; Gupta et al., 1976, 1977, 1978a,b), applying a similar principle, added 10% w/w dextran (M.W. 200,000–400,000) to the physiological saline surrounding the salivary gland epithelium of *Calliphora*, the Malphigian tubules of *Rhodnius*, and rabbit ileum. These latter authors found that dextran at this concentration did not enter cells, reduced ice-crystal damage within a range of plant and animal tissues, and improved the sectioning properties of their specimens (cf. Echlin et al., 1977, 1978; Franks et al., 1977; Skaer et al., 1977). These effects were *not* accompanied by a significant alteration of the functional state of the tissue (as measured by cell agglutination, rate of fluid transport across epithelia, and electrochemical properties). Control experiments show that higher concentrations of dextran (15–20%) can affect the physiological state of the salivary glands of *Calliphora* (Gupta et al., 1978a). In addition to their physiological and cryoprotective effects, the inclusion of albumin or dextran (and also polyvinyl pyrrolidone, polyvinyl alcohol, or polyethylene glycol—Franks and Skaer, 1976) in the fluid medium surrounding the tissue conveniently provides an organic matrix for the extracellular medium so that it can be used as a peripheral analytical standard. This feature is particularly useful in the case of freeze-dried frozen sections for the conversion of element dry weight fractions to wet weight concentration units (Dörge et al., 1975; Rick et al., 1977). In practice it is important to ensure that the thickness of the layer of "physiological standard" encapsulating the mounted specimen does not exceed a few micrometers, otherwise it will seriously affect the quality of freezing in the underlying tissue layers.

Sampling of soft tissues is itself a major potential source of pre-preparative chemical artifact. Unfortunately, in the case of a majority of soft tissues there is no alternative to the sampling of small tissue cubes (Christensen, 1971). Such specimens are surrounded (on all but one side if the outer surface of the organ or tissue is included) by cells severed or otherwise physically damaged during excision. The spilled cytoplasmic contents from these cells will also probably "contaminate" the chemistry of adjacent cells. As mentioned earlier, it is an unhappy coincidence that this damaged superficial layer is included in the narrow vitrified zone of quench-frozen specimens. Cubed samples with an undamaged side should, therefore, be mounted on the cryomicrotome chuck with the undamaged face vertical and arranged so that it is not trimmed (see "Sectioning," below), i.e., with its longitudinal axis perpendicular to the knife edge. In this way all sections throughout the depth of the specimen will include a narrow region of intact cells with minimal freezing damage. Obviously, even this rather limiting arrangement is precluded in the case of samples that have to be taken from deep within an organ or tissue. Furthermore, the lack of adequate inherent differential contrast in fresh-frozen cryosections does not permit the iden-tification of cells damaged during pre-preparation. Given this difficult posi-tion, the microprobe analyst must rely on the results of several individual analyses to counteract the bias of non-biological variabilities stemming from damaged cells. Contrasting of adjacent sections with osmium vapor is a recommended adjunct to the cryosectioning schedule to assist in mor-phological interpretation (Somlyo et al., 1977).

In the specific case of striated muscle, where the length of individual cells greatly exceeds their diameter, Somlyo et al. (1977) overcame the problem of ion loss through cut surfaces by stretching whole (small) muscles over the domed tip of a low-mass steel holder. The muscles were held in this device during freezing and sectioning, and a small dome of tissue (1 mm^3), which did not need trimming, was presented to the knife. An important advantage is that the whole periphery of all sections obtained from the domed region included cells that were not touched during dissection, exci-sion, and mounting, and were also frozen at the fastest rate. Similar holders could easily be constructed for a number of small, soft-bodied tubular struc-tures, e.g., insect Malphigian tubules.

Freezing and Storage   Rapid freezing is regarded as a fixation process. Unlike classical chemical fixation procedures, the aim of the freezing process is to preserve adequate structural detail, but *without* moving the dif-fusible tissue ions and electrolytes from their subcellular and extracellular compartments, so that the microprobe can localize these elements within recognizable structures.

A general discussion of certain practical aspects of tissue freezing has already been presented (see above). However, specific questions will undoubtedly occur to the potential user of the cryomicrotomy technique. For example:

Choice of quenching agent.   The essence of good freezing technique in the context of microprobe requirements is that it provides rapid freezing rates and that the method is simple. Given these dual requirements, a brief literature survey reveals claims and counter-claims regarding the merits (or demerits) of specific freezing agents, which probably suggests that a fairly wide choice is available. Spriggs and Wynne-Evans (1976) favored Freon 22 slush (i.e., Freon partially frozen by liquid N_2 cooling), having also assessed freezing by contact with a polished liquid N_2-cooled copper block, isopentane, and propane. Hodson and Williams (1976) considered melting Freon 22 slush ($-146°C$) superior to melting nitrogen ($-210°C$). Somlyo et al. (1977) also chose Freon 22 supercooled with liquid N_2 (freezing temperature of $-164 \pm 2°C$; solidification temperature, $-175°C$). Others (Appleton, 1978a; Sevéus, 1978; Barnard and Sevéus, 1978) use melting nitrogen for quench freezing. Having previously assessed liquid nitrogen and liquid N_2-cooled copper blocks, we are now using melting nitrogen in our own laboratory. The "freezing shaft" described by Sevéus (1978) for the simultaneous freezing of five individual specimens in nitrogen slush appears to suffer from one serious disadvantage. Although the specimens may well be frozen under identical conditions with such a device, the time interval elapsing during mounting and orientation is quite likely to introduce significant differences in the chemical state of each individual sample *before* freezing. In order to draw maximal advantage from the use of melting N_2, the specimens must be immersed while there is a large proportion of solid nitrogen present, i.e., immediately upon vacuum release (Sevéus, 1978).

Storage of the frozen specimens.   Quench-frozen specimens are usually stored under liquid nitrogen. Appleton (1977) stated that there is no detectable deterioration of morphology or redistribution of diffusible substances in specimens stored in this way for periods up to 3 years.

Cryoprotectants or not?   This question was once considered resolved, but it has re-emerged as a subject of considerable contemporary and future interest. Without doubt, the use of low molecular weight penetrating cryoprotectants must be avoided due to interference with cellular chemistry (Appleton, 1977). But, in the absence of cryoprotectants, the vast bulk of a soft tissue specimen is subject to ice-crystal damage [unless it is either extremely small or is a relatively acellular tissue with a polymeric extracellular matrix whose gel-like characteristics possibly confer natural cryoprotectant properties, e.g., cartilage (Ali et al., 1977)]. Besides physically rupturing cells and organelles, the formation of ice introduces the possibility of excessive osmotic dehydration accompanied by cellular shrinkage (Franks and Skaer, 1976). When ice crystals form, therefore, diffusible substances tend to aggregate around them and are further displaced when the crystals grow. This explains why in the vitreous zone of striated muscle fibers, the elements were retained and distributed in predictable fashion within specific structures, while in the deeper (freeze-damaged) regions of the same fibers, no reproducible specific element localization is found (Sjöström, 1975). Because of this practical encumbrance of ice formation, Barnard and Sevéus (1978) were moved to conclude that "the incompatible requirements of good freeze-drying of sections and apparently well-preserved ultrastructure" suggest to us "the need to abandon attempts to work with uncryoprotected tissue, unless this should consist of a very tiny monolayer of cells." This statement emphasizes the interest generated by the fairly recent introduction of non-penetrating, polymeric cryoprotectants (see above), whose large water-binding capacity and high molecular weight combine to promote the formation of vitreous states and limit the osmotic flow of water (and suspended substances) from cells during shock freezing (Franks and Skaer, 1976). The advantages conferred by these polymers are startling (but see Wilson et al., 1979) but whether they do preserve the chemical integrity of cells at the level of resolution of microprobe analysis requires stringent assessment before cryoprotection by these agents can be routinely considered part of the "dry" cryomicrotomy procedure.

Sectioning   The manipulative aspects of cryosectioning have been described in detail by several workers, but most notably by Appleton (1974a, 1977, 1978a). Although a similar discussion of technical detail is both unnecessary and beyond the scope of the present text, it seems pertinent to briefly answer some of the more general procedural questions that may concern the potential user.

Choice of cryo-ultramicrotome.   Broadly speaking, there are two cryoultramicrotome arrangements commercially available (Sjöström and Squire, 1977). First, there are the cryochamber attachments, which effectively enclose a relatively small volume in the region of the specimen support arm and knife assemblies in otherwise conventional, or only slightly modified, ultramicrotomes, e.g., the LKB CryoKit mounted on LKB Ultrotome III (Sevéus, 1978), the Cambridge Instruments cryo-attachment mounted on a Huxley ultramicrotome Mark II (Hodson and Williams, 1976; Spriggs and Wynne-Evans, 1976), the Reichert OMU 3 ultramicrotome with low-temperature freezing attachment (Geymayer et al., 1978), and the Sorval MT2 microtome with an FTS LTC-2-Frozen thin sectioner and low-temperature controller (Christensen 1969, 1971). Second, a modified microtome can be placed inside a cryostat chamber. A cryostat incorporating the features and specific modifications suggested by Appleton (1974a,b, 1977, 1978a) is available from SLEE Medical Equipment Ltd.

Various workers have produced excellent sections by both types of instruments. The advantages of cryo-attachment systems are that the specimen and knife temperatures can be independently controlled (see below) and, of relatively minor significance, the microtome can be uncoupled for conventional plastic sectioning. The small volume of the attachment chamber, however, interferes with temperature stability and limits the space available for ancillary equipment, such as grid carriers, metal rod for section flattening, and freeze-drying or transfer devices, etc. In comparison, the large volume within the cryostat chamber improves temperature stability, reduces air turbulence in the knife region, and easily accommodates the necessary additional instruments. Sectioning is undoubtedly easier in a cryostat than in an attachment system (Appleton, 1978a), but, although the chamber temperature can be varied in the range $-50°$ to $-140°C$, the specimen and knife temperatures cannot readily be independently controlled. The SLEE cryostat is also permanently committed to cryosectioning, and is normally maintained at a standby temperature of $-25°C$ to $-40°C$. The unmodified LKB is, in our experience, functionally useful only for periods up to 3 h, before ice precipitation immobilizes the specimen arm and covers all internal surfaces.

Sectioning temperatures. Sectioning temperatures are the subject of considerable controversy. Appleton (1974a,b, 1977, 1978a) and Christensen (1971) were able to cut true and complete sections with fairly regular ribbon formation in the temperature range $-70°C$ to $-85°C$, the optimal temperature depending on the chemical and physical constitution of each individual specimen (Appleton, 1978a).

A second group of workers (Hodson and Williams, 1976; Spriggs and Wynne-Evans, 1976) argue that while sectioning at $-80°C$ is certainly easier, the true sections are formed due to a partial and transient thawing phase at its knife edge (Hodson and Marshall, 1972), which probably results in ion and electrolyte redistribution. These authors therefore section at low temperatures by maintaining the specimen at $-135°C$ or lower, the knife at $-125°C$, and the cryochamber at about $-100°C$. Complete sections are not produced under these conditions (Spriggs and Wynne-Evans, 1976); indeed, there is little doubt that the fragments collected are fractures and not true sections. Nevertheless, the appearance of scratch marks and compression lines in the sections cut at these lower temperatures and subsequently freeze-dried indicates that thawing did not occur during sectioning.

It is possible to distinguish a third group of workers, who section at intermediate temperatures and produce true and complete sections (indicated by the appearance of interference colors) that also retain knife scratch marks after freeze-drying. Significantly, these workers maintain the knife at a *higher* temperature than the specimen. Somlyo et al. (1977), using a specially modified and insulated LKB CryoKit (cf. Doty et al., 1974), maintain the specimen at $-110°C$, the knife at $-100°C$, and the cryochamber at $-130°C$. Sevéus (1978) used a combination of specimen temperature

−140°C, knife temperature −80° to −100°C, and a chamber temperature at −100°C. Unlike Appleton, Somlyo et al. (1977) and Sevéus (1978) do not trim their frozen specimens in an attempt to obtain ribbons, since trimming removes a large proportion of the vitrified region of the tissue. In addition, the adhesion of adjacent sections to form ribbons is considered possible only at higher temperatures, where there is a transient thawing during sectioning. Significantly, Sevéus et al. (1978) found that intramitochondrial (Ca + P)-rich particles were present only in brown adipose tissue and liver cryosections cut at −80°C and which showed signs of melting after sectioning. Particles were not found in sections cut at −110°C. The authors concluded that the particles form in the sections cut at the higher temperature due to calcium diffusion from in vivo sites followed by preferential precipitation on mitochondrial nucleating sites.

Drying.   Once the frozen sections have been prepared, they must not be allowed to thaw. Melt drying has been shown to result in a diffusion of elements from their normal locations. Appleton (1974a) tested the reliability of his freeze-drying method on a model system consisting of sodium chloride dissolved in a solution of carboxymethyl cellulose. The sodium chloride assumed elaborate fern-like appearance in freeze-dried sections. Most interestingly, the clear areas adjacent to the crystal "ferns" did not yield significant sodium signals in freeze-dried sections, but the clear areas in melt-dried sections contained large quantities of "spilled" sodium. A recent paper by Barnard and Sevéus (1978) also strongly emphasized the importance of good freeze-drying. They assessed several methods of drying, including air drying at room temperature, drying at −100°C in a jet of extra-dry nitrogen gas, drying at either −100°C or room temperature over molecular sieve, or freeze-drying under a moderate vacuum at −100°C in the cryochamber. They found that, if drying is inadequate, the diffusion of elements is easily seen as an electron-dense "diffusion film," and is especially noticeable in their brown adipose tissue sections over triglyceride droplets and immediately outside the section edges (Figure 19). Barnard and Sevéus (1978) found that the lowest incidence of element diffusion was obtained by drying for at least 2 h either over molecular sieve or under vacuum within the cryochamber of their LKB CryoKit, whose internal temperature was lowered to about −110°C by forced evaporation of liquid N_2 via an extra foil heater.

Other workers have devised slightly different freeze-drying arrangements. For example, Somlyo et al. (1977), Hodson and Williams (1976), and Spriggs and Wynne-Evans (1976) employ special cold-metal transfer devices for carrying the frozen sections to vacuum units situated outside the cryochamber. In contrast, Appleton (1974a, 1977, 1978a) and Sjöström and Thornell (1975) freeze-dry very slowly within the cryochamber and at atmospheric pressure, believing that the surface tension forces involved in the rapid sublimation of ice under vacuum cause a gross disruption of cells.

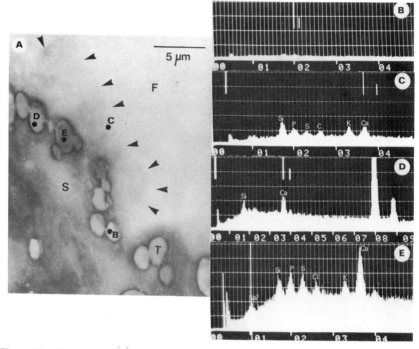

Figure 19. X-ray microanalytical evidence of element diffusion in unfixed cryosections that are inadequately dried. **A**: Micrograph of a cryosection through brown adipose tissue. The flattened section was dried in a gelatine capsule (placed inside the cryochamber) by flushing with extra-dry N_2 gas. The section (S) contains several triglyceride droplets (T); F = support film. Diffusion of material from the section onto the support film (arrows) and triglyceride droplets has clearly occurred. **B–E**: X-ray spectra taken from positions **B–E** depicted in **A**. Spot size = 1 μm; vertical scale = 100 counts per interval. **B**: Spectrum from a triglyceride droplet with a weak or absent diffusion film. Note no peaks and low "Bremsstrahlung." The support film peripheral to the diffusion-film boundary gave a similar spectrum. **C**: Spectrum from the diffusion film on the support film. Note silicon, phosphorous, sulfur, chlorine, potassium, and calcium signals. **D**: Spectrum from triglyceride droplet covered with a moderately electron-dense diffusion film. Note silicon and calcium signals and increased "Bremsstrahlung" (cf. **B**). **E**: Spectrum from triglyceride droplet covered with a strongly electron-dense diffusion film. Note the silicon, phosphorus, sulfur, chlorine, potassium, and calcium signals and the increased "Bremsstrahlung." (From Barnard and Sevéus, 1978.)

The weight of opinion suggests, however, that the lace-like appearance of vacuum dried sections, which Appleton (1978a) would interpret as a freeze-drying artifact, actually represents a tissue region containing holes that were previously occupied by ice crystals and that were faithfully preserved (i.e., the tissue was not allowed to reconstitute) during vacuum-drying (Spriggs and Wynne-Evans, 1976; Barnard and Sevéus, 1978). In summary, therefore, sections containing discrete holes and edges displaying Fresnel fringes, in which knife marks and chatter may also be seen, are well dried.

Finally, and contrary to the observations of Spriggs and Wynne-Evans (1976), it is considered advisable to coat the dry section with carbon to prevent rehydration from atmospheric moisture (Appleton, 1974a, 1978a).

FROZEN-HYDRATED SPECIMEN PREPARATION

All the cryoprocedures discussed so far, and indeed also the non-freezing procedures, grossly destroy the chemical integrity of biological tissues, in that they remove, and in some cases replace, tissue water. We must, therefore, reconsider carefully at this point our definition of "the preservation of tissue chemical integrity." The total retention of those elements detectable by electron probe microanalysis, and the limitation of their movement from in vivo locations over distances less than the limits of the spatial resolution of the analyzer, have hitherto served as our working definition. Accordingly, we may accept that the cryomicrotomy of unfixed fresh tissues, followed by stringently controlled freeze-drying, often meets the specific requirements of this strictly practical definition. Nevertheless, the freeze-dried frozen section suffers certain restrictive disadvantages:

1. The quality of the morphological information obtainable from unfixed, freeze-dried cryosections is generally inferior to that associated with conventional EM techniques (Figure 17). This is not to say that the cryosection does not represent a truer reflection of the disposition of cellular components in vivo (cf. Bacaner et al., 1973), but since the outstanding value of microprobe analysis is the ability to relate chemistry with identifiable morphological structure, this limitation can be restrictive (although perhaps too often exaggerated).
2. Dehydration precludes even qualitative assay of solutes suspended in highly hydrated structural compartments, e.g., intercellular spaces and intracellular vacuoles. The contents of these structures precipitate on the boundary membranes during water withdrawal (Somlyo et al., 1977).
3. Finally, and of considerable physiological consequence, the removal of water precludes a direct quantitative comparison of *solute concentrations* in tissue and cellular compartments having differing and unknown water contents [but see the correction procedures of Dörge et al. (1975) and Appleton (1978a,b)].

The chemical shortcomings of all cryoprocedures are primarily born of the phase changes accompanying tissue freezing. Additional restraints (mentioned above) are introduced during the removal of water, albeit from the solid state. The alternative to drying is to immobilize tissue water and its solutes in deep-frozen tissue throughout all preparative stages, including transference to and analysis in the electron microscope. This technology was pioneered by Echlin et al. (1970) and Nei et al. (1973), and has since steadily commanded the attention of several independent laboratories. Both bulk specimens and sections of plant and animal tissues have been maintained in the frozen hydrated state and chemically analyzed. For example:

1.   *Frozen-hydrated bulk specimens*

[Marshall and Wright (1973), Marshall (1974, 1977), Kaufmann and Gullasch (1972), Gullasch et al. (1973), Gullasch and Kaufmann (1974), Fuchs and Lindemann (1975), Fuchs et al. (1978b), Zierold (1976), Zierold and Schafer (1978), and Zierold et al. (1978)].

2.   *Frozen-hydrated sections*

[Bacaner et al. (1973), Saubermann and Echlin (1975), Gupta et al. (1976, 1977, 1978a,b), and Hutchinson (1977)].

Detailed reviews of the technological aspects of the frozen-hydrated procedure have been presented by Saubermann and Echlin (1975) and Echlin (1978), and a summary of the protocol used by Gupta and his colleagues for the preparation of frozen-hydrated sections is shown in Table 16.

Hydrated Bulk Specimens versus Hydrated Sections

Obviously, the major challenge of the hydration regime is to prevent water evaporation at all manipulative stages. This is more easily achieved in the case of bulk specimens because they can be continuously maintained very near to liquid N_2 temperature; but sections cannot be cut at these ultra-low temperatures. A comparison of the freeze-drying and frozen-hydrated sectioning regimes (Tables 15 and 16) immediately shows that they diverge fundamentally only after section retrieval. Significant freeze-drying of sections can all too easily occur at cryomicrotome chamber temperatures (Gehring et al., 1973; Fuchs and Lindemann, 1975). Gupta et al. (1976) actually exploit this phenomenon to partially dehydrate their sections (10%–20% water loss) before embarking on an elaborate storage and transfer sequence to the microscope cold stage (Table 16). They claim that this loss of water enhances the differential contrast and hence the morphological detail perceivable in their sections. In fact, a major problem with frozen-hydrated specimens (bulk and sections) is that they appear quite featureless in the fully hydrated state (Figures 20 and 21). To improve the quality of the morphological information in bulk specimens so that the electron probe can be localized for correlative analysis, it is also common to sublimate the surface ice by controlled freeze-etching (Fuchs et al., 1978b; Zierold and Schafer, 1977; Zierold et al., 1978; see also the discussion by Echlin, 1978).

Mainly because of the high rate of ice sublimation in the temperature range $-80°C$ to about $-120°C$, i.e., temperatures corresponding to those

Table 16.   Summary of the method used by Gupta and colleagues (see text) for the processing of frozen-hydrated sections of *Calliphora* salivary gland for microprobe analysis

1. *Pre-preparation*	Freshly dissected tissues suspended in saline, pH 7.2 (normal composition mM: Na, 132, K, 20, Ca, 2, Mg, 2; Tris, 10; Cl, 158; phosphate, 4; malate, 2.7; glutamate, 2.7; glucose, 10) containing 10% dextran (MW 230,000). It is essential to establish the concentration of cryoprotectant that is compatible with the normal functioning of each individual system by monitoring physiological parameters.
2. *Freezing*	Small pieces of tissue suspended in drops of Dextran-saline are mounted on metal specimen holders and quench frozen in liquefied Freon 13, cooled to its M.P. ($-181°C$) with liquid N_2. Excess Freon adhering to the blocks must be shaken off.
3. *Sectioning*	Sections cut 1-2 μm thick, either at $-80°C$ with steel knives in a SLEE cryostat, or at $-120°C$ to $-140°C$ with glass knives in an SLEE cryostat, or at $-120°C$ to $-140°C$ with glass knives in a LKB CryoKit. Sections mounted inside cryochamber on pre-cooled duralium collars covered with aluminized nylon films (hydrophilic) and flattened with a cold, polished copper block.
4. *Storage and transport*	The section-bearing collars are immediately transferred to a slotted brass block cooled with liquid N_2 inside a Dewar for storage and transport to the cold stage of the microprobe analyzer. Since the inherent contrast in a fully hydrated section is very low, the structural details are more readily observed in sections allowed to partially dehydrate (10% water loss) under "controlled conditions," i.e., by storage for about 15 min inside the microtome cryochamber before entering the storage Dewar.
5. *Transfer from storage to microprobe cold stage*	The section-bearing collars were transferred to a modified cold stage ($-170°C$) of a JEOL JXA-50A microprobe (Taylor and Burgess, 1977) with a special device that maintained the sections below $-150°C$ to protect them from dehydration and atmospheric contamination. Tissue structure was observed in a scanning transmission image (STEM) and analysis was performed with a stationary focused beam (100–200 nm) or within limited scan rectangular areas.

used for cryosectioning, ultrathin sections are not commonly prepared and maintained in the frozen-hydrated state (but see Bacaner et al., 1973; Hutchinson, 1977). More reliable results have been obtained with sections 1-2 μm thick (Saubermann and Echlin, 1975; Gupta et al., 1976). But, although sections in this thickness range allow electron transmission even at low accelerating voltages of about 15 kV (Rick et al., 1978), and can therefore be quantified by the Hall continuum normalization method (Hall, 1971; 1977b), the analysis of structural features whose dimensions are

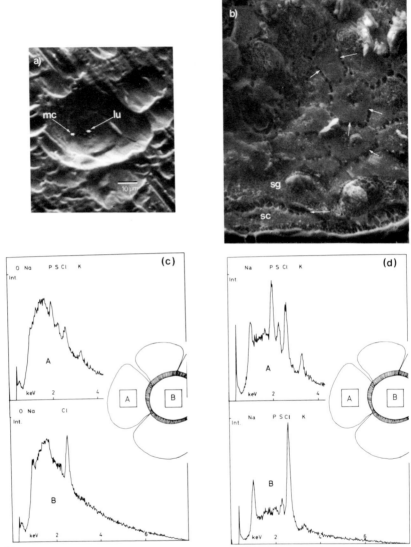

Figure 20. Analysis of frozen-hydrated bulk specimens. **a:** Scanning electron micrograph of the surface of the fully hydrated midgut of *Chironomus thummi* larva fractured in the transverse plane. The fractured surface of the midgut (mc)—a tubular structure consisting of five or six cells enclosing a lumen (lu)—is featureless; positive identification of cellular versus extracellular (lumen) locations could only be achieved by microprobe analysis (see spectra). **b:** Scanning electron micrograph of a superficially freeze-dried frog skin epithelium. Partial dehydration allows a better delineation of gross features, such as cell layers (sc = stratum corneum cell; sg = stratum granulosum cell) and cell boundaries (arrows). This is especially important in complex tissues, but may restrict analysis of intracellular domains. **c:** Energy-dispersive spectra obtained from the analysis of fully hydrated midgut. Analysis was performed by reduced area scanning (2×2 μm) at positions indicated by the stars in **a**. A = cell; B = lumen. **d:** Energy-dispersive spectra obtained from the analysis of areas identical to **c** but where the specimen had been partially freeze-dried prior to analysis. Conditions of analysis were similar to **c**. Note that the P/B ratio is significantly improved in the partially dried specimen, and that the electrolyte distribution patterns remain unaltered. (From Fuchs et al., 1978b. Used by permission.)

142

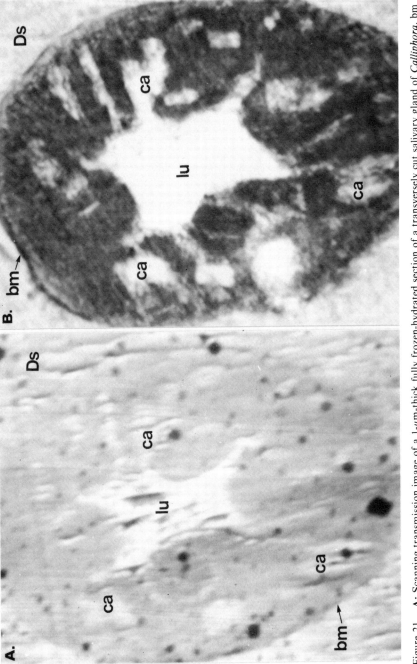

Figure 21. **A**: Scanning transmission image of a 1-μm-thick fully frozen-hydrated section of a transversely cut salivary gland of *Calliphora*. bm = basement membrane; ca = canaliculi; Ds = Dextran + saline; lu = lumen. **B**: Scanning transmission image of the same specimen after freeze-drying within the microanalyzer. Note the improvement in contrast and spatial resolution. (From Gupta et al., 1978a. Used by permission.)

considerably less than the section thickness must inevitably produce variabilities due to sampling error. Similar errors can be produced by the inability to resolve fine structural details in hydrated sections. For example, one explanation of the so-called "chloride anomaly" in *Calliphora* salivary glands (where the chlorine concentration in gland canaliculi measured by microprobe is considerably lower than theoretical values, or of values measured directly by microelectrodes) is that cell processes containing low chlorine levels make variable contributions to "canalicular" measured volumes (Gupta, 1979; Gupta et al., 1977). Sampling errors of both kinds will be more frequently encountered with bulk specimens.

To summarize, the technology of frozen-hydrated specimen preparation finds itself in an unfortunate paradoxical state. On the one hand, bulk specimens, or alternatively 1–2-μm-thick sections, are more likely to preserve tissue water content than ultrathin sections. In the case of bulk specimens, complicated matrix-correction quantitative procedures, which account for internal charging of the specimen, must be adopted (Brombach, 1975; Fuchs et al., 1978a). On the other hand, improved analytical spatial resolution [if not improved structural resolution (Fuchs and Lindemann, 1975)] and simpler quantitative procedures are features of ultrathin specimens. The existing technology, therefore, allows the identification and analysis of only the largest tissue compartments (Figures 20 and 21). Zierold and Schäfer (1978) estimated that the diameter of the smallest analyzable volume in frozen-hydrated bulk specimens is of the order of 1 μm. A recent paper by Talmon et al. (1979) revealed a further possible disadvantage of the frozen-hydrated specimen. They showed, on model specimens consisting of a film of distilled water sandwiched between two polymer films, that samples that are very stable when dry, under electron irradiation are severely etched in the presence of ice. This effect occurs under conditions where local beam heating effects are negligible; the mass loss is caused by radiolysis and may seriously limit the accuracy of quantitative microprobe analyses of biological materials.

Maintenance of Specimen Hydration

All the authors mentioned above described in considerable detail the equipment they use for the transfer of hydrated specimens in a dry, contamination-free, cold environment to the modified cold stages of various SEMs. The basic design features of the storage, transfer, specimen stage, and anticontamination devices are, therefore, very well established. Frozen-hydrated tissue specimens are almost exclusively observed and analyzed in the appropriate optical mode in SEMs, mainly because of the relatively large manipulative volume inside the specimen chamber region of these microscopes. However, very effective cold stages have been developed for TEMs (see Valdre and Horne, 1975). In an x-ray microanalysis context, analysis at low temperatures imparts a considerable improvement in the P/B ratios in freeze-dried cryosections (Appleton, 1977).

Having designed a sequence of devices for the uninterrupted maintenance of the frozen specimen in the hydrated state, great care must be exercised to establish the success of the regime on each individual experimental run. This is especially important in the case of sections, since they are often wrinkled, or otherwise detached from their supporting film and not, therefore, in good thermal continuity with the specimen holder (Gupta et al., 1977). A series of criteria were introduced by Saubermann and Echlin (1975) for assessing whether or not the tissue remains in the fully hydrated state or not. These criteria are equally applicable to bulk specimens and sections. The two most practically significant of these criteria are summarized in simplified form:

1. In order to *prevent* ice sublimation in frozen specimens, the tissue must be kept at all times at temperatures "as low as possible" and in a dry atmosphere. Unless this condition is met, there is a danger that the exposure of the specimen to a vacuum (e.g., in a coating unit or inside the EM column) would immediately create freeze-drying conditions.

2. A good practical method of confirming the hydration of the specimen is to deliberately warm it up, and to record the improvement in P/B ratios that should accompany the loss of water, i.e., of mass (Figure 20). The poor P/B ratios in frozen-hydrated specimens are a considerable encumbrance, and prove especially restrictive in the analysis of sodium and lighter elements by energy-dispersive spectrometry (Appleton, 1978a). However, the loss in mass after freeze-drying (measured by the Hall continuum method) gives a good estimate of the original water content of selected areas of the specimen (Gupta et al., 1976, 1977; Saubermann and Echlin, 1975). This seems to be the only direct method of distinguishing water-filled domains from air spaces in biological tissue (Echlin, 1978).

Utility of Frozen-Hydrated versus Freeze-Dried Cryosection

The advantages of the frozen-hydrated section are:

1. The degree of hydration of tissue components can be measured.
2. It is the only preparative procedure that provides for the analysis of diffusible ions and electrolytes in extracellular spaces and water-filled vacuoles.
3. The concentration of solutes is readily expressed as a function of tissue wet weight (by reference to a hydrated standard).

The advantages of the freeze-dried section are:

1. The post-sectioning protocol is considerably less demanding.
2. The morphological detail it affords is superior.

3. Although elements in the fluid-filled compartments cannot be measured because of migration and precipitation onto boundary walls, the freeze-drying of previously hydrated specimens is *not* accompanied by detectable changes in the distribution of elements within intracellular matrix-rich compartments (Gupta et al., 1976, 1977). Because of advantages 1

Table 17.   Summary of results from measurements with microprobe and microelectrodes in *Calliphora* salivary glands bathed in Ringer's solution

	Microprobe		Microelectrode
	mM kg^{-1} wet wt (measured)	mM kg^{-1} H$_2$O† (estimated)	Concentration mM l^{-1}
Potassium			
Unstimulated			
Basement membrane	76 ± 8 (n = 7)	—	—
Cytoplasm	115 ± 7 (n = 10)	135	143 ± 4 (n = 6)
Canaliculi	160 ± 16 (n = 4)	178	185 ± 6 (n = 6)‖
Lumen	139 ± 5 (n = 5)	139	127 ± 4 (n = 9)
5-HT			
Basement membrane	129 ± 20 (n = 12)	—	—
Cytoplasm	125 ± 3 (n = 42)	161	146 ± 5 (n = 6)
Canaliculi	168 ± 13 (n = 8)	187	199 ± 10 (n = 6)‖
Lumen	161 ± 6 (n = 6)	161	162 ± 3 (n = 9)
Chloride			
Unstimulated			
Basement membrane	134 ± 21 (n = 7)	—	—
Cytoplasm	33 ± 3 (n = 10)	39	
Canaliculi	114 ± 22 (n = 4)	127	185
Lumen	155 ± 7 (n = 5)	155	168 ± 4 (n = 5)
5-HT			
Basement membrane	138 ± 7 (n = 12)	—	—
Cytoplasm	23 ± 2 (n = 42)	30	19 ± 0.5 (n = 6)
Canaliculi	105 ± 12 (n = 8)	117	200
Lumen	168 ± 7 (n = 6)	168	175 ± 4 (n = 5)
Sodium			
Unstimulated			
Basement membrane	76 ± 13 (n = 7)	—	
Cytoplasm	20 ± 3 (n = 10)	24	
Canaliculi	17 ± 10 (n = 4)	19	
Lumen	16 ± 2 (n = 5)	16	
5-HT			Not measured
Basement membrane	80 ± 7 (n = 12)	—	
Cytoplasm	15 ± 1 (n = 42)	19	
Canaliculi	16 ± 5 (n = 8)	18	
Lumen	24 ± 2 (n = 6)	24	

From Gupta et al., 1978a. Used by permission.

† The estimates were made by using the water content of the various regions obtained by comparing frozen-hydrated and frozen-dry sections. The values in this column are based on the assumption that the measured concentrations of potassium, chlorine, and sodium are all in free tissue–H$_2$O.

‖ The tip of the electrode was assumed to be in the canaliculus because of the unusually high potassium values and the much larger changes in transepithelial potential in response to 5-HT as compared with the main lumen.

and 2, the cryosectioning/freeze-drying regime is often the procedure of choice for the microprobe analysis of element distribution *within cells.*

The special advantages offered by the frozen-hydrated specimen are, of course, of unique physiological significance, but the problem that we have frequently encountered throughout this discussion is the ability to assess the viability of data obtained from specimens prepared for microprobe analysis in a given way. Gupta et al. (1978a), besides providing new information about the distribution of ions and water in the salivary glands of *Calliphora,* also correlated their microprobe findings with measurements taken with ion-selective microelectrodes. In addition, they were able to monitor changes induced in the transporting epithelium by the application of stimulatory 5-hydroxytryptamine. The detailed results of this study are of unquestioned physiological interest (Figure 21 and Table 17), but, for our present purposes, there are equally important lessons to be learned from the general experimental approach of these authors. Gupta and his colleagues display how quantitative microprobe data can be accepted and interpreted with greater and well-founded confidence when coupled with biochemical information derived by independent techniques, such as flame spectrophotometry, scintillation counting of radioisotopes, and non-selective microelectrodes. Moreover, the use of drugs with well-established effects on the electrolyte metabolism of a cellular system, e.g., ouabain (Hall, 1977a) can also, in principle, be used to provide a tissue model (whose chemistry has been modified from the norm in a defined way) for the further checking and validation of a preparative method and of the microprobe procedure.

CONCLUDING REMARKS

"When the microscope is improved we shall have the cells analyzed, and all will be electricity or somewhat else."

Emerson

Electron optics and x-ray spectrometry technologies have evolved, and recently fused, to provide an analytical "microscope" that can at least define the inorganic composition of individual cells and their constituent organelles. We have seen that while the analytical instrumentation has reached a state of advanced sophistication, our understanding of the nature of the biological specimen and of the methods of preparing it for microprobe analysis has not yet reached a comparable level of sophistication. The complexity and evanescent nature of the biological matrix, and the often conflicting requirements of the biological microprobe analyst, contribute to the hitherto slow progress of an otherwise exciting technique. These specimen problems achieve some sort of working perspective if we divide biological microprobe applications into two very general categories.

The first category involves the analysis of relatively insoluble products and endogenous compounds, e.g., histochemical reaction products, mineralized bodies in normal or pathological tissues, drug localization in target cells, etc., where a high degree of structural fidelity in the specimen is critically important. Given these requirements, one should perhaps start with a more or less conventional preparative procedure and progress to other procedures that involve fewer chemical constraints but that may still yield the necessary structural detail, e.g., freeze-substitution. Finally and if at all possible, a technique that involves a minimum of chemical interference (i.e., cryomicrotomy) may be undertaken to confirm the previous analytical findings. In other words, there is no virtue in adopting complicated preparative regimes if simpler methods yield satisfactory *answers to the specific questions asked.*

The second category of applications involves the analysis of diffusible substances in cellular and extracellular compartments. The preservation of specimen chemical integrity and fidelity thus becomes the major priority; morphological information is of but secondary importance. Thus, in this context, the microprobe must essentially be considered an analytical instrument, not a microscope with a convenient analytical attachment. With these established priorities, we may more readily discard more traditional methods of specimen preparation in favor of those that, with considerable manipulative care, involve the minimum of chemical compromises. Here again, simplicity is a virtue. For example, the demanding frozen-hydrated preparative procedures offer advantages over cryosectioning/freeze-drying only in cases where water-filled compartments are to be analyzed. Unfortunately, we must learn to accept the problems associated with the poor and restrictive morphological information both these procedures provide. A necessary prelude to the cryosectioning approach is a detailed study of tissue morphology in more conventionally processed specimens; familiarity with the disposition of the tissue eases the imaging and interpretative difficulties encountered in cryosections.

To conclude, no single preparative regime can be used to the exclusion of all others in all circumstances. It is similarly unwise to consider the electron microprobe, even with its unique capabilities, as an analytical instrument divorced from all other analytical techniques.

ADDENDUM

The field of biological x-ray analysis in its many specialized aspects is continuously evolving and expanding; techniques are regularly optimized for specific purposes or attain new practical perspective as advances are made in adjacent areas. In this climate of rapid growth, development, and exploitation of a powerful technology, it is impossible for a single review chapter, however comprehensive and however limited its terms of reference, to accurately capture and reflect the "state of the art." In other words, even the most well-intentioned review is inevitably doomed to be out of date (but hopefully not redundant!) even with the most ambitious publishing schedule. The need to convey definitive new information to an eager population of practitioners is met to some extent by the current trend of publishing collections of papers presented at specialized symposia that are individually exposed to thorough critical scrutiny by independent referees. Attention is drawn to four recent publications of this kind, which appeared either during the preparation of the above chapter (in which case data from relevant papers were incorporated) or immediately after its completion:

Echlin, P., and Kaufmann, R. (eds.) 1978. *Microscopica Acta, Suppl. 2: Microprobe Analysis in Biology and Medicine.* S. Hirzel Verlag, Stuttgart.
(Collection of papers presented at an "International Conference on Microprobe Analysis in Biology and Medicine" in Münster, Germany, 4–8 September, 1977.)

Echlin, P. (ed.) 1978. *Low Temperature Biological Microscopy and Microanalysis.* The Royal Microscopical Society, Oxford, England.
(Collection of papers presented at a meeting entitled "Low Temperature Biological Microscopy" at Cambridge in April 1977, organized on behalf of the R.M.S. by Appleton, Echlin and Robards.)

Lechene, C., and Warner, R. (eds.) 1979. *Microbeam Analysis in Biology.* Academic Press, New York.
(Collection of papers presented at a "Workshop on Biological X-Ray Microanalysis by Electron Beam Excitation," Boston, August 25–26, 1977.)

Johari, O. (ed.) 1979. *Scanning Electron Microscopy (1979) Vols. 1, 2, and 3.* Scanning Electron Microscopy, Inc., AMF. O'Hare, Illinois.
(Collection of papers presented at a conference entitled "International Review of Advances in Techniques and Applications of the Scanning Electron Microscope," April 1979.)

These meetings sponsored by SEM, Inc. are important *annual* international conferences. Details of past and future meetings (the 1980 Conference is to be held on April 21–25 in Chicago) may be obtained from the Editor at SEM Inc., P.O. Box 66507, AMF O'Hare, Il. 60666, U.S.A.).

I would like to thank Miss J. Hughes, Mrs. D. Blake, and Mrs. J. Rees for patiently and expertly typing the manuscript, and also Mr. S. Jones for excellent photographic assistance. I am indebted to all those authors who kindly allowed me to reproduce figures and data from their own original papers, and in a very many cases for providing me with halftone prints of their figures. Finally, I would like to formally acknowledge the invaluable help given by my two colleagues, Dr. David A. Erasmus and Mr. Tudor W. Davies, who were constantly available for discussion and who both read the manuscript and made many suggestions that could only improve the chapter.

LITERATURE CITED

Afzelius, B. A. 1962. Chemical fixatives for electron microscopy. In: *The Interpretation of Ultrastructure* (Harris, R. J. C., ed.), p. 1. Academic Press, New York.

Agostini, B., and Hasselbach, W. 1971. Electron cytochemistry of calcium uptake in the fragmented sarcoplasmic reticulum. Histochemie 28:55.

Ali, S. Y., Craig Gray, J., Wisby, A., and Phillips, M. 1977. Preparation of thin cryo-sections for electron probe analysis of calcifying cartilage. J. Microscopy 111:65.

Andersen, C. A. 1967. An introduction to the electron probe microanalyser and its application to biochemistry. In: *Methods of Biochemical Analysis, Vol. 15* (Glick, D., ed.), p. 147. Intersciences, New York.

Appleton, T. C. 1968. The application of autoradiography to the study of soluble compounds. *Acta Histochem.*, Suppl., 8:115.

Appleton, T. C. 1969. The possibilities of locating soluble labelled compounds by electron microscope autoradiography. In: *Autoradiography of Diffusible Compounds* (Stumpf, W. E., and Roth, L. J., eds.), p. 301. Academic Press, New York.

Appleton, T. C. 1972a. 'Dry' ultrathin frozen sections for electron microscopy and x-ray microanalysis: the cryostat approach. Micron 3:101.

Appleton, T. C. 1972b. 'Dry' ultrathin frozen sections of unfixed and unembedded biological material for x-ray microanalysis of naturally occurring diffusible electrolytes: the cryostat approach. J. Microscopie 13:144.

Appleton, T. C. 1974a. A cryostat approach to ultrathin 'dry' frozen sections for electron microscopy; a morphological and x-ray analytical study. J. Microscopy 100:49.

Appleton, T. C. 1974b. Cryoultramicrotomy: possible applications in cytochemistry. In: *Electron Microscopy and Cytochemistry* (Wisse, E., Daems, W. Th., Molenaar, I., and van Duijn, P., eds.), p. 229. North-Holland Publishing Co., Amsterdam.

Appleton, T. C. 1977. The use of ultrathin frozen sections for x-ray microanalysis of diffusible elements. In: *Analytical and Quantitative Methods in Microscopy* (Meek, G. A. and Elder, H. Y., eds.), p. 247. Cambridge University Press, Cambridge.

Appleton, T. C. 1978a. The contribution of cryoultramicrotomy to x-ray microanalysis in biology. In: *Electron Probe Microanalysis in Biology* (Erasmus, D. A., ed.), p. 148. Chapman and Hall, London.

Appleton, T. C. 1978b. Physiology and pharmacology at a sub-cellular level. In: *Microsc. Acta Suppl. 2. Microprobe Analysis in Biology and Medicine* (Echlin, P., and Kaufmann, R., eds.), p. 1. S. Hirzel Verlag, Stuttgart.

Asahina, E., Shimada, K., and Hisada, J. 1970. A stable state of frozen protoplasm with invisible intracellular ice crystals obtained by rapid cooling. Exp. Cell Res. 59:349.

Bacaner, M., Broadhurst, J., Hutchinson, T. E., and Lilley, J. 1973. Scanning transmission electron microscope studies of deep-frozen unfixed muscle correlated with spatial localization of intracellular elements by fluorescent x-ray analysis. Proc. Nat. Acad. Sci. 70:3423.

Bachmann, L., and Schmitt, W. W. 1971. Improved cryofixation applicable to freeze-etching. Proc. Nat. Acad. Sci. 68:2149.

Baker, J. R. 1960. *Cytological Technique*, 4th Edition, p. 18. Methuen & Co., Ltd., London. John Wiley & Sons Inc., New York.

Bald, W. B., and Robards, A. W. 1978. A device for the rapid freezing of biological specimens under precisely controlled and reproducible conditions. J. Microscopy 112:3.

Bank, H., and Mazur, P. 1973. Vizualization of freezing damage. J. Cell Biol. 57:729.

Barkhaus, R. H., and Höhling, H. J. 1978. Electron microscopical microprobe analysis of freeze-dried and unstained mineralized epiphyseal cartilage. Cell Tiss. Res. 186:541.

Barnard, T., and Sevéus, L. 1978. Preparation of biological material for x-ray microanalysis of diffusible elements. II. Comparison of different methods of drying ultrathin cryosections cut without a trough liquid. J. Microscopy 112:281.

Bauer, H., and Sigarlakie, E. 1973. Cytochemistry on ultrathin frozen sections of yeast cells. Localization of acid and alkaline phosphatase. J. Microscopy 99:205.

Beaman, D. R., Nishiyama, R. H., and Penner, J. A. 1969. The analysis of blood diseases with the electron microprobe. Blood 34:401.

Bell, L. G. E. 1956. Freeze-drying. In: *Physical Techniques in Biological Research* (Oster, G., and Pollister, A. W., eds.), Vol. 3, p. 1. Academic Press, New York.

Bernhard, W. 1971. Improved techniques for the preparation of ultrathin frozen sections. J. Cell Biol. 49:731.

Bernhard, W., and Leduc, E. H. 1967. Ultrathin frozen sections. I. Methods and ultrastructural preservation. J. Cell Biol. 34:757.

Bernhard, W., and Viron, A. 1971. Improved techniques for the preparation of ultrathin frozen sections. J. Cell Biol. 49:731.

Bondareff, W. 1967. Demonstration of an intercellular substance in mouse cerebral cortex. Z. Zellforsch. 81:366.

Boothroyd, B. 1964. The problems of demineralization in thin sections of fully calcified bone. J. Cell Biol. 20:165.

Boothroyd, B. 1968. The adaptation of the technique of micro-incineration to electron microscopy. J. R. Microsc. Soc. 88:529.

Bowen, I. D., and Ryder, T. A. 1977. The application of x-ray microanalysis to enzyme cytochemistry. In: *Electron Microscopy of Enzymes: Principles and Methods* (Hayat, M. A., ed.), Vol. 5, p. 187. Van Nostrand Reinhold, New York.

Bowen, I. D., and Ryder, T. A. 1978. The application of x-ray microanalysis to histochemistry. In: *Electron Probe Microanalysis in Biology* (Erasmus, D. A., ed.), 183. Chapman and Hall, London.

Bowen, I. D., Ryder, T. A., and Downing, N. L. 1976. An x-ray microanalytical azo dye technique for the localization of acid phosphatase activity. Histochemistry 49:43.

Bowen, I. D., Ryder, T. A., and Winters, C. 1975. The distribution of oxidizable mucosubstances and polysaccharides in the planarian *Polycelis tenuis* Iijima. Cell Tiss. Res. 161:263.

Boyde, A., Grainger, F., and James, D. W. 1969. Scanning electron microscopic observations of chick embryo fibroblasts in vitro, with particular reference to the movement of cells under others. Z. Zellforsch. Mikrosk. Anat. 94:46.

Boyde, A., James, D. W., Tresman, R. L., and Willis, R. A. 1968. Outgrowth from chick embryo spinal cord *in vitro*, studied with the scanning electron microscope. Z. Zellforsch. Mikrosk. Anat. 90:1.

Boyde, A., and Williams, J. C. P. 1968. Surface morphology of frog striated muscle as prepared for and examined in the scanning electron microscope. J. Physiol. 197:108.

Boyde, A., and Wood, C. 1969. Preparation of animal tissues for surface-scanning electron microscopy. J. Microscopy 90:221.

Brombach, J. D. 1975. Electron beam x-ray microanalysis of frozen biological bulk specimens below 130 K. II. The electrical charging of the sample in quantitative analysis. J. Microsc. Biol. Cell. 22:233.

Bullivant, S. 1965. Freeze-substitution and supporting techniques. Lab. Invest. 14:1178.

Bullivant, S. 1970. Present status of freezing techniques. In: *Some Biological Techniques in Electron Microscopy* (Parson, D. F., ed.), p. 101. Academic Press, New York.

Burstone, M. S. 1969. Cryobiology techniques in histochemistry, including freeze-drying and cryostat techniques. In: *Physical Techniques in Biological Research* (Pollister, A. W., ed.), 4th ed., Vol. 3, *Cells and Tissues*, p. 1. Academic Press, New York.

Carroll, K. G., and Tullis, J. L. 1968. Observations on the presence of titanium and zinc in human leucocytes. Nature 217:1172.

Castaing, R. 1951. Thesis, University of Paris, ONERA Publ., No. 55.

Chandler, J. A. 1973. Recent developments in analytical electron microscopy. J. Microscopy 98:359.

Chandler, J. A. 1975a. Electron probe microanalysis in cytochemistry. In: *Techniques of Biochemical and Biophysical Morphology*, Vol. 2 (Glick, D., and Rosenbaum, R. M., eds.), p. 307. John Wiley & Sons Inc., New York.

Chandler, J. A. 1975b. Application of x-ray microanalysis to pathology. J. Microsc. Biol. Cell. 22:425.

Chandler, J. A. 1977a. X-ray microanalysis in the electron microscope. In: *Practical Methods in Electron Microscopy*, Vol. 5 (Glauert, A. M., ed.), p. 315. North-Holland Publishing Company, Amsterdam.

Chandler, J. A. 1977b. Wavelength dispersive x-ray microanalysis in biological research. In: *Analytical and Quantitative Methods in Microscopy* (Meek, G. A., and Elder, H. Y., eds.), p. 227. Cambridge University Press, Cambridge.

Chandler, J. A. 1978. The application of X-ray microanalysis in TEM to the study of ultrathin biological specimens—a review. In: *Electron Probe Microanalysis in Biology* (Erasmus, D. A., ed.), p. 37. Chapman and Hall, London.

Chandler, J. A., and Battersby, S. 1976a. X-ray microanalysis of zinc and calcium in ultrathin sections of human sperm cells, using the pyroantimonate technique. J. Histochem. Cytochem. 24:740.

Chandler, J. A., and Battersby, S. 1976b. X-ray microanalysis of ultrathin frozen and freeze-dried sections of human sperm cells. J. Microscopy 107:55.

Christensen, A. K. 1969. A way to prepare thin frozen sections of tissue for electron microscopy. In: *Autoradiography of Diffusible Substances* (Stumpf, W. E., and Roth, R. J., eds.), p. 349. Academic Press, New York.

Christensen, A. K. 1971. Frozen thin sections of fresh tissue for electron microscopy, with a description of pancreas and liver. J. Cell Biol. 51:772.

Coleman, J. R., Nilsson, J. R., Warner, R. R., and Batt, P. 1973a. Electron probe analysis of refractive bodies in *Amoeba proteus*. Exp. Cell Res. 76:31.

Coleman, J. R., Nilsson, J. R., Warner, R. R., and Batt, P. 1973b. Effects of calcium and strontium on divalent ion contents of refractive granules in *Tetrahymena pyriformis*. Exp. Cell Res. 80:1.

Coleman, J. R., and Terepka, A. R. 1972. Electron probe analysis of the calcium distribution in cells of the embryonic chick chorioallantoic membrane. I. A critical evaluation of techniques. J. Histochem. Cytochem. 20:401.

Coleman, J. R., and Terepka, A. R. 1974. Preparatory methods for electron probe analysis. In: *Principles and Techniques of Electron Microscopy: Biological Applications* (Hayat, M. A., ed.), Vol. 4, p. 159. Van Nostrand Reinhold Company, New York.

Colvin, J. R., Sowden, L. C., and Male, R. S. 1975. Variability of the iron, copper and mercury contents of individual red blood cells. J. Histochem. Cytochem. 23:329.

Costa, J. L., Reese, T. S., and Murphy, D. L. 1974. Serotonin storage in platelets: estimation of storage-packet size. Science 183:537.

Costa, J. L., Detwiler, T. C., Feinman, R. D., Murphy, D. L., Patlak, C. S., and Pettigrew, K. D. 1977a. Quantitative evaluation of the loss of human-platelet dense bodies following stimulation by thrombin or A23187. J. Physiol. 264:297.

Costa, J. L., Tanaka, Y., Pettigrew, K., and Cushing, R. J. 1977b. Evaluation of the utility of air-dried whole mounts for quantitative electron microprobe studies of platelet dense bodies. J. Histochem. Cytochem. 25:1079.

Costello, M. J., and Corless, J. M. 1978. The direct measurement of temperature changes within freeze-fractive specimens during rapid quenching in liquid coolants. J. Microscopy 112:17.

Davies, T. W., and Erasmus, D. A. 1973. Cryo-ultramicrotomy and x-ray microanalysis in the transmission electron microscope. Sci. Tools 20:9.

Davies, T. W., and Morgan, A. J. 1976. The application of x-ray analysis in the transmission electron analytical microscope (TEAM) to the quantitative bulk analysis of biological microsamples. J. Microscopy 107:47.

DeFilippis, L. F., and Pallaghy, C. K. 1973. Effect of light on the volume and ion relations of chloroplasts in detached leaves of *Elodea densa*. Australian J. Biol. Sci. 26:1251.

DeFilippis, L. F., and Pallaghy, C. K. 1975. Localization of zinc and mercury in plant cells. Micron 6:111.

Dempsey, G. P., and Bullivant, S. 1976a. A copper block method for freezing non-cryoprotected tissue to produce ice crystal-free regions for electron microscopy. I. Evaluation using freeze-substitution. J. Microscopy 106:251.

Dempsey, G. P., and Bullivant, S. 1976b. A copper block method for freezing non-cryoprotected tissue to produce ice crystal-free regions for electron microscopy. II. Evaluation using freeze fracturing with a cryo-ultramicrotome. J. Microscopy 106:261.

Diculescu, I., Popescu, L. M., Ionescu, N., and Butucescu, N. 1971. Ultrastructural study of calcium distribution in cardiac muscle cells. Z. Zellforsch. Mikrosk. Anat. 121:181.

Dörge, A., Rick, R., Gehring, K., Mason, J., and Thurau, K. 1975. Preparation and applicability of freeze-dried sections in the microprobe analysis of biological soft tissue. J. Microsc. Biol. Cell. 22:205.

Doty, S. B., Lee, C. W., and Banfield, W. G. 1974. A method for obtaining ultrathin frozen sections from fresh or glutaraldehyde-fixed tissues. Histochem. J. 6:383.

Dudley, H. R., and Spiro, D. 1961. The fine structure of bone cells. J. Biophys. Biochem. Cytol. 11:627.

Duprez, A., and Vignes, A. 1967. Détermination ponctuelle du potassium intracellulaire d'hematies humaines. Analyze a la microsonde electronique de Castaing. Comp. Rend. Soc. Biol. 161:1358.

Echlin, P. 1975. The preparation of frozen-hydrated biological material for x-ray microanalysis. J. Microsc. Biol. Cell. 22:215.

Echlin, P. 1978. Low temperature scanning electron microscopy: a review. J. Microscopy 112:47.

Echlin, P., and Moreton, R. 1974. The preparation of biological materials for x-ray microanalysis. In: Microprobe Analysis as Applied to Cells and Tissues (Hall, T., Echlin, P., and Kaufmann, R., eds.), p. 159. Academic Press, New York.

Echlin, P., Saubermann, A., Franks, F., and Skaer, H. leB. 1978. The use of polymeric cryoprotectants in the preparation of biological material for x-ray microanalysis. Microsc. Acta Suppl. 2. Microprobe Analysis in Biology and Medicine (Echlin, P., and Kaufmann, R., eds.), p. 64. S. Hirzel Verlag, Stuttgart.

Echlin, P., Skaer, H. le B., Gardiner, B. O. C., Franks, F., and Asquith, M. H. 1977. Polymeric cryoprotectants in the preservation of biological ultrastructure. II. Physiological effects. J. Microscopy 110:239.

Edelmann, L. 1978a. Visualization and x-ray microanalysis of potassium tracers in freeze-dried and plastic embedded frog muscle. Microsc. Acta Suppl. 2. Microprobe Analysis in Biology and Medicine (Echlin, P., and Kaufmann, R., eds.), p. 166. S. Hirzel Verlag, Stuttgart.

Edelmann, L. 1978b. A simple freeze-drying technique for preparing biological tissue without chemical fixation for electron microscopy. J. Microscopy 112:243.

Elbers, P. F. 1966. Ion permeability of the egg of Limnaea stagnalis L. on fixation for electron microscopy. Biochem. Biophys. Acta 112:318.

Eränkö, O. 1954. Quenching of tissues for freeze-drying. Acta Anatomica 22:331.

Erasmus, D. A. 1974. The application of x-ray microanalysis in the transmission electron microscope to a study of drug distribution in the parasite Schistosoma mansoni (Platyhelminthes). J. Microscopy 102:59.

Erasmus, D. A. 1975. Schistosoma mansoni: Development of the vitelline cell, its role in drug sequestration, and changes induced by Astiban. Exp. Parasitol. 38:240.

Erasmus, D. A., and Davies, T. W. 1979. Schistosoma mansoni and S. haematobium: aspects of the calcium metabolism of the vitelline cell. Expt. Parasitol. 47:91.

Fernández-Morán, H. 1957. Electron microscopy of nervous tissue. In: Metabolism of the Nervous Tissue (Richter, D., ed.), p. 1. Pergamon Press, London.

Fernández-Morán, H. 1960. Low-temperatures preparation techniques for electron microscopy of biological specimens based on rapid freezing with liquid helium. II. Ann. N.Y. Acad. Sci. 85:689.

Fernández-Morán, H. 1961. Lamellar systems in myelin and photoreceptors as revealed by high resolution electron microscopy. In: Macromolecular Complexes (Edds, M. V., ed.), p. 113. Ronald Press, New York.

Fisher, D. B. 1972. Artefacts in the embedment of water-soluble compounds for light microscopy. Plant Physiol. 49:161.

Fisher, D. B., and Housley, T. H. 1972. The retention of water-soluble compounds during freeze-substitution and microautoradiography. Plant Physiol. 49:166.

Fletcher, N. H. 1970. The Chemical Physics of Ice. Cambridge University Press, Cambridge.

Forrest, Q. C., and Marshall, A. T. 1976. Comparative x-ray microanalysis of frozen-hydrated and freeze-substituted specimens. Proc. 6th Europ. Congr. Electron Microscopy, Jerusalem, 1976.

Foster, J. R., and Erasmus, D. A. 1980. *Fasciola hepatica*: elemental changes in rat liver during the acute phase of infection. Exp. Parasitol. (in press).

Franks, F. 1977. Biological freezing and cryofixation. J. Microscopy 111:3.

Franks, F., Asquith, M. H., Hammond, C. C., Skaer, H. leB., and Echlin, P. 1977. Polymeric cryoprotectants in the preservation of biological ultrastructure. I. Low temperature states of aqueous solutions of hydrophilic polymers. J. Microscopy 110:223.

Franks, F., and Skaer, H. leB. 1976. Aqueous glasses as matrices in freeze-fracture electron microscopy. Nature 262:323.

Frederik, P. M., and Klepper, D. 1976. The possibility of electron microscopic auto-radiography of steroids after freeze drying of unfixed testes. J. Microscopy 106:209.

Fuchs, W., Brombach, J. D., and Trösch, W. 1978a. Charging effect in electron-irradiated ice. J. Microscopy 112:63.

Fuchs, W., and Lindemann, B. 1975. Electron beam x-ray microanalysis of frozen biological bulk specimens below 130K. I. Instrumentation and specimen preparation. J. Microsc. Biol. Cell. 22:227.

Fuchs, W., Lindemann, B., Brombach, J. D., and Trösch, W. 1978b. Instrumentation and specimen preparation for electron beam x-ray microanalysis of frozen hydrated bulk specimens. J. Microscopy 112:75.

Furseth, R. 1969. The fine structure of the cellular cementum of young human teeth. Arch. Oral Biol. 14:1147.

Garfield, R. E., Henderson, R. M., and Daniel, E. E. 1972. Evaluation of the pyroantimonate technique for localization of tissue sodium. Tissue Cell 4:575.

Gehring, K., Dórge, A., Wunderlich, P., and Thurau, K. 1973. Uber die Andwendung der elektronenstahmikroanalyse an geftiegetrocknelen Kryoschnitten und an Tiefgefrorenem, wasserhaltigem Gewebe von Rattennieren. In: *Beitrage zur Electronenmikroskofischen Direktabbilgung von Oberflachen Remy RA Verlag Munster* 5:937.

George, S. G., Nott, J. A., Pirie, B. J. S., and Mason, A. Z. 1976. A comparative quantitative study of cadmium retention in tissues of a marine bivalve during different fixation and embedding procedures. Proc. R. Microsc. Soc. Vol. 11, Part 5, Micro 76 Suppl., p. 42.

Gersh, I., and Stephenson, J. L. 1954. Freezing and drying of tissues for morphological and histochemical studies. In: *Biological Applications of Freezing and Drying* (Harris, R. J. C., ed.), p. 329. Academic Press, New York.

Geymayer, W., Grasenick, F., and Hödl, Y. 1978. Stabilizing ultrathin cryo-sections by freeze-drying. J. Microscopy 112:39.

Ghadially, F. N. 1979. Invited review. The technique and scope of electron-probe x-ray analysis in pathology. Pathology 11:95.

Gielink, A. J., Sauer, G., and Ringoet, A. 1966. Histoautoradiographic localization of calcium in oat plant tissues. Stain Technol. 41:281.

Glimcher, M. J., and Krane, S. M. 1968. The organisation and structure of bone, and the mechanism of calcification. In: *Treatise on Collagen. Vol. 2, Part B, Biology of Collagen* (Gould, B. S., ed.), p. 67. Academic Press, New York.

Greenawalt, J. W., and Carafoli, E. 1966. Electron microscope studies on the active accumulation of Sr^{++} by rat-liver mitochondria. J. Cell Biol. 29:37.

Gullasch, J., and Kaufmann, R. 1974. Energy-dispersive x-ray microanalysis in soft biological tissues: relevance and reproducibility of the results as depending on specimen preparation (air drying, cryo-fixation, cold-stage techniques). In:

Microprobe Analysis as Applied to Cells and Tissues (Hall, T., Echlin, P., and Kaufmann, R., eds.), p. 175, Academic Press, New York.

Gullasch, J., Kaufmann, R., and Werner, G. 1973. Electron probe microanalysis in wet biological tissues. Beitr. Electronenmikroskop. Direktabb. Oberfl. 5:937.

Gupta, B. L. 1976. Water movement in cells and tissues. In: *Perspectives in Experimental Biology* (Spencer Davies, P., ed.), Vol. 1, p. 25. Pergamon, Oxford.

Gupta, B. L. 1979. The electron microprobe x-ray analysis of frozen-hydrated sections with new information on fluid transplanting epithelia. In: *Microbeam Analysis in Biology* (Lechene, C., and Warner R., eds.). Academic Press, New York.

Gupta, B. L., Berridge, M. J., Hall, T. A., and Moreton, R. B. 1978a. Electron microprobe and ion-selective microelectrode studies of fluid secretion in the salivary glands of *Calliphora*. J. Exp. Biol. 72:261.

Gupta, B. L., Hall, T. A., Maddrell, S. H. P., and Moreton, R. B. 1976. Distribution of ions in a fluid-transporting epithelium determined by electron-probe x-ray microanalysis. Nature 264:284.

Gupta, B. L., Hall, T. A., and Moreton, R. B. 1977. Electron microprobe x-ray analysis. In: *Transport of Ions and Water in Animals* (Gupta, B. L., Moreton, R. B., Oschman, J. L., and Wall, B. J., eds.), p. 83. Academic Press, New York.

Gupta, B. L., Naftalin, R. J., and Hall, T. A. 1978b. Microprobe measurements of concentrations and gradients of Na, K and Cl in epithelial cells and intercellular spaces of rabbit ileum. Nature 272:70.

Hall, T. A. 1971. The microprobe assay of chemical elements. In: *Physical Techniques in Biological Research, 2nd ed., Vol. 1A* (Oster, G., ed.), p. 157. Academic Press, New York.

Hall, T. A. 1977a. Electron probe x-ray microanalysis. In: *An Introduction to the International Conference on Microprobe Analysis in Biology and Medicine*, Munster/Westphalia (FRG), 4th–8th Sept., 1977.

Hall, T. A. 1977b. Problems of the continuum-normalization method for the quantitative analysis of sections of soft tissue. *Proceedings of the workshop in Biological X-ray Microanalysis by Electron Beam Excitation, Boston*, Aug. 25–26.

Hall, T. A., Clarke Anderson, H., and Appleton, T. 1973. The use of thin specimens for x-ray microanalysis in biology. J. Microscopy 99:177.

Hall, T. A., and Gupta, B. L. 1974. Measurement of mass loss in biological specimens under an electron microbeam. In: *Microprobe Analysis as Applied to Cells and Tissues* (Hall, T., Echlin, P., and Kaufmann, R., eds.), p. 147. Academic Press, New York.

Hall, J. L., Yeo, A. R., and Flowers, T. J. 1974. Uptake and localization of rubidium in the halophyte *Suaeda maritima*. Z. Pflanzenphysiologie 71:200.

Hanzon, V., and Hermodsson, L. H. 1960. Freeze drying of tissues for light and electron microscopy. J. Ultrastruct. Res. 4:332.

Harvey, D. M. R., Flowers, T. J., and Hall, J. L. 1976a. Localization of chloride in leaf cells of the halophyte *Suaeda maritima* by silver precipitation. New Phytol. 77:319.

Harvey, D. M. R., Hall, J. L., and Flowers, T. J. 1976b. The use of freeze-substitution in the preparation of plant tissue for ion localization studies. J. Microscopy 107:189.

Hayat, M. A. 1975. Positive Staining for Electron Microscopy. Van Nostrand Reinhold, New York.

Hayat, M. A. 1980. *Principles and Techniques of Electron Microscopy: Biological Applications, Vol. 1*, 2nd Ed., p. 5. University Park Press, Baltimore.

Hayat, M. A. 1981. Fixation for Electron Microscopy. Academic Press, New York.

Heckman, C. A., and Barrnett, R. 1973. GACH: A water-miscible, lipid-retaining embedding polymer for electron microscopy. J. Ultrastruct. Res. 42:156.

Hereward, F. V., and Northcote, D. H. 1972. A simple freeze-substitution method for the study of ultrastructure of plant tissue. Exp. Cell Res. 70:73.

Hodson, S., and Marshall, J. 1969. A device for cutting ultrathin, unfixed, frozen section for electron microscopy. J. Physiol. 201:63P.

Hodson, S., and Marshall, J. 1970. Ultracryotomy: a technique for cutting ultrathin sections of unfixed, frozen, biological tissues for electron microscopy. J. Microscopy 91:105.

Hodson, S., and Marshall, J. 1972. Evidence against through-section thawing whilst cutting on the ultracryotome. J. Microscopy 95:459.

Hodson, S., and Williams, L. 1976. Ultracryotomy of biological tissues to preserve membrane structure. J. Cell Sci. 20:687.

Hoffman, J. F. 1966. The red cell membrane and the transport of sodium and potassium. Am. J. Med. 41:666.

Höhling, H. J., and Nicholson, W. A. P. 1975. Electron microprobe analysis in hard tissue research: specimen. J. Microsc. Biol. Cell. 22:185.

Höhling, H. J., Steffens, H., Stamm, G., and Mays, U. 1976. Transmission microscopy of freeze dried, unstained epiphyseal cartilage of the guinea pig. Cell Tiss. Res. 167:243.

Howell, S. L., and Tyhurst, M. 1974. Cryo-ultramicrotomy of islets of Langerhans. Some observations on the fine structure of mammalian islets in frozen sections. J. Cell Sci. 15:591.

Howell, S. L., and Tyhurst, M. 1976. 45-Calcium localization in islets of Langerhans, a study by electron-microscope autoradiography. J. Cell Sci. 21:415.

Hutchinson, T. E. 1977. Energy dispersive x-ray microanalysis. In: *Analytical and Quantitative Methods in Microscopy* (Meek, G. A., and Elder, H. Y., eds.), p. 213. Cambridge University Press, Cambridge.

Iglesias, J. R., Bernier, R., and Simard, R. 1971. Ultracryotomy: a routine procedure. J. Ultrastruct. Res. 36:271.

Ingram, F. D., and Ingram, M. J. 1975. Quantitative analysis with the freeze-dried plastic embedded tissue specimen. J. Microsc. Biol. Cell. 22:193.

Ingram, F. D., Ingram, M. J., and Hogben, C. A. M. 1972. Quantitative electron probe analysis of soft biological tissue for electrolytes. J. Histochem. Cytochem. 20:716.

Ingram, F. D., Ingram, M. J., and Hogben, C. A. M. 1974. An analysis of the freeze-dried, plastic embedded electron probe specimen preparation. In: *Microprobe Analysis as Applied to Cells and Tissues* (Hall, T., Echlin, P., and Kaufmann, R., eds.), p. 119. Academic Press, New York.

Ingram, M. J., and Hogben, C. A. M. 1968. Procedures for the study of biological soft tissue with the electron microprobe. In: *Developments in Applied Spectroscopy* (Baer, W. K., Perkins, A. J., and Grove, E. L., eds.), p. 43. Plenum Press, New York.

Jessen, H., Peters, P. D., and Hall, T. A. 1974. Sulphur in different types of keratohyaline granules. A quantitative assay by x-ray microanalysis. J. Cell Sci. 15:359.

Jessen, H., Peters, P. D., and Hall, T. A. 1976. Sulphur in epidermal keratohyaline granules: a quantitative assay by x-ray microanalysis. J. Cell Sci. 22:161.

Kashiwa, H. K. 1966. Calcium in the cells of fresh bone stained with glyoxal bis (2-hydroxyanil). Stain Technol. 41:49.

Kashiwa, H. K. 1968. The glyoxal bis (2-hydroxyanil) method for differential staining of intracellular calcium in bone. In: *Parathyroid Hormone and Thyrocalcitonin (Calcitonin)* (Talmage, R. V., and Belanger, L. F., eds.). Excerpta Medica International Congress Series No. 159.

Kaufmann, R., and Gullasch, J. 1972. Rasterelektronenmikroskopische und mikroanalytische Untersuchungen an wasserhaltigen, tiefgefrorenen Präparaten. Beitr. Elecktronenmikroskop. Direktabb. Oberfl. 5:855.

Kimzey, S. L., and Burns, L. C. 1973. Electron probe microanalysis of cellular potassium distribution. Ann. N.Y. Acad. Sci. 204:486.

Kirk, R. G., Crenshaw, M. A., and Tosteson, D. C. 1974. Potassium content of single human red cells measured with an electron probe. J. Cell. Physiol. 84:29.

Kirk, R. G., and Lee, P. 1978. X-ray microanalysis of cation and hemoglobin contents in red blood cells. In: *Microsc. Acta Suppl. 2. Microprobe Analysis in Biology and Medicine* (Echlin, P., and Kaufmann, R., eds.), p. 102. S. Hirzel Verlag, Stuttgart.

Komnick, H. 1962. Lokalisation von Na^+ und Cl^- in Zellen und Geweben. Protoplasma 55:414.

Krames, B., and Page, E. 1968. Effects of electron microscopic fixatives on cell membranes of the perfused rat heart. Biochim. Biophys. Acta 150:24.

Kuhn,, C. 1972. A comparison of freeze substitution with other methods for preservation of the pulmonary alveolar lining layer. Am. J. Anat. 133:495.

Landis, W. J., Hauschka, B. T., Rogerson, C. A., and Glimcher, M. J. 1977a. Electron microscopic observations of bone tissue prepared by ultracryomicrotomy. J. Ultrastruct. Res. 59:185.

Landis, W. J., Paine, M. C., and Glimcher, M. J. 1977b. Electron microscopic observations of bone tissue prepared anhydrously in organic solvents. J. Ultrastruct. Res. 59:1.

Läuchli, A. 1975. Precipitation technique for diffusible substances. J. Microsc. Biol. Cell. 22:239.

Läuchli, A., Spurr, A. R., and Epstein, E. 1971. Lateral transport of ions into the xylem of corn roots. II. Evaluation of a stellar pump. Plant Physiol. 48:118.

Läuchli, A., Spurr, A., and Wittkopp, R. W. 1970. Electron probe analysis of freeze-substituted, epoxy resin embedded tissue for ion transport studies in plants. Planta 95:341.

Läuchli, A., Stelzer, R., Guggenheim, R., and Henning, L. 1974. Precipitation techniques as a means of intracellular ion localization by use of electron probe analysis. In: *Microprobe Analysis as Applied to Cells and Tissues* (Hall, T., Echlin, P., and Kaufmann, R., eds.), p. 107. Academic Press, New York.

Lechene, C. P., Bronner, C., and Kirk, R. G. 1976. Electron probe microanalysis of chemical elemental content of single human red cells. J. Cell Physiol. 90:117.

Lever, J. D., Santer, R. M., Lu, K. S., and Presley, R. 1977. Electron probe x-ray microanalysis of small granulated cells in rat sympathetic ganglia after sequential aldehyde and dichromate treatment. J. Histochem. Cytochem. 25:295.

Livingston, D. C., Coombs, M. M., Franks, L. M., Maggi, V., and Gahan, P. B. 1969. A lead phthalocyanin method for the demonstration of acid hydrolase in plant and animal tissues. Histochemie 18:48.

Lüttge, U., and Weigl, J. 1965. Zur Mikroautoradiographic wasserloslicher Substanzen. Planta 64:28.

Luyet, B., and Gonzales, F. 1951. Recording ultra rapid changes in temperature. Refrig. Eng. 59:1191.

Mackenzie, A. P. 1965. Factors affecting the mechanism of transformation of ice into water vapour in the freeze-drying process. Ann. N.Y. Acad. Sci. 125:522.

Malhotra, S. K. 1968. Freeze-substitution and freeze-drying in electron microscopy. In: *The Interpretation of Cell Structure* (McGee-Russell, S. M., and Ross, K. F. A., eds.), p. 11. Edward Arnold Ltd., London.

Malhotra, S. K., and Van Harreveld, A. 1965. Some structural features of mitochondria in tissues prepared by freeze-substitution. J. Ultrastruct. Res. 12:473.

Marinozzi, V. 1964. Cytochimie ultrastructurale du nucléole, RNA et protéines intranucléolaires. J. Ultrastruct. Res. 10:433.

Marshall, J. 1972. Ionic analysis of frozen sections in the electron microscope. Micron 3:99.

Marshall, A. T. 1974. X-ray microanalysis of frozen-dried and frozen-hydrated biological specimens. Proc. 8th Int. Cong. Electron Microscopy, Canberra II, 74.

Marshall, A. T. 1977. Electron probe x-ray microanalysis of frozen-hydrated biological specimens. Microsc. Acta 79, 254.

Marshall, A. T., and Wright, A. 1973. Detection of diffusible ions in insect osmoregulatory systems by electron probe x-ray microanalysis, using scanning electron microscopy and a cryogenic technique. Micron 4:31.

Maunder, C. A., Yarom, R., and Dubowitz, V. 1977. Electron-microscopic x-ray microanalysis of normal and diseased human muscle. J. Neurol. Sci. 33:323.

Maynard, P. V., Elstein, M., and Chandler, J. A. 1975. The effect of copper on the distribution of elements in human spermatozoa. J. Reproduct. Fert. 43:41.

McIntyre, I. A., Gilula, N. B., and Karnovsky, M. J. 1974. Cryoprotectant-induced redistribution of intramembranous particles in mouse lymphocytes. J. Cell Biol. 60:192.

Mehard, C. W., and Volcani, B. E. 1975. Evaluation of silicon and germanium retention in rat tissues and diatoms during cell and organelle preparation for electron probe microanalysis. J. Histochem. Cytochem. 23:348.

Mehard, C. W., and Volcani, B. E. 1976. Silicon in rat liver organelles: electron probe microanalysis. Cell Tiss. Res. 166:255.

Meryman, H. T. 1966. Freeze-drying. In: Cryobiology (Meryman, H. T., ed.), p. 609. Academic Press, New York.

Miyamoto, Y., and Moll, W. 1971. Measurements of dimensions and pathway of red cells in rapidly frozen lungs in situ. Resp. Physiol. 12:141.

Mizuhira, V. 1976. Elemental analysis of biological specimens by electron probe microanalysis. Acta Histochem. Cytochem. 9:69.

Monroe, R. G., Gamble, W. J., Lafarge, C. G., Gamboa, R., Morgan, C. L., Rosenthal, A., and Bullivant, S. 1968. Myocardial ultrastructure in systole and diastole, using ballistic cryofixation. J. Ultrastruct. Res. 22:22.

Moor, H. 1964. Die Gefierfixation labender Zellen und Ihre Anwendung in der Electronenmikroskopie. Z. Zellforsch. 62:546.

Moor, H. 1971. Recent progress in the freeze-etching technique. Phil. Trans. R. Soc. Lond. B. Biol. Sci. 261:121.

Moor, H. 1973. Cryotechnology for the structural analysis of biological material. In: *Freeze-etching Techniques and Applications* (Beneditte, E. L., and Favard, P., eds.), p. 11. Societé Francaise de Microscopie Électronique, Paris.

Morgan, A. J. 1980. Strontium metabolism in the earthworm, *Lumbricus terrestris*. I. Morphology of calciferous glands and chloragogenous tissues; preliminary electron microprobe analysis. Cell Tiss. Res. (in press).

Morgan, A. J., and Bellamy, D. 1973. Microanalysis of the elastic fibres of rat aorta. Age Ageing 2:61.

Morgan, A. J., and Davies, T. W. 1976. X-ray analysis in TEM and SEM of calcium metabolism in the earthworm *Lumbricus terrestris*, using strontium as a marker. Proc. Roy. Microsc. Soc. Micro 76 Suppl. Vol. 11, Part 5, p. 33.

Morgan, A. J., Davies, T. W., and Erasmus, D. A. 1975. Changes in the concentration and distribution of elements during electron microscope preparative procedures. Micron 6:11.

Morgan, A. J., Davies, T. W., and Erasmus, D. A. 1978. Specimen preparation. In: *Electron Probe Microanalysis in Biology* (Erasmus, D. A., ed.), p. 94. Chapman and Hall, London.

Müller, H. R. 1957. Gefriertrocknung als Fixierungsmethode an Pflanzenzellen. J. Ultrastruct. Res. 1:109.

Nath, J. 1972. Correlative biochemical and ultrastructural studies on the mechanism of freeze damage to ram semen. Cryobiology 9:240.

Neeracher, H. 1966. Transportuntersuchungen an *Zea mays* mit Hilfe von THO and Mikroautoradiographie. Ber. Schweiz. Bot. Ges. 75:303.

Nei, T. 1978. Structure and function of frozen cells: freezing patterns and post-thaw survival. J. Microscopy 112:197.

Nei, T., Yotsumoto, H., Hasegawa, Y., and Nagasawa, Y. 1973. Direct observation of frozen specimens with a scanning electron microscope. J. Elect. Microsc. 22:185.

Nermut, M. V. 1977. Freeze-drying for electron microscopy. In: *Principles and Techniques of Electron Microscopy: Biological Applications* (Hayat, M. A., ed.), Vol. 7, p. 79. Van Nostrand Reinhold Company, New York.

Neumann, D. 1973. Zur Darstellung pflanzlicher Gewebe nach Gefriersubstitution unter besonderer B'rucksichtigung der Strukturerhaltung der plastiden. Acta Histochem. 47:278.

Neumann, D., Krajewski, Th., and Schumann, K. 1974. Ionenbestimung in Zellkompartimenten durch energiedispersive Röntgenanalyse. Naturwiss. 61:214.

Oschman, J. L., Hall, T. A., Peters, P. D., and Wall, B. J. 1974. Microprobe analysis of membrane-associated calcium deposits in squid giant axon. J. Cell Biol. 61:156.

Oschman, J. L., and Wall, B. J. 1972. Calcium binding to intestinal membranes. J. Cell Biol. 55:58.

Ostrowski, K., Komender, J., Koscianek, H., and Kwarecki, K. 1962a. Quantitative investigation of the P and N loss in the rat liver when using various media in the freeze-substitution technique. Experientia 18:142.

Ostrowski, K., Komander, J., Koscianek, H., and Kwarecki, K. 1962b. Quantitative studies on the influence of the temperature applied in freeze-substitution on P, N, and dry mass losses in fixed tissues. Experientia 18:227.

Padawer, J. 1974. Identification of mast cells in the scanning electron microscope by means of x-ray spectrometry. J. Cell Biol. 61:641.

Pallaghy, C. K. 1973. Electron probe microanalysis of potassium and chloride in freeze-substituted leaf sections of *Zea mays*. Austral. J. Biol. Sci. 25:1015.

Panessa, B. J., and Russ, J. C. 1975. Techniques for practical biological microanalysis. Proc. 8th Ann. SEM Symp., p. 251.

Pautard, F. G. E. 1972. In: *The Comparative Molecular Biology of Extracellular Matrices* (Slavkin, H. C., ed.), p. 440. Academic Press, New York.

Pearse, A. G. E. 1968. *Histochemistry: Theoretical and Applied. 3d ed. Vol. 1, Chap. 3, Freeze-drying of biological tissues*, p. 27. *Chap. 4, Freeze-substitution of tissues and sections*, p. 59. J. & A. Churchill Ltd., London.

Pease, D. C. 1966a. The preservation of unfixed cytological detail by dehydration with "inert" agents. J. Ultrastruct. Res. 14:356.

Pease, D. C. 1966b. Anhydrous ultrathin sectioning and staining for electron microscopy. J. Ultrastruct. Res. 14:379.

Pease, D. C. 1967a. Eutectic ethylene glycol and pure propylene glycol as substituting media for the dehydration of frozen tissue. J. Ultrastruct. Res. 21:75.

Pease, D. C. 1967b. The preservation of tissue fine structure during rapid freezing. J. Ultrastruct. Res. 21:89.

Pease, D. C., and Peterson, R. G. 1972. Polymerizable glutaraldehyde-urea mixtures as polar, water-containing embedding media. J. Ultrastruct. Res. 41:133.

Penttila, A., Kalimo, H., and Trump, B. F. 1974. Influence of glutaraldehyde and/or osmium tetroxide on cell volume, ion content, mechanical stability, and membrane permeability of Ehrlich ascites tumor cells. J. Cell Biol. 63:197.

Pieri, C., Zs.-Nagy, I., Zs.-Nagy, V., Giuli, C., and Bertoni-Freddari, C. 1977. Energy dispersive x-ray microanalysis of the electrolytes in biological bulk specimen. II. Age-dependent alterations in the monovalent ion contents of cell nucleus and cytoplasm in rat liver and brain cells. J. Ultrastruct. Res. 59:320.

Plattner, H., Schmitt-Fumian, W. W., and Bachmann, L. 1973. Cryofixation of single cells by spray freezing. In: Freeze-etching (Benedetti, E. L., and Favard, P., eds.), p. 81. Société Francaise de Microscopie Électronique, Paris.

Posner, A. S. 1972. In: The Comparative Molecular Biology of Extracellular Matrices (Slavkin, H. C., ed.), p. 437. Academic Press, New York.

Rebhun, L. I. 1961. Applications of freeze-substitution to electron microscope studies of invertebrate oocytes. J. Biophys. Biochem. Cytol. 9:785.

Rebhun, L. I. 1965. Freeze-substitution as a function of water concentration in cells. Fed. Proc. 24 (Suppl. 15), S217.

Rebhun, L. I. 1972. Freeze-substitution and freeze-drying. In: Principles and Techniques of Electron Microscopy: Biological Applications (Hayat, M. A., ed.), Vol. 2, p. 1. Van Nostrand Reinhold Company, New York.

Rebhun, L. I., and Sander, G. 1971. Electron microscope studies of frozen-substituted marine eggs. 1. Conditions for avoidance of ultracellular ice crystallization. Am. J. Anat. 130:1.

Renaud, S. 1959. Superiority of alcoholic over aqueous fixation in the histochemical detection of calcium. Stain Technol. 34:267.

Rick, R., Dörge, A., and von Arnim, E. 1978. X-ray microanalysis of frog skin epithelium: evidence for a syncytial Na transport compartment. Microsc. Acta Suppl. 2. Microprobe Analysis in Biology and Medicine (Echlin, P., and Kaufmann, R., eds.), p. 156. S. Hirzel Verlag, Stuttgart.

Riehle, W. 1968. Ueber die Vitrifizierung verduunter wässriger Lösungen. Ph.D. Thesis. Juris Druck and Verlag, Zurich.

Riehle, U., and Hoechli, M. 1973. The theory and technique of high pressure freezing. In: Freeze-etching (Benedetti, E. L., and Favard, P., eds.), p. 31. Société Francaise de Microscopie Électronique, Paris.

Roinel, N., and Morel, F. 1973. Problèmes liés à l'analyse du contenu élémentaire d'hematies isolées à l'aide de la microsonde électronique. J. Physiol. Paris 67:306A.

Roinel, N., and Passow, H. 1975. Analyse quantitative au moyen de la microsonde électronique du contenu en sodium et potassium d'hematies humaines isolées. J. Microsc. Biol. Cell. 22:475.

Roinel, N., Passow, H., and Malorey, P. 1974. The study of the applicability of the electron microprobe to a quantitative analysis of K and Na in single human red blood cells. F.E.B.S. Letters 41:81.

Roomans, G. M., and Boekestein, A. 1978. Distribution of ions in *Neurospora crassa* determined by quantitative electron microprobe analysis. Protoplasma 95:385.

Routledge, L. M., Amos, W. B., Gupta, B. L., Hall, T. A., and Weis-Fogh, T. 1975. Microprobe measurements of calcium binding in the contractible spasmoneme of a Vorticellid. J. Cell Sci. 19:195.

Russ, J. C. 1978. Electron probe x-ray microanalysis—principles. In: *Electron Probe Microanalysis in Biology* (Erasmus, D. A., ed.), p. 5. Chapman and Hall, London.

Ryder, T. A., and Bowen, I. D. 1974. The use of x-ray microanalysis to investigate problems encountered in enzyme cytochemistry. J. Microscopy 101:143.

Ryder, T. A., and Bowen, I. D. 1977. The use of x-ray microanalysis to demonstrate the uptake of the molluscicide copper sulphate by slug eggs. Histochemistry 52:55.

Santer, R. M., Lever, J. D., and Davies, T. W. 1978. X-ray microanalysis in biogenic amine detection in sympatho-chromaffin system. In: *Peripheral Neuroendocrine Interaction* (Coupland, R. E., and Forssman, W. G., eds.), p. 29. Springer-Verlag, Berlin.

Saubermann, A. J., and Echlin, P. 1975. The preparation, examination and analysis of frozen hydrated tissue sections by scanning transmission electron microscopy and x-ray microanalysis. J. Microscopy 105:155.

Schoenberg, C. F., Goodford, P. J., Wolowyk, M. W., and Wootton, G. S. 1973. Ionic changes during smooth muscle fixation for electron microscopy. J. Mechanochem. Cell Motil. 2:69.

Sevéus, L. 1978. Preparation of biological material for x-ray microanalysis of diffusible elements. I. Rapid freezing of biological tissue in nitrogen slush and preparation of ultrathin frozen sections in the absence of trough liquid. J. Microscopy 112:269.

Sevéus, L., Brdiczka, D., and Barnard, T. 1978. On the occurrence and composition of dense particles in mitochondria in ultrathin frozen dry sections. Cell Biol. Int. Reports 2:155.

Simard, R. 1976. Cryoultramicrotomy. In: *Principles and Techniques of Electron Microscopy: Biological Applications* (Hayat, M. A., ed.), Vol. 6. Van Nostrand Reinhold Company, New York.

Simkiss, K. 1976. Intracellular and extracellular routes of biomineralization. Symp. Soc. Exp. Biol. 30:423.

Simson, J. A. V., Bank, H. L., and Spicer, S. S. 1979. X-ray microanalysis of pyroantimonate-precipitable cations. In: *Scanning Electron Microscopy* (Johari, O., ed.), Vol. II, p. 779. Scanning Electron Microscopy Inc., AMF O'Hare, Illinois.

Sjöstrand, F. S. 1967. Freeze-drying preservation. In: *Electron Microscopy of Cells and Tissues. Instrumentation and Techniques.* Vol. 1, p. 188. Academic Press, New York.

Sjöstrand, F. S., and Kretzer, F. 1975. A new freeze-drying technique applied to the analysis of the molecular structure of mitochondrial and chloroplast membranes. J. Ultrastruct. Res. 53:1.

Sjöström, M. 1975. X-ray microanalysis in cell biology. Detection of intracellular elements in the normal muscle cell and in muscle pathology. J. Microsc. Biol. Cell. 22:415.

Sjöström, M., Johansson, R., and Thornell, L. E. 1974. Cryo-ultramicrotomy of muscles in defined state. Methodological aspects. In: *Electron Microscopy and Cytochemistry* (Wiesse, E., Daems, W. Th., Molenaar, J., and van Duijn, P., eds.), p. 387. North-Holland Publishing Co., Amsterdam.

Sjöström, M., and Squire, J. M. 1977. Cryo-ultramicrotomy and myofibrillar fine structure: a review. J. Microscopy 111:239.

Sjöström, M., and Thornell, L. E. 1975. Preparing sections of skeletal muscle for transmission electron analytical microscopy (TEAM) of diffusible elements. J. Microscopy 103:101.

Sjöström, M., Thornell, L. E., and Cedergren, E. 1973. The application of cryo-ultramicrotomy in the study of the fine structure of myofilaments. J. Microscopy 99:193.

Skaer, H. le B., Franks, F., Asquith, M. H., and Echlin, P. 1977. Polymeric cryo-protectants in the preservation of biological ultrastructure. III. Morphological aspects. J. Microscopy 110:257.

Skaer, R. J., Peters, R. D., and Emmines, J. P. 1974. The localization of calcium and phosphorus in human platelets. J. Cell Sci. 15:679.

Skaer, R. J., Peters, P. D., and Emmines, J. P. 1976. Platelet dense bodies: a quantitative microprobe analysis. J. Cell Sci. 20:441.

Sleytr, U. B., and Umrath, W. 1974. A simple device for obtaining complementary fracture planes at liquid helium temperature. J. Microscopy 101:187.

Somlyo, A. P., Somlyo, A. V., Schuman, H., Sloane, B., and Scarpa, A. 1978. Electron probe analysis of calcium compartments in cryosections of smooth and striated muscles. Ann. N.Y. Acad. Sci. 307:523.

Somlyo, A. V. 1976. Ion compartmentalization in frog striated muscle: a study utilizing quantitative electron probe analysis and cryoultramicrotomy. Ph.D. dissertation, University of Pennsylvania No. 77-10223, University Microfilms Inc., Ann Arbor, Michigan.

Somlyo, A. V., Schuman, H., and Somlyo, A. P. 1977. Elemental distribution in striated muscle and the effects of hypertonicity. Electron probe analysis of cryo-sections. J. Cell Biol. 74:828.

Spriggs, T. L. B., and Wynne-Evans, D. 1976. Observations on the production of frozen-dried thin sections for electron microscopy using unfixed fresh liver, fast-frozen without cryoprotectants. J. Microscopy 107:35.

Spurr, A. R. 1969. A low-viscosity epoxy resin embedding medium for electron microscopy. J. Ultrastruct. Res. 26:31.

Spurr, A. R. 1972a. Freeze substitution additives for sodium and calcium retention in cells studied by x-ray analytical electron microscopy. Bot. Gaz. 133:263.

Spurr, A. R. 1972b. Freeze substitution systems in the retention of elements in tissues studied by x-ray analytical electron microscopy. In: *Thin-section Microanalysis* (Russ, J. C., and Panessa, J. B., eds.), p. 49. EDAX Laboratories, Raleigh N.C.

Steinbiss, H., and Schmitz, K. 1973. CO_2-Fixierung und Stofftransport in Benthischen marinen Algen. Planta 112:253.

Takaya, K. 1975a. Energy dispersive x-ray microanalysis of zymogen granules of mouse pancreas using fresh air-dried tissue spread. Arch. Histol. Jap. 37:387.

Takaya, K. 1975b. Electron probe microanalysis of the dense bodies of human blood platelets. Arch. Histol. Jap. 37:333.

Takaya, K. 1975c. Energy dispersive x-ray microanalysis of neurosecretory granules of mouse pituitary using fresh air-dried tissue spreads. Cell Tiss. Res. 159:227.

Takaya, K. 1975d. Intranuclear silicon detection in a subcutaneous connective tissue cell by energy-dispersive x-ray microanalysis using fresh air-dried spread. J. Histochem. Cytochem. 23:681.

Takaya, K. 1977. Electron microscopy of human melanosomes in unstained, fresh air-dried hair bulbs and their examination by electron probe microanalysis. Cell Tiss. Res. 178:169.

Talmon, Y., Davis, H. T., Scriven, L. E., and Thomas, E. L. 1979. Mass loss and etching of frozen hydrated specimens. J. Microscopy 117:321.

Tanaka, Y., Costa, J. L., Pettigrew, K. D., Yoda, Y., and Cushing, R. J. 1975. Quantitative electron microprobe studies of human-platelet dense bodies. *33rd Ann. Proc. Electron Microscopy Soc. Amer.* (Bailey, G. W., ed.), p. 322. Claitor's Publ. Div., Baton Rouge.

Tandler, C. J., Libanati, C. M., and Sauchis, C. A. 1970. The intracellular localization of inorganic cations with potassium pyroantimonate. J. Cell Biol. 45:355.

Tandler, C. J., and Solari, A. J. 1969. Nucleolar orthophosphate ions. Electron microscope and diffraction studies. J. Cell Biol. 41:91.

Taylor, P. G., and Burgess, A. 1977. Cold stage for an electron probe microanalyzer. J. Microscopy 111:51.

Termine, J. D. 1972. In: *The Comparative Molecular Biology of Extracellular Matrices* (Slavkin, H. C., ed.), p. 444. Academic Press, New York.

Tisher, C. C., Weavers, B. A., and Cirksena, W. J. 1972. X-ray microanalysis of pyroantimonate complexes in rat kidney. Am. J. Pathol. 69:255.

Tokuyasu, K. T. 1973. A technique for ultracryotomy of cell suspensions and tissues. J. Cell Biol. 57:551.

Trösch, W., and Lindemann, B. 1975. Electron-beam x-ray microanalysis of isolated, freeze-dried giant chromosomes. J. Microsc. Biol. Cell. 22:487.

Trump, B. F., and Mergner, W. J. 1974. Cell injury. In: *The Inflammatory Process*, 2nd ed., Vol. I (Zweifach, B. W., Grant, L., and McCluskey, R. T., eds.), p. 115. Academic Press, New York.

Umrath, W. 1974. Cooling bath for rapid freezing in electron microscopy. J. Microscopy 101:103.

Umrath, W. 1977. Prinzipien der Kryopräparationsmethoden: Gefriertrocknung und Gefrierätzung. Mikroskopie 33:11.

Valdre, U., and Horne, R. W. 1975. A combined freeze chamber and low temperature stage for an electron microscope. J. Microscopy 103:305.

Van Harreveld, A., and Crowell, J. 1964. Electron microscopy after rapid freezing on a metal surface and substitution fixation. Anat. Rec. 149:381.

Van Harreveld, A., Crowell, J., and Malhotra, S. K. 1965. A study of extracellular space in central nervous tissue by freeze-substitution. J. Cell Biol. 25:117.

Van Harreveld, A., and Steiner, J. 1970. Extracellular space in frozen and ethanol substituted central nervous tissue. Anat. Record 166:117.

Van Harreveld, A., Trubatch, J., and Steiner, J. 1974. Rapid freezing and electron microscopy for the arrest of physiological processes. J. Microscopy 100:189.

Van Steveninck, R. F. M., Chenoweth, A. R. F., and Van Steveninck, M. E. 1973. Ultrastructural localization of ions. In: *Ion Transport in Plants* (Anderson, W. P., ed.), p. 25. Academic Press, New York.

Van Steveninck, R. F. M., Van Steveninck, M. E., Hall, T. A., and Peters, P. D. 1974. A chlorine-free embedding medium for use in x-ray analytical electron microscopic localization of chlorine in biological tissue. Histochimie 38:173.

Van Zyl, J., Forrest, Q. G., Hocking, C., and Pallaghy, C. K. 1976. Freeze substitution of plant and animal tissue for the localization of water-soluble compounds by electron probe microanalysis. Micron 7:213.

Vassar, P. S., Hards, J. M., Brooks, D. E., Hagenberger, B., and Seaman, G. V. F. 1972. Physicochemical effects of aldehydes on the human erythrocyte. J. Cell Biol. 53:809.

Weibel, E. R., and Bolender, R. P. 1973. Stereological techniques for electron microscopic morphometry. In: *Principles and Techniques of Electron Microscopy. Biological Applications*, Vol. 3 (Hayat, M. A., ed.), p. 237. Van Nostrand Reinhold Company, New York.

Werner, G., and Gullasch, J. 1974. Röntgenmikroanalyse an Rattenspermien. Mikroskopie 30:95.

Whittam, R., and Wheeler, K. P. 1970. Transport across cell membranes. Ann. Rev. Physiol. 32:21.

Wilson, A. J., Robards, A. W., Quine, K. H., and Newman, T. 1979. Some experiences in the use of a polymeric cryoprotectant in the freezing of plant tissues. Proc. Roy. Microsc. Soc. Vol. 14, Part 4, p. 224.

Woolley, D. M. 1974. Freeze-substitution: a method for the rapid arrest and chemical fixation of spermatozoa. J. Microscopy 101:245.

Yarom, R., Hall, T. A., and Peters, P. D. 1975a. Calcium in myonuclei: electron microprobe x-ray analysis. Experientia 31:154.

Yarom, R., Maunder, C. A., Scripps, M., Hall, T. A., and Dubowitz, V. 1975b. A simplified method of specimen preparation for x-ray microanalysis of muscle and blood cells. Histochemistry 45:49.

Yarom, R., Peters, P. D., and Hall, T. A. 1974a. Effect of glutaraldehyde and urea embedding on intracellular ionic elements. X-ray microanalysis of skeletal muscle and myocardium. J. Ultrastruct. Res. 49:405.

Yarom, R., Peters, P. D., Scripps, M., and Rogel, S. 1974b. Effect of specimen preparation on intracellular myocardial calcium. Electron microscopic x-ray microanalysis. Histochemistry 38:143.

Zalokar, M. 1966. A simple freeze-substitution method for electron microscopy. J. Ultrastruct. Res. 15:469.

Zierold, K. 1976. X-ray microanalysis of cryofractured tissue specimens of the rat. Proc. 6th Europ. Congr. Electr. Microsc., Jerusalem, Vol. II, p. 223.

Zierold, K., and Schäfer, D. 1978. Quantitative x-ray microanalysis of diffusible ions in skeletal muscle bulk specimen. J. Microscopy 112:89.

Zierold, K., Schäfer, D., and Gullasch, J. 1978. Application of x-ray microanalysis to studies on distribution of diffusible ions. In: *Microsc. Acta Suppl. 2. Microprobe Analysis in Biology and Medicine* (Echlin, P., and Kaufmann, R., eds.), p. 92. S. Hirzel Verlag, Stuttgart.

Zs.-Nagy, I., Pieri, C., Giuli, C., Bertoni-Freddari, C., and Zs.-Nagy, V. 1977. Energy-dispersive x-ray microanalysis of the electrolytes in biological bulk specimen. I. Specimen preparation, beam penetration, and quantitative analysis. J. Ultrastruct. Res. 58:22.

Chapter 3

FROZEN-HYDRATED BULK SPECIMENS

Alan T. Marshall

Zoology Department,
La Trobe University,
Bundoora, Victoria, Australia

INTRODUCTION

Rationale

When biological samples are prepared by conventional procedures of fixation and dehydration for electron microscopy, they can be shown by electron probe x-ray microanalysis to contain few, if any, of the diffusible elements that they are known to contain in the living state. This loss of diffusible elements is well known (Morgan et al., 1978) and has been investigated in some detail by Morgan et al. (1975).

One approach to the problem of preventing loss of diffusible elements from biological specimens that are to be processed for electron probe x-ray microanalysis is to freeze them and maintain them in the frozen-hydrated state during analysis. The technological problems involved at every stage in this process, from freezing to obtaining x-ray spectra, are formidable. Two methods have evolved within the past six or seven years. These are the frozen-hydrated bulk specimen method (Gehring et al., 1973; Marshall and Wright, 1973; Marshall, 1974, 1975a,b, 1977a; Gullasch and Kaufmann, 1974; Lechene et al., 1975; Brombach, 1975; Fuchs and Lindemann, 1975; Forrest and Marshall, 1976; Zierold, 1976; Yeo et al., 1977; Fuchs et al., 1978a,b; Kramer and Preston, 1978; Zierold and Schafer, 1978) and the frozen-hydrated section method (Bacaner et al., 1972, 1973; Hutchinson et al., 1974; Moreton et al., 1974; Saubermann and Echlin, 1975; Gupta et al., 1976, 1978a,b; Saubermann et al., 1977; Saubermann, 1978). Both methods have advantages and disadvantages. The preparation and maintenance of bulk specimens in the frozen-hydrated state are technically easier than preparing and maintaining sections of frozen-hydrated tissues. However, there may be an advantage, in terms of spatial resolution of analysis, to be had in analyzing sections, although that has yet to be conclusively demonstrated. This chapter deals principally with the preparation and analysis of bulk specimens. *It should be stressed that the frozen-hydrated specimens referred to here are fully frozen-hydrated, that is, they are not deliberately or accidentally etched.* This type of specimen is intended to provide *absolute concentrations*, which would not be possible with etched specimens. It is the author's belief, however, that etched specimens will provide perfectly valid qualitative data and may possibly be successfully analyzed in a quantitative way by the ratio model (see Chapter 1).

Freezing

The literature on the freezing of cells and tissues is voluminous. It is divisible into two broad categories: 1) that dealing with slow freezing and the retention of viability and 2) that dealing with fast freezing and the faithful preservation of the living structure. It is with the latter that x-ray microanalysis has so far been concerned. This arises from the generally accepted belief that the less the morphological distortion, the greater will be the chance that solutes have not been redistributed during freezing.

The aim in freezing, therefore, is to eliminate or to reduce the dimensions of intracellular ice crystals. The achievement of perfect vitrification of intracellular ice is prevented by two things. First, the attainment of a high enough freezing rate to produce vitrification in specimens of reasonable size appears to be impossible. Opinions vary about what this rate should be. For vitrification of pure water, a freezing rate of $10^{10}\,°C\ s^{-1}$ has been calculated (Fletcher, 1971), while freezing rates of $2 \times 10^6\,°C\ s^{-1}$ (Riehle, 1968) and $5 \times 10^3\,°C\ s^{-1}$ (Stephenson, 1956) have been calculated to be necessary for biological specimens. Very few measured rates of freezing on tissues appear to be available. However, Glover and Garvitch (1974) found freezing rates between 0 and $-100°C$ to be approximately $1000°C$ s^{-1} and Echlin and Moreton (1976) record similar values. In order to achieve vitrification, the rate of heat removal by conduction must exceed the rate of heat production by crystallization. If heat production overtakes heat loss, then crystallization goes to completion before further cooling can occur (Stephenson, 1956). This is why it is important to reach the temperature at which transition takes place from the vitreous to the crystalline state as quickly as possible. The transition or recrystallization temperature is variously estimated for pure water to be in the region of $-130°C$ (Meryman, 1957) and in tissues to be in the region of $-100°C$ (Stephenson, 1956).

Rapid freezing is commonly achieved by plunging the specimen into a cryogenic fluid. The properties of cryogenic fluids and their rates of freezing have been investigated in detail by several authors (e.g., Stephenson, 1956; Rebhun, 1959; Glover and Garvitch, 1974; Echlin and Moreton, 1976). However, the freezing of specimens that are to be maintained in the frozen-hydrated state in an electron microscope is best carried out in melting nitrogen. This is recommended in order to avoid carrying over potentially volatile hydrocarbons into the electron microscope column, where they will be a potent source of contamination. The production of melting nitrogen is simply carried out by placing a dewar of liquid nitrogen in a vacuum evaporator and evacuating at 10^{-2} torr until the nitrogen solidifies. On removal from the evaporator the nitrogen will melt and remain in the molten state for some hours. Glover and Garvitch (1974) obtained a freezing rate of $1400°C$ s^{-1} with melting nitrogen and it appears to be an effective cryogen (Echlin and Moreton, 1976).

Freezing rates may be improved by the use of a device to promote high-velocity immersion in the cryogen. Such devices have been used by Rebhun (1972), Glover and Garvitch (1974), Echlin and Moreton (1976), Saetersdal et al. (1977), and Costello and Corless (1978). Freezing can also be carried out effectively by pressing the tissue against a block of metal that has been cooled by a cryogenic fluid (Van Harreveld et al., 1974; Christensen, 1972; Sjöström and Thornell, 1975; Dempsey and Bullivant, 1976a), and Bald and Robards (1978) have described a method of ultra-rapid freezing in pressurized cryogenic fluids. Once the specimen has been frozen, it can be stored under liquid nitrogen until further treatment.

Freeze-Fracturing

For analysis, an internal surface of the specimen must be exposed. This can be done in a variety of ways that may produce different morphological appearances at the surface.

The methods of exposing an internal surface may be broadly categorized as freeze-fracturing and freeze-cleaving. They both have close affinities to freeze-sectioning. The processes and conditions involved appear to be ill-defined and the literature is controversial. However, it is possible to see some trends by examining the literature on freeze-fracturing, freeze-cleaving, and freeze-sectioning. What emerges is that, on the one hand, freeze-fracturing is an impact-fracture carried out at very low temperatures and results in a surface that has a distinct topography *without* the necessity of etching, i.e., subliming ice. On the other hand, freeze-cleaving is the result of a cutting type of operation at rather higher temperatures and results in a smooth, featureless surface in which topography can only be revealed by etching. A correlation with temperature is not invariable and it may well be that the nature of the blade and the velocity at which the cleaving or fracturing stroke is made strongly influence the type of surface produced.

Dempsey and Bullivant (1976b) fracture under liquid nitrogen or on a cryo-ultramicrotome at a specimen temperature of $-180°C$ to produce a fracture surface with detailed topography. In contrast, fracture surfaces produced under high vacuum by freeze-cleaving at $-100°C$ are smooth (Southworth et al., 1975). This type of effect was related primarily to differences in temperature by Kirk and Dobbs (1976), who found that replicas of surfaces produced on a cryo-ultramicrotome were smooth at $-65°C$ and had a detailed topography at $-110°C$.

In freeze-sectioning it is now evident that, at low temperatures, fracturing occurs, whereas at higher temperatures cutting occurs (Saubermann et al., 1977). However, there are curious differences in the temperatures at which different authors claim to have sectioned without difficulty and to have observed cutting and fracturing phenomena. Thus, Appleton (1974) and Davies and Erasmus (1973) failed to obtain sections using specimen temperatures below $-85°C$ and $-100°C$, respectively, while Hodson and Marshall (1970), Doty et al. (1974), Roomans and Sevéus (1976), Spriggs and Wynne-Evans (1976), and Hodson and Williams (1976) were able to obtain sections at much lower temperatures. It seems very probable that sections "cut" at temperatures below $-100°C$ are in fact produced by a fracturing process, whereas at higher temperatures, i.e., above $-100°C$, true cutting is the process involved. The latter probably involves a momentary surface thawing as the sections are cut.

Freeze-fracturing has been reviewed by Sleytr and Robards (1977). These authors conclude that freeze-fracturing, freeze-cleaving, and freeze-sectioning will yield a similar gross topography if done at low enough temperatures. At higher temperatures there is a form of local heating during freeze-cleaving and freeze-sectioning that gives rise to smooth surfaces. This distinction is not recognized by Echlin (1978), who maintains that images with good topographical detail must be a consequence of etching.

In the present author's experience, freeze-fracturing at low temperatures produces a fracture surface with sufficient topographic detail to permit identification of structures for analysis (Marshall, 1977a). The thermal histories of these specimens are such that etching is a very unlikely possibility.

Low-Temperature Stages

Low-temperature stages can vary from the very crude to the very sophisticated. If analyses are qualitative and can be carried out rapidly, and if the probable growth of ice crystals together with some sublimation are acceptable, then all that is necessary is to place a liquid nitrogen–cooled substage carrying the specimen into a scanning electron microscope. Marshall and Wright (1973) used only a slight elaboration of this method to obtain very useful analyses of frozen-hydrated specimens. They kept the substage insertion rod in place and cooled it with liquid nitrogen, thereby slowing down the rate of temperature rise of the substage. In this way specimens could be examined and analyzed for periods up to approximately 30 min.

For prolonged examination of specimens without ice-crystal growth and sublimation occurring, more sophisticated arrangements are required. A variety of low-temperature stages for both SEMs and TEMs have been described (e.g., Gehring et al., 1973; Hörl, 1972; Nei et al., 1973; Griffiths and Venables, 1972; Cort and Steeds, 1972; Bacaner et al., 1973; Echlin and Moreton, 1973, 1974; Gullasch and Kaufmann, 1974; Hutchinson et al., 1974; Lechner, 1974; Dorge et al., 1975; Fuchs and Lindemann, 1975; Koch, 1975; Saubermann and Echlin, 1975; Marshall, 1977a; Taylor and Burgess, 1977; Varriano-Marston et al., 1977; Fuchs et al., 1978b). Many, but not all, of these stages have been capable of maintaining the specimen below the recrystallization temperature of ice, i.e., −130°C.

If commercially available low-temperature stages are used, then it seems that, without exception, they are modified to provide temperatures lower than those that they were designed to provide. These stages are usually designed for morphological work and without modification are not suitable for x-ray microanalysis.

Specimen Transfer

After freeze-fracturing, the exposed surface of the specimen must be protected from contamination by ice condensing from the atmosphere. The specimen should also be maintained at as low a temperature as possible, preferably below −130°C, which is the recrystallization temperature of pure water. A number of ingenious devices have been devised to accomplish this. The whole process of specimen transfer is considerably easier if the microscope has a reasonably large airlock. If an airlock is not provided by the manufacturer it is usually necessary to devise one.

Transfer of bulk specimens into the microscope for x-ray microanalysis has been accomplished by immersion in a container of liquid nitrogen (Gullasch and Kaufmann, 1974), by transporting in a portable airlock (Fuchs and Lindemann, 1975; Zierold, 1976; Fuchs et al., 1978b), or by means of a capping device (Echlin and Moreton, 1973; Marshall, 1977a). Capping devices for transporting frozen-hydrated sections have also been employed by Echlin and Moreton (1974) and protection of sections by means of a cold shroud has been used by Hutchinson et al. (1974), Lechner (1974), and Saubermann and Echlin (1975).

Specimen Coating

Frozen-hydrated specimens that have not been etched, charge in an electron beam. The degree of charging is dependent upon beam current and the temperature of the specimen (Marshall, 1975a; 1977a). The conductivity of ice decreases markedly below $-70°C$ (Durand et al., 1967) and charging of solutions of frozen gelatine commences at about $-80°C$ (Marshall, 1977a). Charging leads to the development of high surface potentials, which result in deflection of the primary electron beam (Marshall, 1975a) and a reduction in the energies of the injected electrons (Brombach, 1975; Marshall, 1977a). This situation can be dealt with by coating the specimen with an effective electrical conductor (Marshall, 1975a, 1977a; Fuchs and Lindemann, 1975; Fuchs et al., 1978b).

Echlin (1978) states that charging occurs only in etched specimens and not in uncoated fully frozen-hydrated specimens, when examined with low beam currents and high accelerating voltages. This is not the case, however, under the conditions used for microanalysis, i.e., relatively low accelerating voltages and high beam currents.

Brombach (1975) and Fuchs et al. (1978a) have shown that in uncoated specimens examined at low accelerating voltages, a space charge develops that drastically alters the shape of the electron interaction volume. The interaction volume, which is the source of the emitted x-rays, is calculated to be "pancake shaped" and is probably of extremely small dimensions. Thus the spatial resolution of analysis is potentially very high. A space charge still develops in coated specimens, but it is probably more limited than in uncoated specimens. Fuchs and Lindemann (1975) and Fuchs et al. (1978b) used carbon coating and low accelerating voltages (8 kV), a procedure that gives rise to a weak x-ray signal. These authors have suggested that controlled etching could be used to improve the signal. The removal of ice could possibly be quantified by observing the change in the oxygen peak with a thin window detector.

When specimens that have been coated with an efficient electrical conductor, such as chromium, are analyzed at relatively high kV, a good x-ray spectrum can be obtained (Marshall, 1977a). The spatial resolution still appears to be reasonably high, as established by empirical means (Marshall, 1977a) (Table 1).

An advantage of the use of chromium for coating, together with higher accelerating voltage, is that a larger secondary electron yield, and therefore

Table 1.   Analyses by static beam at two positions (separated by 2 μm) in the apical region of a cell in the hindgut of an insect

Element	Peak integrals	
	Cuticle	Apex of cell
K	123	444
Cl	1084	657
S	294	337
P	16	527
Mg	16	72
Na	165	47

better-quality secondary electron images, are obtained. The specimen is also able to withstand higher beam currents, since the thermal conductivity is greater.

Conductive coatings can be produced by evaporating carbon, aluminum, chromium (Marshall, 1977a), and beryllium (Marshall, 1977b). However, only chromium and beryllium coatings are consistently reproducible. Carbon requires a high evaporating temperature, with a risk of radiative heat damage to the specimen (Marshall, 1977a), and it is not a good electrical conductor. Aluminum almost invariably forms a black oxide (even though evaporation is carried out in a nitrogen atmosphere), which is a poor conductor. Aluminum coating has been advocated by Echlin and Moreton (1973, 1974), although they admit (Echlin and Moreton, 1974) that "our attempts at evaporating aluminum onto specimens have so far resulted only in a thin black (oxide?) deposit." The formation of this black oxide film seems to be virtually unavoidable.

Specimen Temperature

If fully frozen-hydrated specimens are to be analyzed it is important that specimen temperature is kept as low as possible during preparation and analysis in order to avoid sublimation. Below $-130°C$ the rate of sublimation of ice is very low (0.001 nm s^{-1}) even in a perfect vacuum (Koehler, 1968). It is also desirable to keep the specimen temperature below the ice recrystallization temperature in order to avoid any possibility of ice-crystal growth.

The accurate measurement of specimen temperature at every stage in the process is difficult. Marshall (1977a) estimated that specimen temperature did not rise above $-120°C$ at any stage prior to electron irradiation, while Fuchs et al. (1978b) estimate that their specimens do not rise in temperature above $-148°C$. During electron irradiation, specimen temperature may rise considerably in a localized region. It is certainly easy to drill a hole in a frozen-hydrated specimen with a static beam and a modest beam current. Beam damage has been noted by Fuchs and Lindemann (1975) and Marshall (1977a).

INSTRUMENTATION

The following section describes the instrumentation and instrument modifications used and developed by the author for the analysis of frozen-hydrated bulk specimens.

The microscope (Figure 1) is a JEOL JSM 35R scanning electron microscope that was modified in several ways. In addition, cold-cathode ionization gauges were fitted to facilitate monitoring gun and specimen chamber pressures. The microscope was always back-filled with dry nitrogen and the roughing pump was fitted with a molecular sieve trap to prevent back-streaming of rotary pump oil.

The standard cryostage was fitted with additional copper braid to increase thermal conduction, and it was also fitted with a heater operated by direct current. The thermocouple thermometer provided to monitor stage temperature was made more accurate by providing facilities for an external reference. (It should be noted that this thermometer is not good enough for

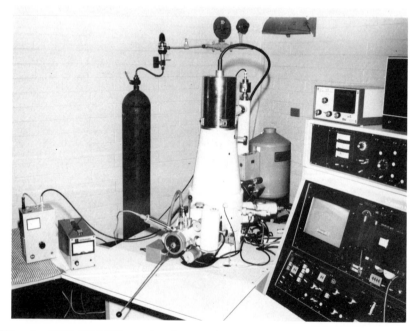

Figure 1.   JEOL JSM 35R scanning electron microscope equipped for low-temperature x-ray analysis.

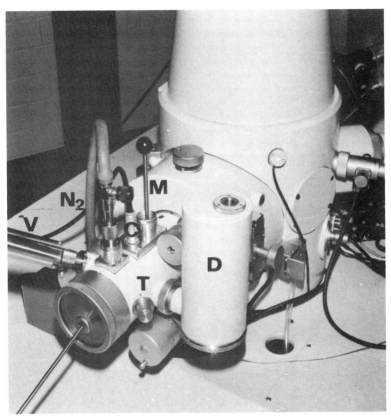

Figure 2. Airlock fitted with cryostage (D, cryostage dewar), manipulator (M), ion-sputtering cathode (C), nitrogen gas inlet (N_2), vacuum gauge (V), and a port for a thermocouple lead (T).

establishing absolute values of stage temperature, and these measurements must be made with a considerably more accurate thermometer and checked at intervals.) A Faraday cup for measuring beam current was mounted close to the specimen position. It consisted of an aluminum block that was thermally and electrically insulated and contained a deep, narrow-bore hole. Beam current is measured with a digital specimen-current amplifier (GW Electronics Type 9DM). Beam current is also monitored by means of the final lens aperture current. The thin foil aperture was insulated from the aperture holder, and the lead, which was normally used for applying a heating current to the aperture for cleaning, is used to monitor aperture current by means of the specimen-current amplifier.

The cryostage airlock was fitted with an inlet for dry nitrogen gas, a special manipulator for uncapping the specimen holder, a vacuum gauge, an ion-sputtering cathode, and a port for inserting a thermocouple (Figure 2).

The standard specimen holder was discarded and a new one was fabricated that allowed the specimen stub to be capped with a threaded airtight cap (Figure 3). A variety of specimen stubs may be used, but the most useful one is a vice-type.

Prior to insertion into the microscope, specimens are coated in a modified JEOL JEE 4 vacuum evaporator. The evaporator was provided with a cold stage that may be rotated 90° in the vertical plane, a port for inserting the specimen, a manipulator for uncapping the specimen holder, and a thermocouple for reading stage temperature. A single-edge razor blade can be fitted to the cold stage for fracturing specimens. The evaporating filament was enclosed in a slotted Pyrex shield and a slotted aluminum plate separated the filament from the specimen (Figure 4). Dry nitrogen gas, which is chilled by liquid nitrogen, can be bled into the bell jar.

Analyses are made with an EDAX energy-dispersive x-ray spectrometer (183 preamplifier/amplifier and 707 A multichannel analyzer). The resolution of the 10 mm² detector is 165 keV at Mn Kα, take-off angle is 40° at a working distance of 35 mm with zero specimen tilt, and the solid angle is less than 0.0040 steradians.

METHODOLOGY

Procedure

Tissue samples are removed from animals and quickly mounted in the vice-type stub in the specimen holder. The specimen holder is plunged into melting nitrogen so that the specimen comes into contact with the nitrogen first.

Figure 3.   Vice-type specimen holder.

Figure 4.   Modified vacuum evaporator with capped specimen holder in place on cold stage.

While under liquid nitrogen, the specimen stub is capped and the whole is transferred in liquid nitrogen to the vacuum-evaporator.

The cold stage of the vacuum-evaporator is precooled to below $-170°C$. Dry nitrogen gas (chilled with liquid nitrogen) is bled into the bell jar. The insertion port is opened and the specimen holder rapidly transferred, by means of a rod, from liquid nitrogen to the cold stage. The rotary pump valve is then opened and the bell jar is rough pumped while the gas supply is reduced and cut off. There is thus a positive pressure in the bell jar while the insertion port is open. This effectively prevents the entry of atmospheric air and the deposition of frost. The stage temperature is continuously monitored during the transfer process by means of a

Figure 5.   Specimen holder, uncapped and with insertion rod engaged.

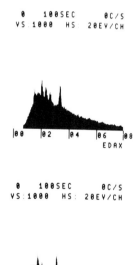

Figure 6.   Frozen-hydrated specimens. Upper spectrum from a beryllium-coated specimen, lower spectrum from a chromium-coated specimen (chromium Kα peak is at 5.4 keV).

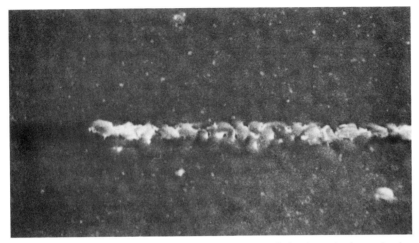

Figure 7. Thermal damage inflicted on a frozen-gelatine solution by a slowly moving beam $(1 \times 10^{-9} \text{ A})$. $\times$ 1760

Figure 8. Frozen-hydrated specimen showing an atypical fractured surface that is smooth and shows knife marks. $\times$ 1120

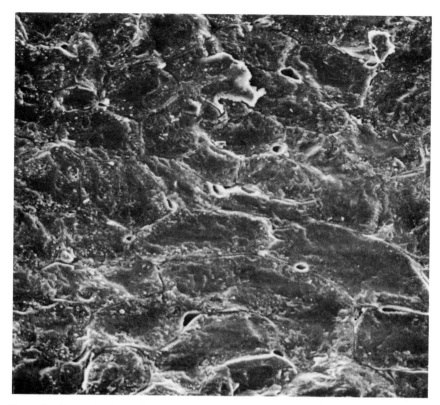

Figure 9.   Frozen-hydrated insect Malpighian tubules (*Teleogryllus oceanicus*), fractured in cross-section. × 200

thermocouple thermometer. Pumping is carried out until a pressure better than 2×10^{-5} torr is attained. At this pressure the specimen is rotated 90°, uncapped, and returned to the horizontal position (Figure 5). By re-engaging the specimen insertion rod the specimen holder can be moved forward in the cold stage so that the specimen is fractured by the cold blade. Following fracturing, the specimen is coated. During coating the specimen is rocked ±20°. After coating, the specimen is recapped and withdrawn from the evaporator and stored under liquid nitrogen until insertion into the SEM. With practice, the specimen temperature does not rise above −135°C during transfers in and out of the vacuum evaporator.

With the cold stage of the SEM airlock at −170°C, the airlock is leaked with dry nitrogen gas and the specimen holder rapidly transferred to the cold stage. After repeated flushing with nitrogen and evacuation, the airlock door leading into the column is opened, thus ensuring that the specimen cap is removed under the highest vacuum attainable. The specimen holder is inserted into the cryostage in the microscope column, where the stage temperature is -160 to $-165°C$. The specimen is now ready for analysis.

Coating

Coating is done either with 20 nm of chromium or 10 nm of berryllium. Coating with chromium is straightforward and presents no difficulties with respect to either application or analysis (Marshall, 1977a,b). Beryllium, however, is highly toxic (Schubert, 1958) and must be handled with extreme caution. The vacuum evaporator must exhaust directly to the exterior of the building and the bell jar should be repeatedly flushed with nitrogen after coating and before opening. The evaporating source should be well colli- mated with shields and apertures so that the deposition of beryllium on the equipment within the bell jar is restricted to as small an area as possible. All

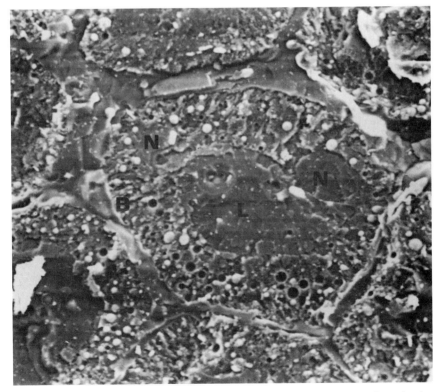

Figure 10. Frozen-hydrated insect Malpighian tubule fractured in cross-section, showing lumen (L), nuclei (N), and infoldings of basal plasma membrane (B). The white spheres are spherites of calcium phosphate and the tubule is surrounded by frozen hemolymph. × 1790.

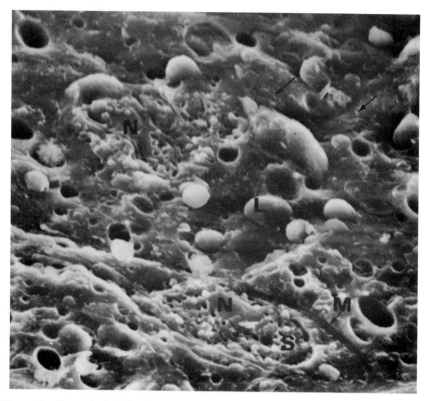

Figure 11.   Frozen-hydrated cockroach fat body showing lipid droplets (L) in fat storage cells, mitochondria (arrows), cell membrane (M), and nuclei (N) of mycetocyte cells containing symbionts (S). × 1680

coated equipment must be scrupulously cleaned. Beryllium is soluble in hydrochloric and sulfuric acid (McQuillan and Farthing, 1961) and coated Pyrex shields, etc., can be cleaned in these acids. Cleaning of the evaporator components should be carried out in a well-exhausted fume cupboard and protective gloves, mask, and goggles should be worn. Waste beryllium should be disposed of in accordance with the local health regulations.

Coating with beryllium is not recommended as a routine procedure, although this metal is probably the ideal coating material, having high electrical and thermal conductivity, Kα emission at 0.110 keV, and very low continuum production. The health hazard is considerable, and the precautions that must be observed make beryllium coating a very tedious and time-consuming process.

Similar specimens that have been coated with chromium and beryllium have essentially similar x-ray spectra (Figure 6). Chromium coating does not appear to result in significant absorption of sodium or magnesium x-rays (Marshall, 1977b,c), despite the high mass absorption coefficients, nor does it reduce the energies of injected electrons (Marshall, 1977a). The chromium $L\alpha$ peak would interfere with the oxygen $K\alpha$ peak if oxygen were being analyzed, as advocated by Fuchs et al. (1978b).

Beam Heating

Thermal damage by beam heating is incurred by a stationary or slowly moving beam with beam currents of 1×10^{-9} Å (Figure 7). Considerable beam current ($1-5 \times 10^{-9}$ A) can be tolerated by some specimens if the beam moves in a fast-scanning raster, preferably at television scanning speeds.

Fracture Surface

Freeze-fracturing under the conditions previously specified usually, but not always, yields a surface that exhibits considerable topography. Occasionally, a smooth, featureless surface showing the presence of knife marks is obtained (Figure 8). This type of surface must be interpreted, in accordance with Sleytr and Robards (1977), as a consequence of local heating.

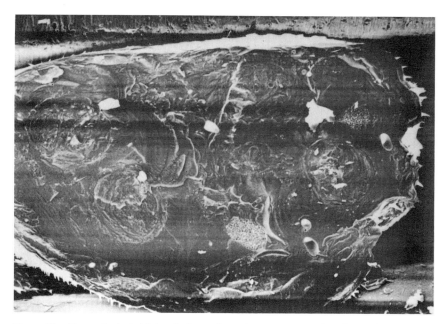

Figure 12.   Frozen-hydrated brine-fly larva shown in transverse section. Internal organs, surrounded by hemolymph, are clearly visible. Charge lines are due to debris on the fracture surface. × 70

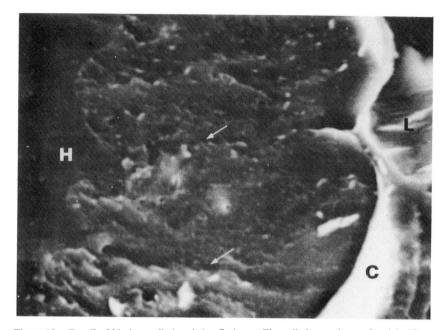

Figure 13. Detail of hindgut cells in a brine-fly larva. The cells have a layer of cuticle (C) on the apical surface, separating the cell cytoplasm from the frozen contents of the lumen (L). Intercellular spaces (arrows) are evident and channels produced by infoldings of the basal plasma membrane and mitochondria can be identified. A basement membrane separates the cytoplasm from the hemolymph (H). × 1790

Examples of surfaces, showing considerable morphological detail, are given in Figures 9–13.

QUANTITATIVE ANALYSIS

Pulse Pile-up and Live-Time Correction

Because of the extremely large continuum signal that is obtained at the low-energy end of the spectrum, problems arise from inaccuracies in live-time correction and pulse pile-up. The reasons for these problems have been discussed by the author in Chapter 1.

With the fast discriminator of the EDAX 183 preamplifier/amplifier (fast channel time constant = 0.5 μs) set just above the noise level (20 cps with no input signal) the output rate exceeds the input rate (e.g., output 680 cps, input 400 cps). This arises because the fast discriminator is set too high and the fast inspection channel misses the large number of low-energy pulses. Thus in 100 s of live-time, approximately 40,000 counts are obtained (Figure 14). When the fast discriminator is set to a noise level of 700 cps,

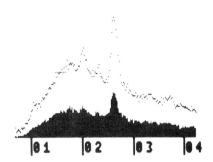

0   100SEC   0C/S
VS:2500   HS: 20EV/CH

Figure 14.   Spectra from a frozen-hydrated specimen. Fast discriminator set at 20 cps with no input signal (bars) and at 700 cps with no input signal (dots). The latter spectrum has more counts and is "cleaner."

0   100SEC 40112INT
VS:1000   HS: 20EV/CH

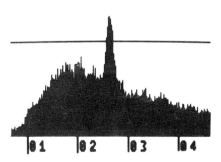

Figure 15.   Spectrum from a frozen-hydrated specimen with fast discriminator set at a noise level of 20 cps.

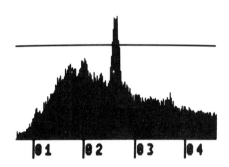

● 38SEC 40070INT
VS:1000   HS: 20EV/CH

Figure 16.   Spectrum from a frozen-hydrated specimen with fast discriminator set at a noise level of 700 cps. When compared with Figure 15, the spectrum is "cleaner." Pulse pile-up rejection has been more effective and the peaks are more evident.

the output rate remains approximately the same (650 cps) but the input rate is now 1200 cps and the total counts in 100 s live-time becomes 120,000 (Figure 14). When spectra that have the same number of counts are compared, it can be seen that pulse pile-up rejection is much more efficient with the fast discriminator set at 700 cps noise level and a much "cleaner" spectrum is obtained (Figures 15 and 16).

Selecting a Region for Analysis

To select an area of the fractured surface that is suitable for analysis it is necessary to choose a part of the specimen that is flat and that is in proper electrical contact with the cryostage.

A flat specimen surface is not always easy to recognize by eye. However, changes in local tilt-angle will result in changes in electron backscattering, which will be discernible by changes in the level of the continuum signal from a slowly scanning beam. Figure 17 shows an example of a line scan across a frozen salt solution in a polyethylene tube of 0.8-mm internal bore. The trace indicates that the surface is reasonably flat and that there is little variation in local tilt angle. If a scan rotation device is fitted to the microscope it is very useful for scanning the specimen in several different directions to assess the overall flatness.

Good electrical contact of the selected area should be checked by observing the energy at which the continuum falls to zero. If contact is not perfect, then the continuum will fall to zero at energies below the accelerating voltage. This is due to the slight charging that occurs in regions of this type.

The energy at which the continuum falls to zero should be carefully checked on specimens and standards on each occasion that an analysis is performed. Failure to standardize the injection energies will lead to erroneous results. This can be seen by considering the formula for x-ray intensity developed by Green and Cosslett (1961) (see Chapter 1):

$$I \simeq (U - 1)^{1.63}$$

Using this relationship for K Kα x-rays, a fall in injection voltage from 15 to 13.5 kV predicts a reduction in intensity of 20%. Since this expression was developed for pure elements, it may not be strictly applicable to specimens in an ice matrix; however, it does indicate a need for the utmost caution.

The area selected for analysis must be free from particles of fracture debris, since these will strongly influence the analysis.

Reproducibility

Reproducible analyses, as assessed on standards, can be made with fast rasters of reduced area. The degree of reproducibility declines as the area is reduced below 25 μm^2. The acceptable limit is about 4 μm^2. Analyses with static beams are not reproducible at the beam currents previously cited, i.e.,

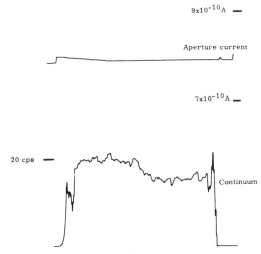

Figure 17.   Continuum recorded between 6200–7880 eV during a line scan across a frozen standard (approx. 1 mm diameter). Aperture current was simultaneously recorded.

Table 2.   Analysis of a 200 mmol l^{-1} KCl solution at 165°C, 15 kV, 0.5×10^{-10} A, 200 s

Element	Peak integrals*		
	100 μm^2 raster	25 μm^2 raster	Static spot
K	1089 ± 133 (5)	1095 ± 133 (4)	839 ± 214 (3)
Cl	1438 ± 247 (5)	1664 ± 178 (4)	1382 ± 312 (3)

* Mean ± SD, number of observations in parentheses.

1×10^{-9} A. Reproducibility with static beams can be improved (Table 2) by substantial reduction in beam current, but this will extend the analysis time. The cause of the decline in reproducibility with static beams is probably, in the case of high beam currents, the thermal damage. In the case of low beam currents, the limitation may be due to ice crystal size or possibly ion migration, as is found in glasses (Varshneya et al., 1966; Goodhew, 1975). Further investigations of this problem are being done by this author.

Standards

Quantitative analysis can be carried out by the ratio model (Russ, 1973) or by the use of matching standards and calibration curves. The ratio model

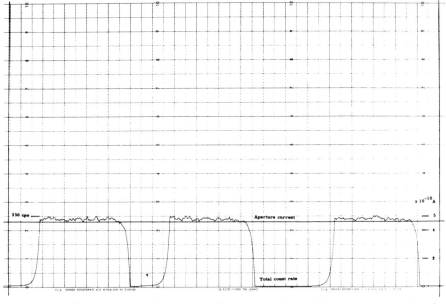

Figure 18.   Showing the typical constancy of beam current (recorded as aperture current) and reproducibility (recorded as total count rate) for three 200-s analyses of a frozen solution of KCl.

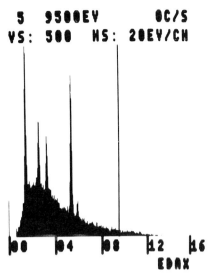

5   9500EV   0C/S
VS: 500   HS: 20EV/CH

0   4   8   12   16
EDAX

Figure 19.   Spectrum from a frozen-hydrated standard. The analysis was made at a nominal accelerating potential of 15 kV. The continuum cut-off is 13.5 kV. For atomic ratios calculated by computer, the best results are given by using an assumed injection potential of 9.5 kV. This figure is derived by taking a tangent from the midpoint of the continuum curve and noting the value at which it intersects the energy scale (indicated by position of cursor).

has been tested by Marshall (1977a) and both methods are discussed in detail in Chapter 1.

The matching standard method is used to obtain absolute concentrations (this is the major reason for using fully frozen-hydrated rather than etched specimens) and for this purpose it is necessary to have a means of monitoring beam current during the analysis. As already described, the author records aperture current as a means of monitoring beam current. Beam current must be kept constant during an analysis and from one analysis to another. The results can then be expressed comparatively in terms of mass per unit volume.

The reproducibility with constant beam current can be seen in Figure 18, where several analyses of the same potassium chloride standard are shown. The total count rates are identical for every analysis. It is also essential to monitor the energy at which cut-off of the continuum occurs, since this may vary, as previously discussed.

Although surface charging potentials are eliminated by coating samples with chromium, it seems probable that an internal space charge can still develop in the ice matrix and that this can cause retardation of the injected electrons. The practical consequence of this seems to be that a high-energy tail is present in the spectrum. This means that if the continuum cut-off

Table 3. Analysis of 200 mmol l⁻¹
KCl at − 165° C using a nominal accelerating voltage of 15 kV and an
observed continuum cut-off at 13.5
kV

Element	Atomic ratios	
	13.5 kV	9.5 kV
K	1.00	1.00
Cl	1.18	1.08

value is used for calculating atomic ratios, erroneous results may occur, since it cannot be assumed that the effective electron energy has a "delta-function-like distribution" (see Brombach, 1975). Some estimate of the effective injection voltage can be made by taking a tangent to the continuum curve from about the midpoint of the spectrum. An example is shown in Figure 19, and the effect on the calculation of atomic ratios is shown in Table 3.

For frozen-hydrated specimens, matching standards can be prepared from fixed and leached tissue samples that are equilibrated in salt solutions before freezing or from protein-containing salt solutions. However, there is no discernible advantage in using these instead of simple aqueous salt solutions as standards; indeed, there is some evidence that selective ion binding by fixed proteins may occur and give misleading results (Marshall, unpublished data). An example of quantitative analysis carried out by means of standards composed of simple salt solutions follows.

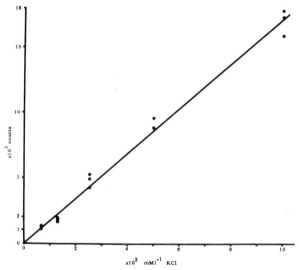

Figure 20. Calibration curve for potassium derived from frozen solutions of potassium chloride.

EXAMPLE OF AN ANALYSIS

Figure 13 shows a transverse fracture through ileal cells in the posterior region of a frozen-hydrated brine-fly larva (*Ephydrella* sp.). Areas from this type of specimen were analyzed using a $10 \times 10\,\mu m$ rapid scanning raster at a magnification of $\times 1000$ and a beam current of 1×10^{-9} A.

Standards were analyzed under exactly the same conditions as the specimens. In order to obtain absolute ion concentrations, a calibration line was fitted by the least-squares method to peak-minus-background counts from a range of potassium chloride solutions that were frozen and fractured in small-bore polyethylene tubes. The concentrations of elements other than potassium and chlorine in the specimens were obtained from their atomic ratios relative to chlorine. These were calculated by the ratio model. A typical calibration graph is shown in Figure 20.

Brine-fly larvae experience rapid and large fluctuations in habitat salinity (normally $300-1200\,mmol\;l^{-1}Na^{+}$) and the hindgut appears to be involved in regulating the sodium and chlorine composition of the hemolymph. The ileum consists of two cell types (large and small cells), which, on the basis of their ultrastructures, may be supposed to function in a differential manner (Marshall and Wright, 1974). The results of an

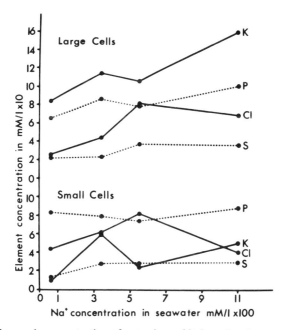

Figure 21.   Changes in concentration of potassium, chlorine, phosphorus, and sulfur in the hindgut cells of brine-fly larvae adapted to different salinities.

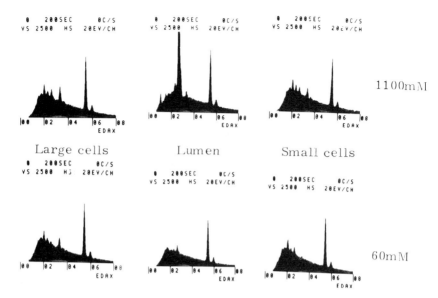

Figure 22. Typical spectra recorded from different cell types and lumina in the hindgut of brine-fly larvae adapted to different salinities.

analysis of these two cell types in larvae adapted to different salinities are shown in Figure 21 and typical spectra are shown in Figure 22.

The limit of detection for the elements examined here is about 2–5 mmol l^{-1}. However, counting statistics, which are a function of peak-to-background ratios, will preclude the observation of the minimum detection limit in every analysis. The detection of sodium below concentrations of approximately 50 mmol l^{-1} was not possible. This appears to be a consequence of the very steep background profile at the low-energy end of the spectrum and inadequate pulse pile-up rejection in this region. Coating specimens with beryllium resulted in no improvement; inability to detect sodium is thus not a consequence of the absorption of sodium x-rays by the chromium coat.

These preliminary results demonstrate the feasibility of the direct measurement of elemental concentrations at the cellular level.

LITERATURE CITED

Appleton, T. C. 1974. A cryostat approach to ultrathin 'dry' frozen sections for electron microscopy: a morphological and x-ray analytical study. J. Microscopy 100:49.

Bacaner, M., Broadhurst, J., Hutchinson, T., and Lilley, J. 1972. High resolution localisation of ions correlated with electron optical image of muscle sarcomeres by analysis of non-dispersive x-rays generated by electron bombardment in scanning electron microscope. Fed. Proc. 31:324.

Bacaner, M., Broadhurst, J., Hutchinson, T., and Lilley, J. 1973. Scanning transmission electron microscope studies of deep-frozen unfixed muscle correlated with spatial localisation of intracellular elements by fluorescent x-ray analysis. Proc. Nat. Acad. Sci. 70:3423.

Bald, W. B., and Robards, A. W. 1978. A device for the rapid freezing of biological specimens under precisely controlled and reproducible conditions. J. Microscopy 112:3.

Brombach, J. D. 1975. Electron-beam x-ray microanalysis of frozen biological bulk specimens below 130 K. 2. The electrical charging of the sample in quantitative analysis. J. Microscopie Biol. Cell. 22:233.

Christensen, A. 1972. Frozen thin sections of fresh tissue for electron microscopy, with a description of pancreas and liver. J. Cell Biol. 51:772.

Cort, D. M., and Steeds, J. W. 1972. A liquid helium cooled stage for the scanning electron microscope. Proc. Fifth Europ. Cong. Electron Microscopy, p. 376.

Costello, M. J., and Corless, J. M. 1978. The direct measurement of temperature changes within freeze-fracture specimens during rapid quenching in liquid coolants. J. Microscopy 112:17.

Davies, T. W., and Erasmus, D. A. 1973. Cryo-ultramicrotomy and x-ray microanalysis in the transmission electron microscope. Sci. Tools 20:9.

Dempsey, G. P., and Bullivant, S. 1976a. A copper block method for freezing non-cryoprotected tissue to produce ice-crystal-free regions for electron microscopy. 1. Evaluation using freeze-substitution. J. Microscopy 106:251.

Dempsey, G. P., and Bullivant, S. 1976b. A copper block method for freezing non-cryoprotected tissue to produce ice-crystal-free regions for electron microscopy. 2. Evaluation using freeze-fracturing with a cryo-ultramicrotome. J. Microscopy 106:261.

Dorge, A., Rick, R., Gehring, K., Mason, J., and Thuran, K. 1975. Preparation and applicability of freeze-dried sections in the microprobe analysis of biological soft tissue. J. Microscopie Biol. Cell. 22:205.

Doty, S. B., Lee, C. W., and Banfield, B. (1974). A method of obtaining ultrathin frozen sections from fresh or glutaraldehyde fixed tissue. Histochem. J. 6:383.

Durand, M., Deleplanque, M., and Kahane, A. 1967. Bulk conductivity of ice between −25°C and −100°C with ion exchange membranes. Sol. Stat. Comm. 5:759.

Echlin, P. 1978. Low temperature scanning electron microscopy: a review. J. Microscopy 112:47.

Echlin, P., and Moreton, R. 1973. The preparation, coating and examination of frozen biological materials in the SEM. Scanning Electron Microscopy 1973:325.

Echlin, P., and Moreton, R. 1974. The preparation of biological materials for x-ray microanalysis. In: *Microprobe Analysis as Applied to Cells and Tissues* (Hall, T., Echlin, P., and Kaufmann, R., eds.), p. 159. Academic Press, London.

Echlin, P., and Moreton, R. 1976. Low temperature techniques for scanning electron microscopy. Scanning Electron Microscopy 1976:753.

Fletcher, N. H. 1971. Structural aspects of the ice-water system. Rep. Prog. Phys. 34:913.

Forrest, Q. G., and Marshall, A. T. 1976. Comparative x-ray microanalysis of frozen-hydrated and freeze-substituted specimens. Proc. Sixth Europ. Cong. Electron Microscopy, p. 218.

Fuchs, W., Brombach, J. D., and Trosch, W. 1978a. Charging effect in electron-irradiated ice. J. Microscopy 112:63.

Fuchs, W., and Lindemann, B. 1975. Electron beam x-ray microanalysis of frozen biological bulk specimen below 130 K. 1. Instrumentation and specimen preparation. J. Microscopie Biol. Cell. 22:227.

Fuchs, W., Lindemann, B., Brombach, J. D., and Trosch, W. 1978b. Instrumentation and specimen preparation for electron beam x-ray microanalysis of frozen hydrated bulk specimens. J. Microscopy 112:75.

Gehring, K., Dorge, A., Wunderlich, P., and Thurau, K. 1973. On the application of the electron microprobe for the analysis of freeze-dried cryo-sections and deep frozen, water containing tissue of the rat kidney. Bietr. Electronenmikroskop. Direktabb. Oberfl. 5:937.

Glover, A. J., and Garvitch, Z. S. 1974. The freezing rate of freeze-etch specimens for electron microscopy. Cryobiology 11:248.

Goodhew, P. J. 1975. Electron probe microanalysis of glasses containing alkali metals. Microstruct. Sci. 3:631.

Green, M., and Cosslett, V. E. 1961. The efficiency of production of characteristic x-radiation in thick targets of a pure element. Proc. Phys. Soc. 78:1206.

Griffiths, B. W., and Venables, J. A. 1972. Scanning electron microscopy at liquid helium temperatures. Scanning Electron Microscopy 1972:9.

Gullasch, J., and Kaufmann, R. 1974. Energy-dispersive x-ray microanalysis in soft biological tissues: relevance and reproducibility of the results as depending on specimen preparation (air drying, cryofixation, cool-stage techniques). In: *Microprobe Analysis as Applied to Cells and Tissues* (Hall, T., Echlin, P., and Kaufmann, R., eds.), p. 175. Academic Press, London.

Gupta, B. L., Hall, T. A., Maddrell, S. H. P., and Moreton, R. B. 1976. Distribution of ions in a fluid-transporting epithelium determined by electron-probe x-ray microanalysis. Nature 264:284.

Gupta, B. L., Berridge, M. J., Hall, T. A., and Moreton, R. B. 1978a. Electron microprobe and ion-selective microelectrode studies of fluid secretion in the salivary glands of *Calliphora*. J. Exp. Biol. 72:201.

Gupta, B. L., Hall, T. A., and Naftalin, R. J. 1978b. Microprobe measurement of Na, K and Cl concentration profiles in epithelial cells and intercellular spaces of rabbit ileum. Nature 272:70.

Hodson, S., and Marshall, J. 1970. Ultracryotomy: a technique for cutting ultrathin sections of unfixed frozen biological tissues for electron microscopy. J. Microscopy 91:105.

Hodson, S., and Williams, L. 1976. Ultracryotomy of biological tissues to preserve membrane structure. J. Cell Sci. 20:687.

Hörl, E. M. 1972. Liquid helium stage for the scanning electron microscope. Proc. Fifth Europ. Cong. Electron Microscopy, p. 374.

Hutchinson, T. E., Bacaner, M., Broadhurst, J., and Lilley, J. 1974. Elemental microanalysis of frozen biological thin sections by scanning electron microscopy and energy selective x-ray analysis. In: *Microprobe Analysis as Applied to Cells and Tissues* (Hall, T., Echlin, P., and Kaufmann, R., eds.), p. 101. Academic Press, London.

Kirk, R. G., and Dobbs, G. H. 1976. Freeze fracturing with a modified cryo-ultramicrotome to prepare large intact replicas and samples for x-ray microanalysis. Sci. Tools 23:28.

Koch, G. R. 1975. Preparation and examination of specimens at low temperatures. In: *Principles and Techniques of Scanning Electron Microscopy*, Vol. 4 (Hayat, M. A., ed.). Van Nostrand Reinhold, New York.

Koehler, J. K. 1968. The technique and application of freeze-etching in ultrastructure research. Adv. Biol. Med. Phys. 12:1.

Kramer, D., and Preston, J. 1978. A modified method of x-ray microanalysis of bulk frozen plant tissue and its application to the problem of salt exclusion in mangrove roots. Microsc. Acta Suppl. 2:193.

Lechene, C., Strunk, T., and Warner, R. 1975. Perspectives in electron probe microanalysis of biological samples kept frozen. Proc. 10th Ann. Conf. Microbeam Analysis Soc., p. 49A.

Lechner, G. 1974. Experiences with a cooled transfer stage between the cryo-ultramicrotome and the SEM. Eighth Intern. Cong. Electron Microscopy, Canberra 2:58.

Marshall, A. T. 1974. X-ray microanalysis of frozen dried and frozen hydrated biological specimens. Eighth Intern. Cong. Electron Microscopy, Canberra 2:74.

Marshall, A. T. 1975a. X-ray microanalysis of frozen hydrated biological specimens: the effect of charging. Micron 5:275.

Marshall, A. T. 1975b. Electron probe x-ray microanalysis. In: *Principles and Techniques of Scanning Electron Microscopy*, Vol. 4 (Hayat, M. A., ed.), p. 103. Van Nostrand Reinhold, New York.

Marshall, A. T. 1977a. Electron probe x-ray microanalysis of frozen hydrated biological specimens. Microsc. Acta 79:254.

Marshall, A. T. 1977b. X-ray microanalysis of frozen-hydrated and freeze-substituted specimens. Proc. Intern. Union of Physiol. Sci. 12:143.

Marshall, A. T. 1977c. Iso-atomic droplets as models for the investigation of parameters affecting x-ray microanalysis of biological specimens. Micron 8:193.

Marshall, A. T., and Wright, A. 1973. Detection of ions in insect osmoregulatory systems by electron probe x-ray microanalysis using scanning electron microscopy and a cryoscopic technique. Micron 4:31.

Marshall, A. T., and Wright, A. 1974. Ultrastructural changes associated with osmoregulation in the hindgut cells of a saltwater insect *Ephydrella* sp. (Ephydridae: Diptera). Tissue and Cell 6:301.

McQuillan, M. K., and Farthing, T. W. 1961. Beryllium. Endeavour 20:11.

Meryman, H. T. 1957. Physical limitations of the rapid freezing method. Proc. Roy. Soc. Lond. B, 147:452.

Moreton, R. B., Echlin, P., Gupta, B. L., Hall, T. A., and Weis-Fogh, T. 1974. Preparation of frozen hydrated tissue sections for x-ray microanalysis in the scanning electron microscope. Nature 247:113.

Morgan, A. J., Davies, T. W., and Erasmus, D. A. 1975. Changes in the concentration and distribution of elements during electron microscope preparative procedures. Micron 6:11.

Morgan, A. J., Davies, T. W., and Erasmus, D. A. 1978. Specimen preparation. In: *Electron Probe Microanalysis in Biology* (Erasmus, D. A., ed.), p. 94. Chapman & Hall, London.

Nei, T., Yosumoto, H., Hasegawa, Y., and Nagasawa, Y. 1973. Direct observation of frozen specimens with a scanning electron microscope. J. Electron Microscopy 22:185.

Rebhun, L. I. 1959. Freeze-substitution: fine structure as a function of water concentration in cells. Fed. Proc. 24 (Suppl. 15):217.

Rebhun, L. I. 1972. Freeze-substitution and freeze drying. In: *Principles and Techniques of Electron Microscopy*, Vol. 2 (Hayat, M. A., ed.), p. 3. Van Nostrand Reinhold, New York.

Riehle, U. (1968). *Über die Vitrifizierung verdünnter wassriger Lösungen*. Juris Druck and Verlag, Zurich.

Roomans, G. M., and Sevéus, L. A. 1976. Subcellular localisation of diffusible ions in the yeast *Saccharomyces cerevisiae*: quantitative microprobe analysis of thin freeze-dried sections. J. Cell Sci. 21:119.

Russ, J. C. 1973. X-ray microanalysis in the biological sciences. J. Submicrosc. Cytol. 6:55.

Saetersdal, T. S., Roli, J., Myklebust, R., and Engedal, H. 1977. Preservation of shock-frozen myocardial tissue as shown by cryo-ultramicrotomy and freeze-fracture studies. J. Microscopy 111:297.

Saubermann, A. J. 1978. X-ray microanalysis of frozen hydrated tissue sections as a physiological tool. Microsc. Acta Suppl. 2:130.

Saubermann, A. J., and Echlin, P. 1975. The preparation, examination and analysis of frozen hydrated tissue sections by scanning transmission electron microscopy and x-ray microanalysis. J. Microscopy 105:155.

Saubermann, A. J., Riley, W. D., and Beeuwkes, R. 1977. Cutting work in thick section cryomicrotomy. J. Microscopy 111:39.

Schubert, J. 1958. Beryllium and berylliosis. Sci. Amer. 199:27.

Sjöström, M., and Thornell, L. E. 1975. Preparing sections of skeletal muscle for transmission electron analytical microscopy (TEAM) of diffusible elements. J. Microscopy 103:101.

Sleytr, U. B., and Robards, A. W. 1977. Freeze-fracturing: a review of methods and results. J. Microscopy 111:77.

Southworth, D., Fisher, K., and Branton, D. 1975. Principles of freeze-fracturing and etching. Tech. Biochem. Biophys. Morphol. 2:247.

Spriggs, T. L. B., and Wynne-Evans, D. 1976. Observations on the production of frozen-dried thin sections for electron microscopy using unfixed fresh liver, fast-frozen without cryoprotection. J. Microscopy 107:35.

Stephenson, J. L. 1956. Ice crystal growth during the rapid freezing of tissues. J. Cell Biol. 2: Suppl. 45.

Taylor, P. G., and Burgess, A. 1977. Cold stage for electron probe microanalyser. J. Microscopy 111:51.

Van Harreveld, A., Trubatch, J., and Steiner, J. 1974. Rapid freezing in electron microscopy for the arrest of physiological processes. J. Microscopy 100:189.

Varriano-Marston, E., Gordon, J., Davis, E. A., and Hutchinson, T. E. 1977. Cryomicrotomy applied to the preparation of frozen hydrated muscle tissue for transmission electron microscopy. J. Microscopy 109:193.

Varshneya, A. K., Cooper, A. R., and Cable, M. 1966. Changes in composition during electron microprobe analysis of K_2O-SrO-SiO_2 glass. J. Appl Phys. 37:2199.

Yeo, A. R., Läuchli, A., Kramer, D., and Gullasch, J. 1977. Ion measurements by x-ray microanalysis in unfixed, frozen, hydrated plant cells of species differing in salt tolerance. Planta 134:35.

Zierold, K. 1976. X-ray microanalysis of cryofractured tissue specimens of the rat. Proc. Sixth Europ. Cong. Electron Microscopy 2:223.

Zierold, K., and Schafer, D. 1978. Quantitative x-ray microanalysis of diffusible ions in the skeletal muscle bulk specimen. J. Microscopy 112:89.

Chapter 4

FROZEN-HYDRATED SECTIONS

Alan T. Marshall

Zoology Department,
La Trobe University,
Bundoora, Victoria, Australia

RATIONALE

The rationale for using frozen-hydrated sections for analysis rather than frozen-hydrated bulk specimens is that both optical and analytical resolution should be better in the former than in the latter. It is extremely doubtful that more ultrastructural detail can be seen in a frozen-hydrated section than in a fractured frozen-hydrated bulk specimen with the present state of the technology. Analytical resolution, however, may be slightly better in frozen-hydrated sections, but this has yet to be conclusively demonstrated.

RESOLUTION

Estimates of analytical resolution in sections are of the order of 2–300 nm (Gupta et al., 1977b, 1978a) while for bulk specimens resolution is estimated to be 1–2 μm (Marshall, 1977; Chapter 3) and may be considerably less (see Fuchs et al., 1978).

A major problem in the analysis of frozen-hydrated sections is that optical resolution is generally very poor. This may be because frozen-hydrated sections have so far been examined only in transmission scanning electron microscopes (TSEM) at low accelerating voltages (<50 kv)

(Bacaner et al., 1973; Moreton et al., 1974; Saubermann and Echlin, 1975; Gupta et al., 1976, 1978a,b; Saubermann, 1978) and transmission electron microscopes (Varriano-Marston et al., 1977; Appleton, T. C., personal communication). It may well be that resolution will be improved by examination in scanning transmission electron microscopes at 100 kV or higher. Some improvement in the TSEM can be obtained by using television scanning (Marshall, unpublished data). Contrast and optical resolution are greatly improved by slight (10%) dehydration of the sections (Gupta et al., 1978a). Slight dehydration does not affect quantitation if a standard is present in the same section and suffers a similar amount of dehydration (Gupta and Hall, 1979).

HYDRATION

A considerable amount of effort has been expended to demonstrate that sections do not dehydrate in an uncontrolled fashion. The evidence that has been obtained includes monitoring water vapor content of the column as ion current by quadruple mass spectrometer, measuring chamber pressure and assuming this to be proportional to water vapor pressure, observing the continuum count before and after warm up, and observing the increase in contrast in the scanning transmission electron image as a consequence of warm up (Saubermann and Echlin, 1975; Echlin, 1978). The first two methods are of doubtful validity unless all frost can be excluded from the substage prior to insertion on the cryostage. This can rarely be accomplished whatever the precautious taken. If frost is present on the substage, it can of course produce a spurious signal when the section is warmed.

PREPARATION OF SECTIONS

The principles of freezing specimens and transferring sections to the electron microscope are essentially the same as for bulk specimens, and were dealt with in Chapter 3. The preparation of sections for x-ray microanalysis is still a matter of some controversy. It seems clear from a number of studies on sectioning that sections cut or cleave at relatively high temperatures ($> -80°C$) and fracture continuously at lower temperatures ($< -100°C$) (see Kirk and Dobbs, 1976). In principle it seems logical to attempt to section at low temperatures ($< -130°C$) to avoid ice recrystallization and to reduce the possibility of local thawing occurring when the work done in sectioning is partially converted to heat. In practical terms, however, it is easier to cut sections at temperatures above $-80°C$, since they will adhere to form a ribbon (Appleton, 1974, 1977, 1978); moreover, they will tend to be of uniform thickness. The question then is: Does thawing occur when sections are cut at higher temperatures, and, if so, is this thawing sufficient to redistribute diffusible elements and invalidate analyses? Thornburg and Mengers (1957) have calculated that the heat dissipated during sectioning would be sufficient to melt a 60-nm layer in a 100-nm section. Hodson and Marshall (1970, 1972) suggest that the specimen temperature may rise locally by 100°C during sectioning and the opinion of Hodson and Williams (1976) seems to be that thawing will certainly occur at temperatures as high as −80°C. Saubermann et al. (1977) have shown that sufficient heat is produced during sectioning at −80°C to briefly melt a 0.5-μm section if *all* the heat goes into the section. It has not, however, been established that all the heat does indeed go into the section. Sevéus (1978) believes that ribboning of frozen sections is an indication of thawing and recommends a low specimen temperature (< −140°C) and a somewhat higher knife temperature (−100°C) to obtain flat, transparent single sections.

From the foregoing it seems probable that some transitory thawing does occur when sections are cut at temperatures higher than −80°C. It remains something of an open question whether this thawing is sufficient to produce a redistribution of diffusible elements. Appleton (1974, 1977) has shown that in the case of a model system (NaCl in carboxymethyl cellu-

Figure 1.   Modified LKB cryokit chamber. A substage for holding grids is shown (arrow) mounted on a cold stage attached to the knife holder.

lose), redistribution does not appear to occur; he also argues that the retention of compartmental differences in ionic concentrations (as measured by x-ray microanalysis) within cells in mouse pancreas suggests that redistribution has not occurred.

The question of whether recrystallization occurs at temperatures around $-80°C$ has not received much attention. Sjöström (1975) found that no crystal growth occurred in sections of glycerol-protected muscle at $-100°C$ after 30 min, and he suggested that the recrystallization temperature in such specimens must be above $-70°C$. Whether this can be taken as representative of untreated muscle is of course open to question. This work, however, is in agreement with that of Dempsey and Bullivant (1976a,b), who compared material by freeze-fracturing at temperatures below

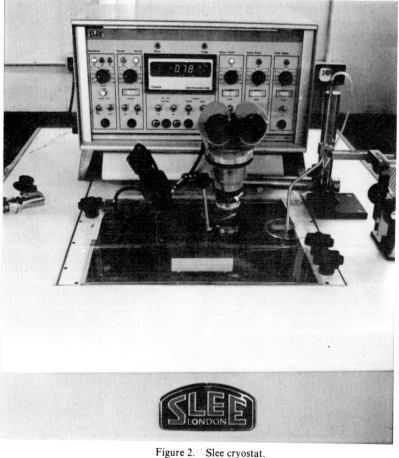

Figure 2.   Slee cryostat.

Figure 3.   Chamber of Slee cryostat showing vacuum device for "drawing" a ribbon of sections.

−140°C and by freeze-substitution at −70°C and found no evidence of ice recrystallization.

MICROTOMES

In general, sections prepared at low temperatures ($< -100°C$) are dry cut as single sections and maneuvered onto a grid or other specimen holder by means of a mounted hair. This type of section can be produced in one of the microtome attachments commonly referred to as "cryokits" (Figure 1). In order to obtain ribbons of sections at higher temperatures ($> -100°C$), however, it is necessary to have a turbulence-free system and a chamber that is in thermal equilibrium with knife and specimen (Appleton, 1978). These requirements are easily met only in a deep refrigerated chamber containing the microtome. Such an instrument is known as a "cryostat," and there appears to be only one commercially available instrument at the present time (Figures 2 and 3).

COATING

Frozen-hydrated sections are usually mounted on metal (aluminum) coated support films (Echlin and Moreton, 1974) in order to improve thermal and electrical conductivity. The application of a further coat of metal onto the section has been prescribed (Echlin and Moreton, 1974) but was later found to be unnecessary (Echlin, 1975). It has been shown by Fuchs et al. (1978) that sections mounted on a conducting film will not charge appreciably if the electron range is larger than about half the section thickness.

CRYOPROTECTANTS

Low molecular weight cryoprotectants, such as glycerol and dimethyl-sulfoxide, probably lead to ionic shifts and ion loss (Sjöström and Thornell, 1975; Franks, 1977; Schlaffer, 1977). However, high molecular weight polymers, such as polyvinyl pyrrolidone, dextran, and hydroxyethyl starch, appear to be effective in preventing ice-crystal formation (Franks and Skaer, 1976; Franks, 1977; Franks et al., 1977) and preserving cell ultrastructure (Skaer et al., 1977) when used at high concentrations (25% w/w). Furthermore, these substances appear to be frequently, but not always (Gupta et al., 1977b), physiologically inert. They have the supplementary benefit of being good encapsulating substances, which facilitates sectioning (Skaer et al., 1977).

QUANTITATIVE ANALYSIS

Quantitative analysis of frozen-hydrated sections has so far been done by the Hall peak-to-background ratio method (see Gupta et al., 1977a,b, 1978a).

The basis of this method of analysis for obtaining absolute concentrations in terms of millimoles per kilogram is the provision of a matching standard adjacent to the specimen within the same section. This means that the specimens are frozen within a physiological solution containing an organic polymer, usually dextran (Gupta et al., 1977b). Such a standard is assumed to have the same mean atomic number as the specimen and thus the same quanta of continuum radiation are assumed to be generated per unit mass. The latter is a vital assumption, since a band of continuum radiation is taken as a measure of the total mass of the analyzed microvolume.

The basic equation is:

$$C_x^{sp} = \frac{(P/W)^{sp}}{(P/W)^{st}} \cdot C_x^{st}$$

(where C_x is elemental concentration in m mol kg^{-1}, P is peak minus background counts, W is counts from a band of continuum, and sp and st refer to specimen and standard, respectively). In the above equation W must be attributed solely to the specimen. It is therefore necessary to assess and to deduct the continuum that may be generated by the specimen support (grid or holder and film) (Gupta et al., 1978a).

Since it appears to have been extremely difficult so far to resolve structure in fully frozen-hydrated sections (Gupta et al., 1977a,b), analyses are frequently carried out on partially (10%) dehydrated sections. It becomes necessary then to assume that specimen and standard dehydrate to the same extent if the above formula is to remain valid. This difficulty has been discussed in detail by Gupta et al. (1977b).

If sections are cut at high temperatures ($-80°$C) and are therefore of uniform thickness, then the mass per unit area will be approximately the same at all points of the specimen and standard. The equation then becomes:

$$C_x^{sp} = \frac{P^{sp}}{P^{st}} \cdot C_x^{st}$$

This equation cannot be applied to sections that are cut at low temperatures ($< -100°$C), since they are produced by continuous fracturing and are not of uniform thickness.

The determination of atomic ratios by means of the ratio model (Russ, 1973) has none of the problems associated with the peak-to-background method, since the continuum count is not required and the degree of dehydration is inconsequential.

LITERATURE CITED

Appleton, T. C. 1974. A cryostat approach to ultrathin 'dry' frozen sections for electron microscopy: a morphological and x-ray analytical study. J. Microscopy 100:49.

Appleton, T. C. 1977. The use of ultrathin frozen sections for x-ray microanalysis of diffusible elements. In: Analytical and Quantitative Methods in Microscopy (Meek, G. A., and Elder, H. Y., eds.), p. 247. Cambridge University Press, London.

Appleton, T. C. 1978. The contribution of cryo-ultramicrotomy to x-ray microanalysis in biology. In: Electron Probe Microanalysis in Biology (Erasmus, D. A., ed.), p. 148. Chapman & Hall, London.

Bacaner, M., Broadhurst, J., Hutchinson, T., and Lilley, J. 1973. Scanning transmission electron microscopy studies of deep-frozen unfixed muscle correlated with spatial localisation of intracellular elements by fluorescent x-ray analysis. Proc. Nat. Acad. Sci. 70:3423.

Dempsey, G. P., and Bullivant, S. 1976a. A copper block method for freezing non-cryoprotected tissue to produce ice-crystal-free regions for electron microscopy. 1. Evaluation using freeze-substitution. J. Microscopy 106:251.

Dempsey, G. P., and Bullivant, S. 1976b. A copper block method for freezing non-cryoprotected tissue to produce ice-crystal-free regions for electron microscopy. 2. Evaluation using freeze fracturing with a cryo-ultramicrotome. J. Microscopy 106:261.

Echlin, P. 1975. The preparation of frozen-hydrated biological material for x-ray microanalysis. J. Microscopie Biol. Cell. 22:215.

Echlin, P. 1978. Low-temperature scanning electron microscopy: a review. J. Microscopy 112:47.

Echlin, P., and Moreton, R. 1974. The preparation of biological materials for x-ray microanalysis. In: *Microprobe Analysis as Applied to Cells and Tissues* (Hall, T., Echlin, P., and Kaufmann, R., eds.), p. 159. Academic Press, London.

Franks, F. 1977. Biological freezing and cryofixation. J. Microscopy 111:3.

Franks, F., Asquith, M. H., Hammond, C. C., Skaer, H. B., and Echlin, P. 1977. Polymeric cryoprotectants in the preservation of biological ultrastructure. 1. Low temperature states of aqueous solutions of hydrophilic polymers. J. Microscopy 110:223.

Franks, F., and Skaer, H. B. 1976. Aqueous glasses as matrices in freeze-fracture electron microscopy. Nature 262:323.

Fuchs, W., Brombach, J. D., and Trösch, W. 1978. Charging effects in electron-irradiated ice. J. Microscopy 112:63.

Gupta, B. L., Berridge, M. J., Hall, T. A., and Moreton, R. B. 1978a. Electron microprobe and ion-selective microelectrode studies of fluid secretion in the salivary glands of *Calliphora*. J. Exp. Biol. 72:201.

Gupta, B. L., and Hall, T. A. 1979. Quantitative electron probe x-ray microanalysis of electrolyte elements within epithelial tissue compartments. Fed. Proc. 38:144.

Gupta, B. L., Hall, T. A., Maddrell, S. H. P., and Moreton, R. B. 1976. Distribution of ions in a fluid-transporting epithelium determined by electron-probe x-ray microanalysis. Nature 264:284.

Gupta, B. L., Hall, T. A., and Moreton, R. B. 1977a. Some aspects of the microanalysis of frozen-hydrated tissue sections. Eighth Intern. Conf. X-ray Optics and Microanalysis, p. 110A.

Gupta, B. L., Hall, T. A., and Moreton, R. B. 1977b. Electron probe x-ray microanalysis. In: *Transport of Ions and Water in Animals* (Gupta, B. L., Moreton, R. B., Oschman, J. L., and Wall, B. J., eds.), p. 83. Academic Press, London.

Gupta, B. L., Hall, T. A., and Naftalin, R. J. 1978b. Microprobe measurement of Na, K and Cl concentration profiles in epithelial cells and intercellular spaces of rabbit ileum. Nature 272:70.

Hodson, S., and Marshall, J. 1970. Ultracryotomy: a technique for cutting ultrathin frozen sections of unfixed biological tissues for electron microscopy. J. Microscopy 91:105.

Hodson, S., and Marshall, J. 1972. Evidence against through-section thawing whilst cutting on the ultracryotome. J. Microscopy 95:459.

Hodson, S., and Williams, L. 1976. Ultracryotomy of biological tissues to preserve membrane structure. J. Cell Sci. 20:687.

Kirk, R. G., and Dobbs, G. H. 1976. Freeze fracturing with a modified cryo-ultra-microtome to prepare large intact replicas and samples for x-ray microanalysis. Sci. Tools 23:28.

Marshall, A. T. 1977. Electron probe x-ray microanalysis of frozen-hydrated biological specimens. Microsc. Acta 79:254.

Moreton, R. B., Echlin, P., Gupta, B. L., Hall, T. A., and Weis-Fogh, T. 1974.

Preparation of frozen hydrated tissue sections for x-ray microanalysis in the scanning electron microscope. Nature 247:113.

Russ, J. C. 1973. X-ray microanalysis in the biological sciences. J. Submicrosc. Cytol. 6:55.

Saubermann, A. J. 1978. X-ray microanalysis of frozen-hydrated tissue sections as a physiological tool. Microsc. Acta, Suppl. 2:130.

Saubermann, A. J., and Echlin, P. 1975. The preparation, examination and analysis of frozen hydrated tissue sections by scanning transmission electron microscopy and x-ray microanalysis. J. Microscopy 105:155.

Saubermann, A. J., Riley, W. D., and Beeuwkes, R. 1977. Cutting work in thick section cryomicrotomy. J. Microscopy 111:39.

Schlafer, M. 1977. Drugs that modify cellular responses to low temperature (cryoprotectants): a pharmacological perspective of preserving biological systems by freezing. Fed. Proc. 36:2590.

Sevéus, L. 1978. Preparation of biological material for x-ray microanalysis of diffusible elements. 1. Rapid freezing of biological tissue in nitrogen slush and preparation of ultrathin frozen sections in the absence of trough liquid. J. Microscopy 112:269.

Sjöström, M. 1975. Ice crystal growth in skeletal muscle fibres. J. Microscopy 105:67.

Sjöström, M., and Thornell, L. E. 1975. Preparing sections of skeletal muscle for transmission electron analytical microscopy (TEAM) of diffusible elements. J. Microscopy 103:101.

Skaer, H. B., Franks, F., Asquith, M. H., and Echlin, P. 1977. Polymeric cryoprotectants in the preservation of biological ultrastructure. 3. Morphological aspects. J. Microscopy 110:257.

Thornburg, W., and Mengers, P. E. 1957. An analysis of frozen section techniques; sectioning of fresh-frozen tissue. J. Histochem. Cytochem. 5:47.

Varriano-Marston, E., Gordon, J., Davis, E. A., and Hutchinson, T. E. 1977. Cryomicrotomy applied to the preparation of frozen hydrated muscle tissue for transmission electron microscopy. J. Microscopy 109:193.

Chapter 5

SECTIONS OF FREEZE-SUBSTITUTED SPECIMENS

Alan T. Marshall

Zoology Department,
La Trobe University,
Bundoora, Victoria, Australia

INTRODUCTION

Freeze-substitution involves rapid freezing of a tissue specimen and then dissolving the ice in the specimen by means of a liquid dehydrating agent at low temperatures. Following the removal or "substitution" of the ice matrix by the dehydrating agent, the specimen is brought to room temperature and embedded in a plastic embedding medium. The method offers the major advantage of relatively easy sectioning on conventional ultramicrotomes, compared to the sectioning of frozen-hydrated specimens.

While there is still much critical investigation to be made before freeze-substitution is unequivocally accepted as a preparation method for x-ray

Table 1.   Ion losses during freeze-substitution in ether or Acrolein/ether and infiltration in modified Spurr's resin*

Specimen	Ion loss as a percentage of total ion content					
	Substitution				Infiltration	
	Na	Cl	Rb	K	Na	Cl
Leaf (*Suaeda maritima L.*)	0.28	0.30	1.28	0.83	0.29	0.45
Barley root	0.5	4.0	+1.0**	—	—	—
Cockroach nerve	2.4	8.0	+4.0	—	—	—

* Abstracted from Harvey et al. (1976) and Van Zyl et al. (1976).
**Plus symbol indicates a percentage gain.

microanalysis, some general statements and comparisons with other methods can be made:

1.   At least some highly diffusible elements are retained in cells with virtually no loss.
2.   Diffusible elements are probably not translocated within cells.
3.   Within lumina and extracellular spaces, diffusible elements are retained in situ to a far greater extent than is the case with preparation methods involving freeze-drying. Retention, however, is possibly not as good as in frozen-hydrated sections.

The rather considerable early literature on freeze-substitution has been reviewed by Pearse (1968) and more recent literature relevant to x-ray microanalysis by Morgan et al. (1978). As far as the x-ray microanalysis of diffusible elements is concerned, there are four papers of major importance (Läuchli et al., 1970; Harvey et al., 1976; Van Zyl et al., 1976; Forrest and Marshall, 1976). These papers deal with attempts to determine element losses during preparation and the extent of element redistribution. It is shown by the use of radionuclides or flame photometry that the loss of sodium, rubidium, chlorine, and potassium is minimized during the substitution phase if the freeze-substitution medium is anhydrous ether or a 20% (v/v) mixture of anhydrous acrolein in ether. Ether was compared to acetone by Harvey et al. (1976) and to a variety of other solvents by Van Zyl et al. (1976). Harvey et al. (1976) also determined losses during infiltration in modified Spurr's resin (Pallaghy, 1973) and found these to be minimal. These data are summarized in Table 1. Preliminary investigations by Forrest and Marshall (1976) suggest that translocation of diffusible elements within freeze-substituted specimens is not great (see "Examples" later in this chapter).

The papers cited in the foregoing clearly establish the possibility of using freeze-substitution as a preparative method for x-ray microanalysis. It should be noted, however, that none of these authors considers the possibility [first raised by Pallaghy (1973)] that water production during polymerization of epoxy resins might result in ion loss or movement. This question is considered in detail later.

FREEZING

The problems associated with freezing living tissues have already been discussed in Chapters 2 and 3. The question of ice-crystal formation and its suppression, however, is particularly pertinent to freeze-substituted preparations, because ice-crystal damage is a more obvious limitation in terms of the resolution of cell structure in resin-embedded sections than it is in frozen bulk specimens or frozen-hydrated sections.

It is important to point out immediately that good preservation of cellular ultrastructure by freeze substitution (comparable to that seen in chemically fixed tissue) is generally obtainable only in the outermost layers of the specimen in a zone some 10–20 μm wide if the sample is untreated with cryoprotective agents or is not deliberately dehydrated. The presence and extent of this zone are critically dependent on the method of freezing and more precisely on the freezing rate. Thus, Dempsey and Bullivant (1976) attained a 12-μm-wide zone by freezing tissue on a copper block cooled with liquid nitrogen, while Rebhun (1959, 1972) could not obtain an ice-free zone without dehydrating the tissues (by air-drying or osmotically) during freezing in Freon 12 or Freon 22 (Rebhun and Sander, 1971; Rebhun, 1972). Similarly, Harvey et al. (1976) found no difference in preservation from the outside to the inside of a block after freezing in acetone at $-90°$C or methylbutane. Williams and Hodson (1978) also found no ice-crystal–free zone when tissues were frozen in boiling liquid nitrogen. It must be concluded that in these latter cases the freezing rate was too low.

An apparent contradiction to this theory is seen in the results of Hereward and Northcote (1972), who used Freon 22 for quenching and obtained a 100-μm-wide zone that was free of ice-crystal damage. Dempsey and Bullivant (1976) suggest that this could be attributed to the short substitution time (3 days) used by Hereward and Northcote. They point out that in tissue frozen in Freon 13 and substituted in the same solvent (acetone) for 14 days it is impossible to obtain ice-crystal–free preparations. It is implied that, in the short substitution period, ice is not removed and that it melts on warming, thereby permitting the still plastic cell components to regain the positions from which they were displaced by ice crystals on freezing.

In the author's experience, regions relatively free of ice-crystal damage can be obtained by freezing in propane. Propane has been shown by Rebhun (1972) to give considerably faster cooling rates than Freon 13 and Freon 22. Boiling liquid nitrogen is a very poor quenching fluid, with a very low cool-

Figure 1.   Quenching equipment for freeze-substitution. Liquid propane thimble on the left and tubes of substitution fluid on the right.

ing rate, by virtue of the insulating gas jacket that forms around any specimen placed in it. Propane, then, is recommended as a quenching fluid.

Many authors advise against the use of propane on the grounds of safety, since it forms explosive mixtures with liquid oxygen that may condense into it from the atmosphere. Only very small quantities are necessary, however, and the following method has been used routinely in the author's laboratory for some years without problems.

High-purity propane is condensed from a low-pressure gas cylinder, by leading the gas into a small thimble (Balzars) that is cooled by liquid nitrogen (Figure 1). The level of the nitrogen in the dewar determines the temperature of the propane, and the flow of nitrogen gas from the narrow-necked dewar prevents the condensation of oxygen.

The small capacity of the thimble, however, does make it difficult to use devices for rapid injection of specimens into the quenching fluid. Experiments with injection devices have not resulted in better preservation than that obtained by hand, in the author's experience. The smallness of the dewar and thimble and the closeness of the latter to the top of the dewar do allow a very close approach of the specimen to the quenching fluid without passing through cold nitrogen gas, which may be the reason why injection devices have not so far produced improved preservation compared to hand quenching.

The quenching technique is simply to dissect and excise the required sample as quickly as possible and to transfer it rapidly to the propane with stainless steel forceps. The sample should be small, to ensure rapid freezing and subsequently complete substitution.

PREPARATION OF SUBSTITUTION FLUID

The recommended substitution fluid is a 20% (v/v) mixture of acrolein in diethyl ether. Some caution must be exercised in using acrolein, since it is a highly explosive and toxic chemical (Hayat, 1980). It must be used only in scrupulously clean glassware, since contaminants can initiate rapid and highly exothermic polymerization. It is a highly volatile liquid with a flash point below 0°F. Acrolein is highly toxic, but it can be detected by olfaction at a concentration at or below the maximum allowable vapor concentration of 0.5 ppm.

Figure 2. Dry box for freeze-substitution, enclosed in a fume hood. Dry box contains substitution fluids, tubes, dessicator, and a hair hygrometer.

Mixing of acrolein and ether must be carried out in a fume cupboard and the mixture stored in a brown glass bottle. It is advisable to wear protective goggles, since liquid acrolein in contact with the eyes can produce severe damage. Spills can be neutralized with a 10% solution of sodium bisulfite, and such a solution should be available.

The mixture must be dried by storing over activated molecular sieve (Union Carbide Linde pellets 3A or 4A) for at least 12 h. Drierite (anhydrous calcium sulfate) should not be used because it is known to remove the hydroquinone present in acrolein as a polymerization inhibitor. A convenient way to activate the molecular sieve is to heat it to 400°C for 24 h and allow it to cool in a dessicator. A considerable vacuum should be produced in the dessicator. The requirement for rigorously anhydrous conditions is crucial to the success of freeze-substitution.

The stock solution of substitution fluid should now be opened only in a dry box. The author uses a "perspex" box with a fume cupboard (Figure 2). Drying of the box atmosphere is accomplished by means of activated molecular sieve and the box is suitable for use when the relative humidity falls to 5–10% at 20°C. Overnight drying is usually sufficient.

Polyethylene tubes with well-sealing caps are filled in the dry box with activated molecular sieve and substitution fluid in a 1:1 ratio. The capacity of the tubes should be 5–10 ml. The tubes are transferred to a low-temperature refrigerator at −90°C until required.

SUBSTITUTION AND EMBEDDING

In order to facilitate a rapid transfer of quenched tissue from propane to substitution tubes, the tubes are removed from the low-temperature refrigerator and stored in a dewar containing a CO_2/acetone slush at −80°C. The tubes can then be kept next to the quenching fluid and momentarily opened to allow transfer of tissue from the propane thimble with precooled forceps.

After transfer, the tubes are replaced in the low-temperature refrigerator and substitution is allowed to proceed at −90°C for 14–20 days. This period seems to be adequate on an empirical basis and is not far removed from that used by Harvey et al. (1976). These authors measured the rate of dye removal from agar blocks in ether at −72°C and on this basis chose 21 days as adequate for substitution. The temperature is then slowly raised, typically one day each at −60°, −40°, −20°, −10°C, and then to room temperature.

The rationale of this protocol is that 1) −90°C is probably well below the eutectic point (see Pearse, 1968) and as near to the presumed recrystallization temperature (−100°C) as possible and 2) slow rewarming up to the eutectic point (−40° to −20°C) allows complete substitution of any ice still remaining in the center of the block and holding at $-10°C$ facilitates complete fixation. Slow rewarming was shown by Sasse and Matthaei (1977) to give improved fixation at the light microscope level. The tubes are transferred in a polyethylene bag from the refrigerator (at $-10°C$) to the dry box. After sufficient time has elapsed for them to thermally equilibrate, the specimens are removed from the tubes and infiltrated with increasing concentrations of dried resin monomer and diethyl ether. The specimens are finally embedded in "Beem" capsules and transferred to an oven for polymerization. After polymerization, the blocks are stored in dessicators over molecular sieve.

SECTIONING

In general, sections must be cut dry. This means that it is extremely difficult to obtain sections thinner than 0.5 μm. Thinner sections compress and crumple on the knife edge and cannot be flattened. Sectioning onto fluids is hazardous, since inorganic ions will diffuse out of the section with great rapidity. Harvey et al. (1976) have measured a loss of 87% of Na^{22} and 58% of Cl^{36} from labeled plant specimens that were freeze-substituted and sectioned onto water. Anhydrous solvents, such as ethylene glycol, hexylene glycol, glycerol, and dibutyl phthallate, have been tried as trough fluids by the author, but a marked change in the x-ray spectrum compared with dry cut sections has been observed in each case. Solvents such as ethylene glycol absorb water from the atmosphere so rapidly that they cannot in fact be regarded as dry after more than a very short time after being placed in the trough. The only fluid so far found to be of any use by the author is liquid paraffin. Thin sections can be cut into it (as with all these solvents, the surface tension properties are such that the sections invariably sink and have to be fished out of the trough), but there is then the added complication of removing the paraffin. Removal can be done by flooding the grid with Freon TF, thereby washing off the paraffin, and allowing the Freon to evaporate.

For most purposes, sections of 0.5 to 1.0 μm will have to be used. Sections are picked up off the knife edge with a hair or fine tungsten needle and transferred to a filmed grid on the knife. The section is then gently flattened and firmly pressed down onto the film. This requires a little practice but it is possible to obtain reasonably flat sections that adhere firmly to the film. The author uses nylon grids from which a square (of 1–2 mm side) is cut out of the central area with a scalpel. A tough film is required to support most, if not the entire area, of the section in the central square (Figure 3).

Plastics used as support films vary considerably in their mechanical strength, resistance to electron radiation, and shrinkage in the electron beam. The properties of various materials for support films have been

Figure 3.   Carbon-coated Nylon grid with a central square of mesh cut out, and a freeze-substituted section mounted in the square on a supporting film.

extensively discussed by Baumeister and Hahn (1978). The most suitable plastic known to the author is Pioloform, which is a polyvinylformal. It should be noted that crosslinking may occur between a section and the support film and that irradiation-induced contraction of the film will lead to shrinkage of the section (Baumeister and Hahn, 1978). It has been shown by Kolbel (1972) that methacrylate sections shrink by 20–25% on nitrocellulose films, less on polyvinylformal films, and only 5% on carbon films.

When the section is firmly on the grid, it is coated with 20 nm of aluminum, chromium, or beryllium. If the section is not adhering well it can be sandwiched by applying a second film on top of the section. This can be readily done by placing the grid on a peg and bringing the film, cast onto a ring, over the grid. A piece of photographic film with holes cut into it with a cork borer makes a good support for the sandwiching film. A metal coat is then applied to the top film. The finished grids should be stored over molecular sieve until required.

RESINS

The resins that have so far been used for freeze-substituted specimens for x-ray microanalysis have been Spurr's resin (Läuchli et al., 1970) and modified Spurr's resin (Pallaghy, 1973). The formulation of the latter is:

	Weight
VCD (Vinylcyclohexene dioxide)	10 g
NSA (Nonenyl succinic anhydride)	26 g
Dibutyl phthalate	12 g
DMP-30 (Tridimethyl amino methyl phenol)	1.6 g

This modification was developed because Spurr's medium (Spurr, 1969) contains appreciable quantities of chlorine, particularly in the flexibilizer DER 736 (diglycidyl ether of polypropylene glycol). The resulting chlorine-free resin has a low viscosity and reasonable sectioning qualities. The hardness of the blocks can be changed by varying the dibutyl phthalate and DMP-30 content. This resin would be very satisfactory except that it is an epoxy resin, and, as Pallaghy (1973) pointed out, the possibility exists that free water is formed during the polymerization of epoxy resins. This could, of course, affect the distribution of diffusible ions in specimens contained within the resin.

The reactions that occur during anhydride hardening of epoxy resins are complex, but the general mechanism of hardening is believed to be as follows (modified from Brydson, 1975):

1. Opening of the anhydride ring by an alcoholic hydroxyl group

$$R^1\!-\!CO \diagdown \atop \diagup \atop R^1\!-\!CO \quad O + HOR^{11} \rightarrow$$

2. Reaction of the carboxylic group with the epoxy group

3. Etherification of the epoxy group by hydroxyl groups

4. Reaction of the monoester with a hydroxyl group

5. Hydrolysis of the anhydride to acid by the water released in reaction 4.
6. Hydrolysis of the monoester with water to give acid and alcohol.

There appears to be no information available that would make it possible to estimate the amount of water produced in reaction 4 nor whether it could exist for any appreciable length of time before participating in reactions 5 and 6. Nevertheless, it is clear that the possibility does exist that ions could be leached from, or translocated within, embedded tissue during polymerization. How serious this might be has yet to be assessed. Recent unpublished work carried out in the author's laboratory by Q. G. Forrest showed that ~85% of Na^{22} was retained in isotope-loaded samples of cockroach nerve cord after polymerization in the chlorine-free modification of Spurr's resin.

A loss of about 15% of diffusible sodium from cockroach nerve cord to the surrounding polymer would not be readily detected by x-ray microanalysis. The concentration of this element is very low and a 15% change in concentration would probably be within the error due to counting statistics. Certainly no diffusible elements of any kind can be detected at distances greater than the resolution limits of the analysis from the periphery of any specimen embedded in epoxy resin ($\simeq 1 \ \mu$m). If it is assumed that the diffusion distances are small, then it might be expected that a 15% loss of potassium from most specimens would result in a detectable amount of potassium in the polymerized resin surrounding the specimens. This expectation has not so far been fulfilled.

The possibility of ion loss or translocation due to water release during polymerization can be circumvented by using an acrylic resin, i.e., methacrylate. The polymerization of methacrylate proceeds without the production of water. For example, the polymerization of methyl methacrylate is as follows (Saunders, 1973):

$$nH_2C{=}\underset{\underset{COOCH_3}{|}}{\overset{\overset{CH_3}{|}}{C}} \quad \rightarrow \quad \left[-H_2C{-}\underset{\underset{COOCH_3}{|}}{\overset{\overset{CH_3}{|}}{C}}- \right]_n$$

A suitable embedding medium can be prepared from a mixture of n-butyl and methyl methacrylate in the ratio of 80:20 by weight. The inhibitor (hydroquinone) need not be removed if sufficient catalyst (benzoyl peroxide) is added (~2–2.5%) (Hayat, 1980).

It has been shown again by Forrest (unpublished work) that Na^{22} is totally retained in cockroach nerve cord during methacrylate embedding and polymerization. Further, the present author has used the dye merocyanin, which is completely insoluble in methacrylate monomer but is extremely soluble in the slightest trace of water, to show that water can be completely excluded from methacrylate during infiltration and polymerization. The advantages and disadvantages of these two resins must now be considered.

Epoxy resins, when compared with methacrylates, have the advantages of little polymerization damage to the specimen and relatively greater resistance to radiation damage in the electron beam. They are consequently much easier to use than methacrylates, which require considerable attention to detail if polymerization and radiation damage are to be minimized.

Mass loss during conventional electron microscopy from epoxy resin sections can be as much as 20%, and for methacrylate sections it can be 50% (Isaacson, 1977). In the case of epoxy resin sections, most, if not all, of the mass loss may be attributed to radiation damage if beam current densities are low. Methacrylate resins, however, are extremely sensitive to heating damage (Luft, 1973). Heating damage may be a significant factor in the relatively thick (0.5–1.0 μm) sections used for x-ray microanalysis. Mass loss during x-ray microanalysis is reduced markedly by coating the sections with aluminum or chromium. This is presumably due to increased heat conduction through the metal coat. Reimer (1959) noted a small decrease in mass loss if plastic sections were coated with carbon. Heating damage is probably also reduced by mounting the section on a film support over a large hole cut in a nylon grid. It is known that irradiation of metal grid bars can raise specimen temperatures considerably (Isaacson, 1977). Mass loss can be reduced in methacrylate sections quite considerably by the aforementioned procedures; however, it is unlikely that it can be reduced to the same extent as for epoxy resin sections.

Polymerization damage can be severe in methacrylate-embedded specimens. Luft (1973) ascribes this to the restriction of growing polymer chains in tissue, which gives the effect of increased viscosity. Increased viscosity leads in turn to osmotic swelling by monomer diffusing across tissue boundaries that are impermeable to the polymer. The end result is a swelling of tissue components as monomer continues to diffuse into and dissolve in the polymer.

The effect can be diminished in several ways, but only two are acceptable for specimens destined for x-ray microanalysis. The use of uranyl nitrate in the resin mixture (Ward, 1958) as a means of reducing polymerization damage, for example, is not desirable because the M-line emission energy of uranium is very close to the potassium K-line emission energy.

The acceptable methods for reducing polymerization damage are: 1) to increase polymerization temperature to 90°–100°C, and 2) to prepolymerize the embedding medium. An increase in polymerization temperature ideally keeps the polymerized resin in the specimen less viscose and thereby avoids the osmotic effect. Prepolymerized resin, obtained either by heating the monomer (Borysko and Sapranouskas, 1954) or by adding polymer powder to make a slurry, matches the viscosities within and outside the tissue and again avoids the osmotic effect. It should be obvious that the two methods are not compatible and should not be used simultaneously. It should also be obvious that, in the second case, tissue should be infiltrated in low-viscosity monomer, not in high-viscosity prepolymerized resin.

The epoxy resins previously discussed and methacrylate resins have densities close to 1.0, which is an advantage in terms of specimen contrast (the lower the density of the embedding medium, the greater the specimen contrast). This is an important consideration when viewing sections for analysis, since they are, of course, unstained. The attainable contrast varies with the specimen and there is sometimes better contrast with one resin than with the other. It is, however, usually mandatory to examine ultrathin sections (cut wet) to estimate the extent of ice-crystal damage and the state of preservation of tissue components. Ultrathin sections of epoxy-embedded freeze-substituted sections are extremely hard to stain. Staining times must frequently be quadrupled to achieve even weak contrast. In this respect ultrathin methacrylate sections offer a distinct advantage: they are frequently very easy to stain, and, if necessary, the section can be deliberately "thinned" in the electron beam (i.e., mass loss by radiation damage is encouraged), thereby considerably enhancing contrast. The latter technique can also be employed in difficult cases with uncoated thick (0.5–1.0 μm) sections used for analysis. This will result in increased contrast, reduced continuum, and enhanced P/B ratios. As will be seen later, however, the use of uncoated sections for analysis leads to other problems and thinning sections in this way means that analyses in terms of elemental mass per unit volume are no longer valid. There may also be a risk of mass loss from the tissue itself. This procedure is recommended only when all else fails. A density approaching 1.0 is also important if estimates of section thickness are to be made using latex spheres, as in the technique described in Chapter 1.

The author has made preliminary examinations of several tissues that have been embedded in both epoxy and methacrylate resins and he has not so far found any differences that cannot be ascribed to either biological variation or analytical error. In addition, a small number of comparisons between tissues that have been analyzed as frozen-hydrated bulk specimens and freeze-substituted epoxy-embedded specimens have not revealed any apparently significant differences (see section on "Examples").

When using freeze-substituted sections it is advisable to take as much care as possible that they are not exposed to moist air for more than short periods of time. This is particularly advisable with methacrylate sections, in view of the findings of Fisher (1972), who found that, in methacrylate sections of freeze-substituted red beet cells, pigment from pigment vacuoles formed droplets on the *outside* surface of sections on exposure to water vapor. He also suggested that the absorption of water vapor by freeze-substituted tissues in methacrylate could induce shrinkage.

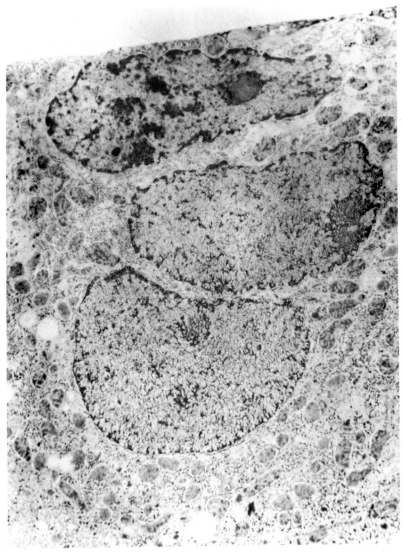

Figure 4.   Showing gradient of ice-crystal damage across a section of freeze-substituted mouse liver. Damage increases from outside (top) to inside (bottom). × 7000

It is probable that the absorption of water vapor may be reduced or entirely prevented by coating sections with hydrophobic carbon. Hydrophobic, as opposed to hydrophilic, carbon is produced under certain conditions of evaporative coating. The principal requirement is that the electrodes are > 12 cm from the specimen (Mahl and Moldner, 1972). Appleton (1978) has found carbon coating to be effective in preventing water uptake by

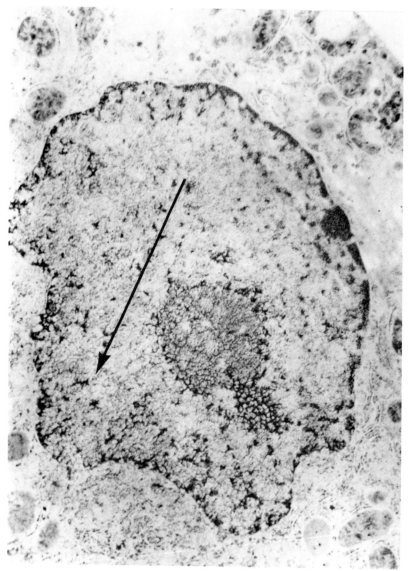

Figure 5. Showing gradient (arrow) of ice-crystal damage in a nucleus of a cell in freeze-substituted mouse liver. × 14,580

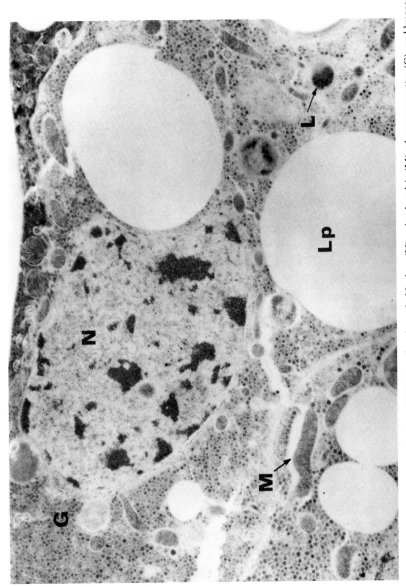

Figure 6.   Trophocyte cells from freeze-substituted fat body of cockroach. Nucleus (N), mitochondria (M), glycogen rosettes (G), and lysosomes (L) are well preserved. Lipid droplets (Lp) may have been dissolved. × 11,650

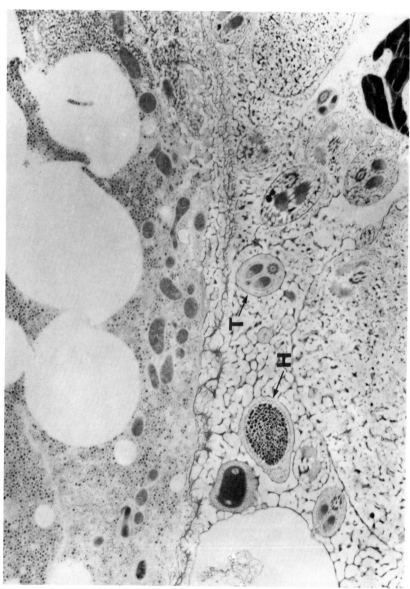

Figure 7. Comparison of ice-crystal damage in a tissue of low water content (trophocyte cell of cockroach fat body) and an adjacent tissue of high water content (testis). The trophocyte in the upper part of the micrograph contains well-preserved organelles. In the testis in the lower part of the micrograph, extensive ice-crystal damage is present. Note sperm tails (T) and

freeze-dried sections. In the author's view, aluminum coating is also effective for this purpose.

MORPHOLOGICAL PRESERVATION

Ultrathin, Stained Sections

The standard of morphological preservation is readily seen in ultrathin "wet cut" sections that have been stained. Figure 4 is a sample of freeze-substituted mouse liver. The increasing size of ice-crystal damage can be seen, particularly in the nuclei, as the distance from the outside edge of the specimen increases. The same effect can be seen across a single nucleus in Figure 5.

As might be expected, morphological preservation is seen at its best in tissues with low water content. Such a tissue is insect fat body, and Figure 6 shows an example of excellent preservation in cockroach fat body. The lipid

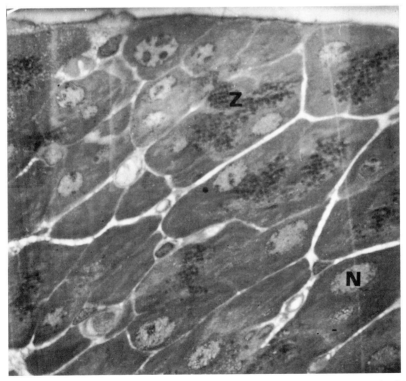

Figure 8.   Low-magnification scanning transmission electron micrograph of an unstained 1.0-μm methacrylate section of mouse pancreas. Nuclei (N) and zymogen granules (Z) are clearly seen. $\times$ 1,500

droplets may have been dissolved by the freeze-substitution solvent, but this has not been determined with certainty. In Figure 7 the section passes through a piece of cockroach fat body, with low water content, adjacent to a piece of testis of relatively high water content. The difference in preservation between the two tissues is striking.

Thick, Unstained Sections

It is easy to have good morphological preservation and to be able to identify cell organelles in ultrathin stained sections, but these are difficult in 1.0-μm unstained sections that are to be analyzed. Contrast is inherently weak, particularly in specimens with good preservation. Greatly improved images are obtained in a STEM. Tissues vary in the degree of contrast obtainable in epoxy and methacrylate resins, presumably because of slight differences in both resin and tissue densities. Particularly good examples of both image quality and morphological preservation are shown in Figures 8–10 (showing a 1.0-μm section of mouse pancreas in methacrylate photographed in JEOL 100C STEM at 100 kv). In the author's experience these examples represent the best obtainable by present methods.

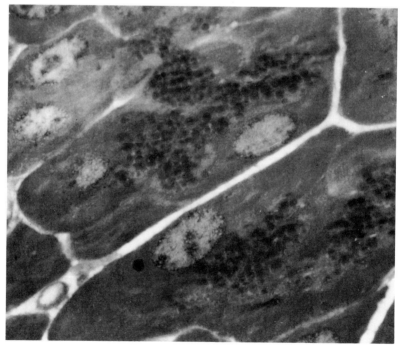

Figure 9.   Higher-magnification micrograph of the same region that was shown in Figure 8. × 3,000

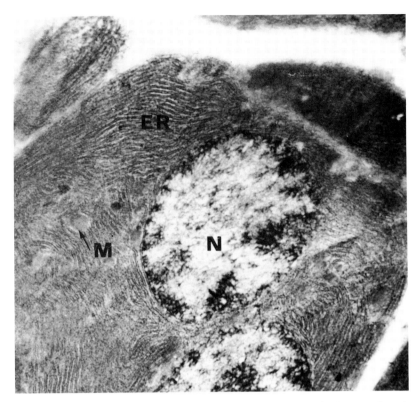

Figure 10.   Scanning transmission electron micrograph of an unstained 1.0-μm methacrylate section of mouse pancreas. Nucleus (N), mitochondria (M), and rough endoplasmic reticulum (ER) are clearly visible. × 12,000

ANALYTICAL PROCEDURES

The analysis of sections is not straightforward. There are a number of practical considerations that will have a profound effect upon the analysis if they are not dealt with. This applies equally well to the analysis of conventionally fixed and embedded sections and freeze-dried sections. (It is assumed that extraneous x-rays have been eliminated from the system and that the detector is well collimated. These precautions have been discussed in Chapter 1.)

Beam Current

Beam current must remain constant during an analysis and between analyses if an analysis in terms of mass per unit volume is to be attempted. The measurement of beam current has been discussed in Chapter 1), where

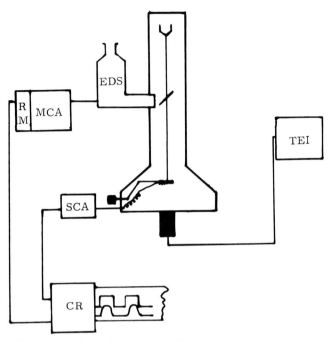

Figure 11.   Diagram showing method of measuring transmitted electron currents in a STEM or TEM. Abbreviations: energy-dispersive detector—EDS; multichannel analyzer—MCA; rate meter—RM; specimen current amplifier—SCA; chart recorder—CR; and transmitted electron image—TEI.

attention was drawn to the difficulties of both measuring and stabilizing beam current in a TEM or STEM *during* an analysis. The best method that the author has found is to collect the transmitted beam current on a polished copper disc mounted in the viewing chamber, i.e., above the fluorescent screen in a TEM (Figure 11). Replacement of the diffraction beam stop with a similar rod on which the disc is mounted has proved to be a very satisfactory arrangement. The disc may have concentric grooves machined into the upper surface that are filled with phosphor. These will facilitate reproducible positioning of the disc in the beam if the disc is moved in and out of the beam to facilitate image viewing. In the STEM mode, an image can be obtained without moving the collecting disc if a central hole of suitable diameter is drilled through the disc. The electron current is led out to a specimen current amplifier via shielded cable.

This device allows beam current stability to be easily monitored during an analysis, and by shifting the grid so that the beam passes through a suitable hole or a reference area in the section, beam current can be monitored between analyses (see Figure 13). A means of monitoring transmitted electron beam current provides a very convenient means of monitoring other parameters that affect an analysis, i.e., charging, specimen drift, contamination, and mass loss.

Charging

Charging of sections can occur if the conductive film is discontinuous or if the sections are uncoated. The occurrence of charging can be readily detected as excursions of the transmitted beam current. This can be readily seen in Figure 12. In this case, charging has resulted in momentary deflections of the beam from a relatively dense subcellular structure to less dense cytosol; consequently, the transmitted electron beam shows sporadic increases in value. Total count rate was recorded simultaneously and the upward deflections of beam current are mirrored by downward deflections of count rate. The latter is a consequence of the beam moving to less dense areas of the specimen, which yielded a lower count rate.

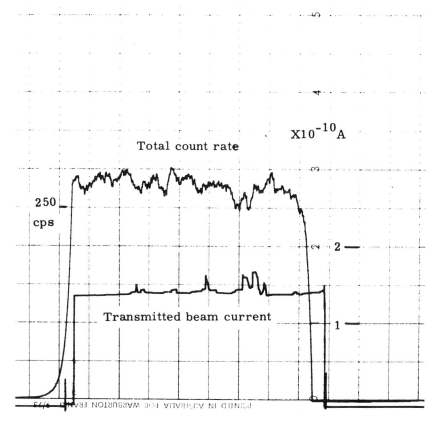

Figure 12. Effect of specimen charging on transmitted beam current and total count rate. Charging is indicated by the short-term deviations in the traces.

Drift

Specimen drift can be easily recognized as long-term excursions of transmitted beam current. Figure 13 shows an example of drift superimposed on charging. Drift has occurred in that the beam has moved from a dense to a less dense region of the specimen. The increase in transmitted beam current is paralleled by a fall in total count rate. Note the stable beam current recording when the beam was positioned in a hole in the grid. Figure 14 shows a further example, in which the beam position has gradually shifted from a less dense to a denser region of the specimen.

Contamination

If contamination occurs during an analysis it can be readily recognized by an exponential decrease in transmitted beam current and a simultaneous linear increase in continuum counts. This is shown in Figure 15. If contamination is a problem, it can be reduced by analyzing a larger area. The effect of changing from a 20-nm static spot to a 0.25-μm^2 raster and then a 1.0-μm^2 raster is shown in Figure 16. The rate of decrease in transmitted beam current is reduced, as is the rate of increase in continuum count.

Mass Loss

Mass loss can be recognized by an increasing transmitted beam current and a falling continuum count rate. Mass loss can be restricted to ~20% using alu-

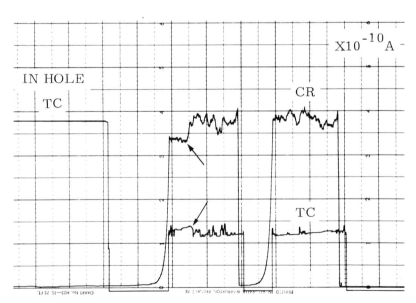

Figure 13. Effect of drift on transmitted beam current (TC) and count rate (CR). Drift is indicated by the long-term deviations in the traces (arrows). Short-term deviations indicate charging. Chart speed was 0.5 cm/min.

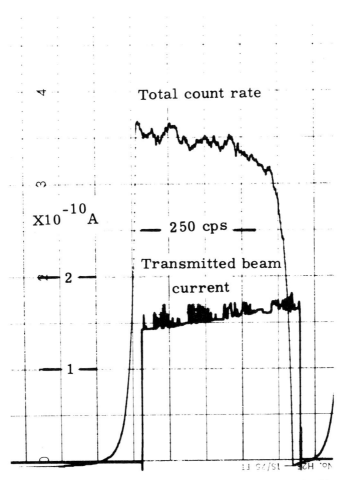

Figure 14.   Effect of gradual drift is shown as a linear decrease in transmitted beam current and an increase in count rate. Short-term deviations in transmitted beam current are due to charging. Chart speed was 0.5 cm/min.

minum-coated sections even with a high total radiation dose of 30,000–40,000 C/cm^2.

QUANTITATIVE ANALYSIS

Methods of quantitative analysis have been outlined in Chapter 1. Plastic sections can be cut with moderate ease, and their thickness can easily be checked by the method described in Chapter 1. Furthermore, standards can be readily made (Spurr, 1974; Chandler, 1976). It is therefore feasible to analyze freeze-substituted sections in terms of mass per unit volume. As

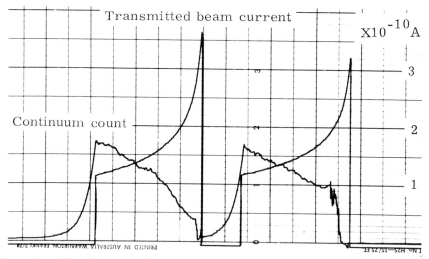

Figure 15.   Effect of contamination on count rate and transmitted beam current. Transmitted beam current decreases exponentially and count rate increases linearly. Chart speed was 0.5 cm/min.

previously mentioned, beam current must be stabilized and reproduced from one analysis to another. In these circumstances, the simple equation

$$C_x = \frac{P_x}{P_{kx}} \cdot C_{kx}$$

(where C_x and C_{kx} are the unknown and known concentrations and P_x and P_{kx} are the observed peak intensities of the unknown and known concentrations, respectively) can be applied. The possible changes in volume of specimens and resins that have been previously discussed must, however, be borne in mind.

The advantage of analyzing in terms of mass per unit volume is that the results can be directly related to the concentrations prevailing in life if it is assumed that the resin in the tissue specimen replaces water without a gross redistribution of diffusible elements.

In an analysis of this type it may be useful to have information about the mass of the analyzed volume. This can be obtained from the continuum count, but it can also be obtained from the transmitted beam current (Figure 17). The advantages of using transmitted beam current are that the instrument contribution to the continuum count does not have to be determined and that this measure of the total mass is more sensitive than the continuum method (Halloran et al., 1978). The relative mass fraction can therefore be obtained as the ratio of peak intensity to transmitted beam current, and by the use of standards, absolute densities and mass fractions can be obtained.

This is similar to the method proposed by Halloran et al. (1978), but is in essence much simpler. If transmitted beam current cannot be directly measured as in a STEM, then the photomultiplier response of the transmitted electron detector can be used as described by Halloran et al. (1978).

Halloran et al. (1978) have shown that a measurement of beam attenuation through a section in a STEM by means of the transmitted electron detector can be used as a measure of total mass per unit area:

$$C_x \text{ (concentration of element x)} = \frac{M_x \text{ (elemental mass/unit area)}}{M \text{ (total mass/unit area)}} \qquad \text{(Hall, 1971)}$$

M is also defined as the mass thickness ρt where ρ is specimen density and t is section thickness.

Concentration (Cx) can be expressed in terms of peak counts (Px) and mass thickness (ρt) as:

$$Cx = \alpha \frac{Px}{\rho t}$$

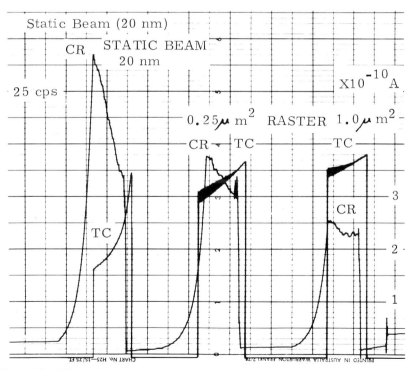

Figure 16. Effect of beam current density on contamination rate. There is a decrease in decline and growth of transmitted beam current (TC) and continuum count rate (CR), respectively, as current density decreases. Chart speed was 0.5 cm/min.

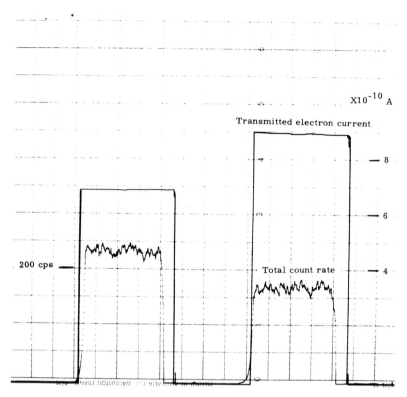

Figure 17. The difference in transmitted beam current and total count rate between consecutive analyses of two different zymogen granules in an acinar cell of mouse pancreas is directly related to their differing densities. Chart speed was 1.0 cm/min.

(where α is a proportionality constant relating characteristic x-ray intensity to concentration and is found from calibration curves obtained from standards for any particular element). Peak counts for an element per unit mass thickness are plotted against concentration of the element; α is then found as the slope of the curve.

As discussed by Halloran et al. (1978), the transmitted electron intensity I is related to the initial intensity I_0 by an expression similar to Beer's Law:

$$I = I_0 e^{-S_t \rho t}$$

(where S_t is the total mass scattering cross-section). If S_t can be shown to be constant, then the expression reduces to

$$\ln (I_0/I) = \rho t$$

In the case of resin-embedded sections analyzed in a STEM at 100 kV, it has already been shown in Chapter 1 that $\ln (I_0/I) = \rho t$ is a linear relationship for thicknesses up to 1 μm.

All that is necessary to do then is to measure peak intensity (P_x) and I_0 and I (directly as transmitted electron currents) on standards with differing concentrations of element x. A calibration curve is constructed. Measurement of P_x, I_0, and I on the section of unknown composition can then be directly translated into concentration from the calibration curve.

A possible source of error in this method is if the major contribution to beam attenuation in the section of unknown composition is protein, i.e., if a

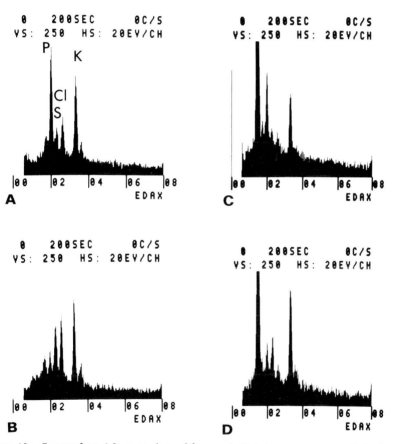

Figure 18.   Spectra from 1.0-μm sections of freeze-substituted mouse pancreas. Endoplasmic reticulum, **A** and **C**; zymogen granules, **B** and **D**. Spectra **A** and **B** are from uncoated methacrylate section and **B** and **D** are from aluminum-coated methacrylate section.

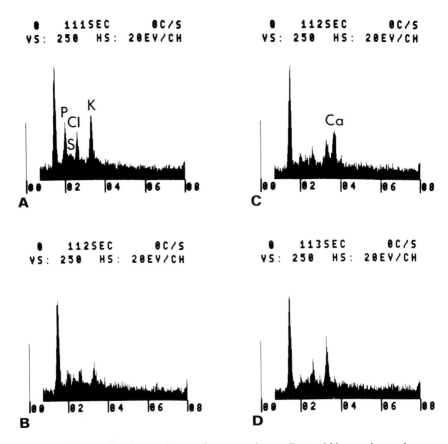

Figure 19. Spectra showing persistence of concentration gradients within a nucleus and across the nuclear membrane in a 1.0-μm section of freeze-substituted cockroach fat body. Heterochromatin, **A**; euchromatin, **B**; electron-dense body in nucleus, **C**; and cytoplasm adjacent to nucleus, **D**. Spectra are from aluminum-coated section of chlorine-free epoxy resin.

very dense protein granule is being analyzed for example. In this case ρ will be different from the ρ of the resin standards.

If beam attenuation is being used for density measurement, then the overall section thickness t must be measured as in Chapter 1. The relative density of any structure that extends through the specimen is then proportional to the ratio $\ln(I_0/I)/t$. If absolute densities are required, then protein standards would have to be used.

RESOLUTION

The analytical resolution can be determined experimentally. This is done by determining the minimum distance between two analytical sites, analyzed with a static probe, from which distinctly different spectra can be obtained across a phase boundary. It must be borne in mind that electron scatter within the section will be the limiting factor only if extraneous x-rays generated by the microscope and scattered electrons have been eliminated. In the author's experience a resolution at least 250 nm should be obtainable in a 1.0-μm section.

EXAMPLES

1. Spectra typical of those obtainable from methacrylate sections.
    Mouse pancreas was freeze-substituted and embedded in methacrylate. Analyses were made in a JEM 100B STEM at 100 kV with a static beam. Section thickness was 1.0 μm. Spectra A and B are from an uncoated section and spectra C and D are from an aluminum-coated section (Figure 18). Spectra A and C are from endoplasmic reticulum and spectra B and D are from zymogen granules.
2. Spectra showing persistence of concentration differences within a nucleus after freeze-substitution.
    Cockroach fat body was freeze-substituted and embedded in chlorine-free epoxy resin. Analyses were made in a JEM 100B STEM at 100 kV with a static beam. Sections were coated with aluminum and were 1.0 μm thick. Spectra are from the nucleus of a trophocyte cell (Figure 19). Spectrum A is from heterochromatin, and B is from euchromatin, C is from a small electron-dense body in the nucleus, and D is from cytoplasm immediately adjacent to the nucleus.
3. Comparison of spectra from a freeze-substituted specimen and a frozen-hydrated bulk specimen.

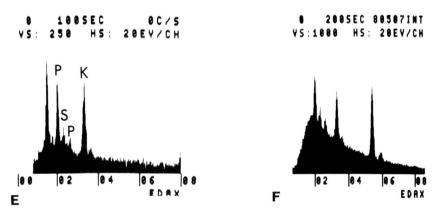

Figure 20. Comparison of freeze-substituted and bulk frozen-hydrated specimens. Spectrum E is from cytoplasm of freeze-substituted cockroach fat body. Spectrum F is from cytoplasm of bulk frozen-hydrated fat body.

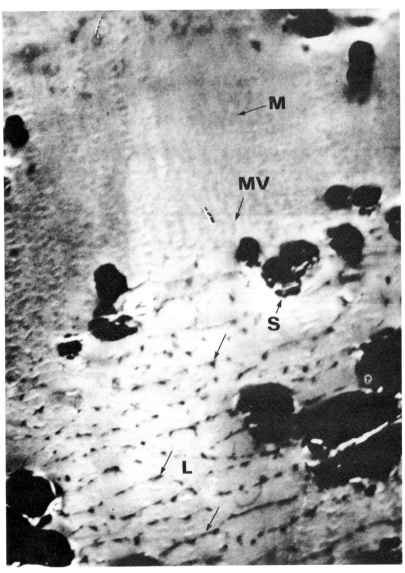

Figure 21.   Scanning transmission electron micrograph of part of a 0.5-μm thick section of freeze-substituted Malpighian tubule from the cricket *Teleogryllus oceanicus*. Deposits of potassium chloride (arrows) are present in the lumen (L). Other structures are spherites (S), microvilli (MV), and mitochondria (M). × 5400

Table 2. Comparison of axoplasmic atomic ratios from frozen-hydrated (bulk specimen) and freeze-substituted (sections in chlorine-free epoxy resin) cockroach axons

Element	Frozen-hydrated				Freeze-substituted			
	K	P	S	Cl	K	P	S	Cl
Axoplasm								
1	1.00	0.09	0.05	0.17	1.00	0.14	0.07	0.14
2	1.00	0.05	0.06	0.66	1.00	0.18	0.00	0.07
3	1.00	0.04	0.01	0.04	1.00	0.15	0.07	0.16
4	1.00	0.04	0.01	0.05	1.00	0.04	0.02	0.07
5	1.00	0.03	0.01	0.01	1.00	0.01	0.07	0.07
6	1.00	0.03	0.01	0.02	1.00	0.14	0.07	0.12
7	1.00	0.04	0.11	0.11	1.00	0.12	0.06	0.07
8	1.00	0.03	0.01	0.04				
9	1.00	0.09	0.04	0.10				

a. Analyses of the cytoplasm of trophocyte cells in cockroach fat body are shown in spectra E and F (Figure 20). Spectrum E is from freeze-substituted fat body embedded in chlorine-free epoxy resin. An aluminum-coated 1.0-μm section was analyzed in a JEM 100B STEM at 100 kV with a static spot 20 nm in diameter. Spectrum B is from a frozen-hydrated bulk specimen of fat body, coated with chromium and analyzed at $-160°C$ in a JSM 35 SEM at 15 kV with a static beam.

b. The semiquantitative similarity between freeze-substituted and frozen-hydrated specimens is shown in Table 2 (from Forrest and Marshall, 1976). These analyses represent the axoplasmic compositions of giant axons in cockroach nerve cord, expressed as elemental atomic ratios.

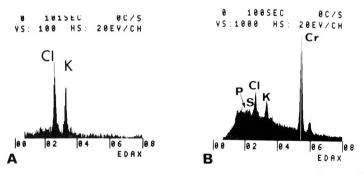

Figure 22. Spectra from lumen of freeze-substituted cricket Malpighian tubule (A) and lumen of frozen-hydrated bulk specimen of cricket Malpighian tubule (B).

4. Comparison of spectra from extracellular space in a freeze-substituted specimen and a frozen-hydrated bulk-specimen.

Cricket Malpighian tubules were freeze-substituted and embedded in epoxy resin. A 0.5-μm thick section is shown in Figure 21. The lumen of the tubule contains large electron dense spherites and smaller electron-dense particles. The latter represent KCl present in the original luminal solution and deposited around the ice crystals during freezing. The KCl retains its original position in the lumen during the freeze-substitution process. A spectrum from an analysis of the luminal contents using a small raster (JEM 100B STEM at 100 kV) is shown in Figure 22a. This should be compared with Figure 22b, which is spectrum from the luminal contents of a frozen-hydrated bulk specimen of Malpighian tubule, as is seen in Figure 10 in Chapter 3 (analyzed in a JSM 35 at 15 kV with a small raster).

LITERATURE CITED

Appleton, T. A. 1978. The contribution of cryoultramicrotomy to x-ray microanalysis in biology. In: *Electron Probe Microanalysis in Biology* (Erasmus, D. A., ed.), p. 148. Chapman & Hall, London.

Baumeister, W., and Hahn, M. 1978. Specimen supports. In: *Principles and Techniques of Electron Microscopy*, Vol. 8 (Hayat, M. A., ed.), p. 1. Van Nostrand Reinhold, New York.

Borysko, E., and Sapranouskas, P. 1954. A new technique for comparative phase-contrast and electron microscope studies of cells grown in tissue culture, with an evaluation of the technique by means of time-lapse cinemicrographs. Bull. Johns Hopkins Hosp. 95:68.

Brydson, J. A. 1975. *Plastics Materials*. Newnes-Butterworths, London.

Chandler, J. A. 1976. A method for preparing absolute standards for calibration and measurement of section thickness with x-ray microanalysis of biological ultrathin specimens in EMMA. J. Microscopy 106:391.

Dempsey, G. P., and Bullivant, S. 1976. A copper block method for freezing non-cryoprotected tissue to produce ice-crystal-free regions for electron microscopy. 1. Evaluation using freeze-substitution. J. Microscopy 106:251.

Fisher, D. B. 1972. Artifacts in the embedment of water-soluble compounds for light microscopy. Plant Physiol. 49:161.

Forrest, Q. G., and Marshall, A. T. 1976. Comparative x-ray microanalysis of frozen-hydrated and freeze-substituted specimens. Proc. Sixth Europ. Cong. Electron Microscopy, Jerusalem, p. 218.

Hall, T. A. 1977. The microprobe assay of chemical elements. In: *Physical Techniques in Biological Research*, Vol. 1(c) (Oster, G., ed.), p. 157. Academic Press, New York.

Halloran, B. P., Kirk, R. G., and Spurr, A. R. 1978. Quantitative electron probe microanalysis of biological thin sections: the use of STEM for measurement of local mass thickness. Ultramicroscopy 3:175.

Harvey, D. M. R., Hall, J. L., and Flowers, T. J. 1976. The use of freeze-substitution in the preparation of plant tissue for ion localization studies. J. Microscopy 107:189.

Hayat, M. A. 1980. *Principles and Techniques of Electron Microscopy*, Vol. 1. 2nd Ed. University Park Press, Baltimore, Maryland.

Hayat, M. A. 1981. *Fixation for Electron Microscopy*. Academic Press, New York.

Hereward, F. V., and Northcote, D. H. 1972. A simple freeze-substitution method for the study of ultrastructure of plant tissues. Exp. Cell Res. 70:73.

Isaacson, M. S. 1977. Specimen damage in the electron microscope. In: *Principles and Techniques of Electron Microscopy*, Vol. 7 (Hayat, M. A., ed.), p. 1. Van Nostrand Reinhold, New York.

Kolbel, Von H. K. 1972. Influence of various support films on image size and contrast of thin-sectioned biological objects in electron microscopy. Mikroskopie 28:202.

Läuchli, A., Spurr, A. R., and Wittkopp, R. W. 1970. Electron probe analysis of freeze-substituted, epoxy resin embedded tissue for ion transport studies in plants. Planta 95:341.

Luft, J. H. 1973. Embedding media—old and new. In: *Advanced Techniques in Biological Electron Microscopy* (Koehler, J. K., ed.), p. 1. Springer Verlag, Berlin.

Mahl, H. Von, and Moldner, K. 1972. Observations on evaporated carbon coatings. Mikroskopie 28:139.

Morgan, A. J., Davies, T. W., and Erasmus, D. A. 1978. Specimen preparation. In: *Electron Probe Microanalysis in Biology* (Erasmus, D. A., ed.), p. 94. Chapman & Hall, London.

Pallaghy, C. K. 1973. Electron probe microanalysis of potassium and chloride in freeze-substituted leaf sections of *Zea mays*. Aust. J. Biol. Sci. 26:1015.

Pearse, A. G. E. 1968. *Histochemistry, Theoretical and Applied*. Churchill, London.

Rebhun, L. I. 1959. Freeze-substitution: Fine structure as a function of water concentration in cells. Fed. Proc. 24 (Suppl. 15):217.

Rebhun, L. I. 1972. Freeze-substitution and freeze-drying. In: *Principles and Techniques of Electron Microscopy*, Vol. 2 (Hayat, M. A., ed.), p. 2. Van Nostrand Reinhold, New York.

Rebhun, L. I., and Sander, G. 1971. Electron microscope studies of frozen-substituted marine eggs. 1. Conditions for avoidance of intracellular ice crystallization. Am. J. Anat. 130:1.

Reimer, L. 1959. Quantitative Untersuchungen zur Massenabnahme von Embettungsmitteln (Methacrylat, Vestopal und Araldit) unter Elektronenbeschuss. Z. Naturforsch. 14B:566.

Sasse, D., and Matthaei, C. R. 1977. Improvement of freeze-substitution by programmed rewarming. Stain Technol. 52:299.

Saunders, K. J. 1973. *Organic Polymer Chemistry*. Chapman & Hall, London.

Spurr, A. R. 1969. A low-viscosity epoxy resin embedding medium for electron microscopy. J. Ultrastruct. Res. 26:31.

Spurr, A. R. 1974. Macrocyclic polyether complexes with alkali elements in epoxy resin as standards for x-ray analysis of biological tissues. In: *Microprobe Analysis as Applied to Cells and Tissues* (Hall, T. A., Echlin, P., and Kaufmann, R., eds.), p. 213. Academic Press, London.

Van Zyl, J., Forrest, Q. G., Hocking, C., and Pallaghy, C. K. 1976. Freeze-substitution of plant and animal tissue for the localization of water-soluble compounds by electron probe microanalysis. Micron 7:213.

Ward, R. T. 1958. Prevention of polymerization damage in methacrylate embedding media. J. Histochem. Cytochem. 6:398.

Williams, L., and Hodson, S. 1978. Quench cooling and ice crystal formation in biological tissues. Cryobiology 15:323.

Chapter 6

INFLUENCE OF SPECIMEN TOPOGRAPHY ON MICROANALYSIS

F. D. Hess

Department of Botany and Plant Pathology,
Purdue University,
West Lafayette, Indiana

INTRODUCTION

In the preface of Hall's (1968) paper published in the National Bureau of Standards special publication, "Quantitative Electron Probe Micro-analysis," the author remarks that there is a large gap between x-ray microanalysis "specialists" and biologists who apply x-ray microanalysis in their research. Hall states, "Experience has shown that probe specialists pay no attention to biological papers at probe meetings." This was particularly evident when problems related to biological applications of x-ray microanalysis were discussed. The purpose of this chapter, therefore, is to describe, in detail, a common problem in x-ray microanalysis that is often overlooked or disregarded in biological applications. That problem is the influence of specimen topography on bulk sample x-ray microanalysis. Other problems related to x-ray microanalysis (e.g., in-specimen flu-orescence, specimen charging, thermal and radiation damage to the specimen, specimen density variations, and contamination of the specimen surface) have been discussed elsewhere (see Marshall, 1975; Lechene and Warner, 1977).

Failure to understand the effects of specimen topography during x-ray microanalysis can lead to considerable error in both qualitative and quantitative analysis. Bomback (1973) concluded that inaccuracies in x-ray microanalysis arising from analysis of rough surfaces can make detected x-ray intensities "more dependent on surface morphology than on local elemental concentration." The topics related to x-ray microanalysis of rough specimens discussed in this chapter are applicable to all disciplines using this analytical method.

RECOGNITION OF SPECIMEN TOPOGRAPHY

The electron microprobe (EMP) usually uses an electron beam with a diameter of 200 nm to 1000 nm. Such large beam diameters do not provide high-resolution primary and secondary electron mode images of surfaces. If the surface of the sample being analyzed in an EMP has been carefully cut and polished, there is no need for evaluating surface morphology. If, however, the specimen being analyzed has not been cut and polished, as with many thick biological specimens, a careful examination of the surface is necessary to determine if problems induced by topography are likely to be encountered. The scanning electron microscope (SEM), with its 10-nm electron beam diameter, provides substantial resolution in the secondary electron image mode of operation and allows detailed evaluation of a sample's surface morphology. If the EMP is used for x-ray microanalysis, the specimen should first be studied by scanning electron microscopy to determine its topography in the area to be analyzed (Anderson and Leitner, 1969; Hess et al., 1975). The disadvantage here is that two different instruments must be used, and the researcher must find the same area of the specimen for analysis in the SEM and in the EMP (Anderson and Leitner, 1969). An alternative is to equip the SEM with x-ray microanalysis instrumentation. This allows morphological evaluation and x-ray microanalysis with one instrument. The SEM was designed primarily for secondary electron mode imaging, hence the potential for x-ray microanalysis may be less than that of the EMP (which was designed specifically for x-ray microanalysis).

The specific advantages of using the SEM or EMP equipped with energy-dispersive or wavelength-dispersive spectrometers has recently been discussed by Lechene and Warner (1977) and Walinga (1977).

EFFECTS OF SPECIMEN TOPOGRAPHY ON X-RAY MICROANALYSIS

There are three predominate instances where specimen topography has a significant influence on the results obtained during elemental analysis. The first is that an irregular specimen surface will cause local changes in the angle of electron beam incidence (the angle between the plane of the specimen at the incident point and the electron beam), the x-ray take-off angle (the angle between the plane of the specimen and a line from the elicited x-ray to the detector), and the size and shape of the excited volume. The second is that the specimen shape, or a remote area on the specimen, may protrude from the surface and absorb emitted x-rays, thus preventing their reaching the detector. A third instance is that emitted x-rays and back-scattered electrons can strike remote areas of an irregular specimen and cause the emission of detectable x-rays not associated with the area of interest. Figure 1 graphically summarizes these three types of problems on specimens with nonuniform topography.

The three instances described above as adversely affecting x-ray microanalysis occur in both the SEM energy-dispersive system and in the EMP wavelength-dispersive system. When the EMP system is used, an additional topographical effect is introduced because of the geometric focus-ing requirement. The x-ray source must be on the Rowland circle (focusing circle) of the spectrometer if the x-rays are to be detected. Russ (1970) reported that a vertical displacement of 40 μm (caused by topographic variation) can cause "defocusing" of the x-ray optics that will result in a change in count rate of 50%. The energy-dispersive spectrometers used in SEM x-ray microanalysis have no focusing requirement (Woldseth, 1973), hence there is no loss in intensity due to the defocusing phenomenon described for wavelength-dispersive spectrometers. This property of energy-dispersive spectrometers makes them insensitive to changes in vertical dis-placement of the specimen because of specimen roughness (Russ, 1970).

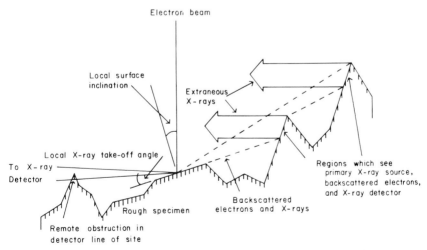

Figure 1.   The influences of specimen topography on x-ray microanalysis. Redrawn by per-mission from Bomback (1973)

Angle of Electron Beam Incidence and Take-Off Angle of X-Rays

In most SEM applications the specimen stage is tilted during morphological observations and x-ray analysis. Tilting the specimen results in an effectively larger cross-sectional beam area incident to the specimen surface (Woldseth, 1973). The effective cross-sectional area of the beam may also be increased when specimens having a rough surface are analyzed. Rough specimen surfaces will influence the local angle of electron beam incidence (Beaman and Isasi, 1972; Bomback, 1973), which will affect the depth of electron beam penetration (Berkey and Whitlow, 1971). These effects will change the intensity and distribution of the primary x-ray production (Birks, 1963; Berkey and Whitlow, 1972; Bomback, 1973), which if unrecognized will result in an inaccurate analysis. Russ (1973) reported that the local specimen orientation must be known accurately in quantitative x-ray microanalysis, because an error in tilt angle of only a few degrees can introduce a "several percent" error into the results.

Whenever x-rays pass through matter they are absorbed exponentially with distance: $I/I_0 = e^{\mu\rho t}$, where I_0 is the original x-ray intensity, μ is the mass absorption coefficient, ρ is the density of the material, and t is the distance the x-ray photons must travel through the substrate (Russ, 1974). The take-off angle from a flat polished specimen is an important consideration because it determines the maximum distance x-ray photons must travel through the sample. Because x-rays are emitted in all directions, the distance (t) through the sample that the x-rays travel is greater for detectors with a low take-off angle than for those with a high take-off angle (Läuchli, 1972). An example of this is presented in Figure 2. If the detector has a high take-off angle (θ_1), the distances the x-rays travel in the specimen is t_1. If the take-off angle of the detector is lowered to an angle θ_2, the distance the x-rays travel in the sample is $t_1 + t_2$ (where t_2 is the additional distance the x-rays travel to escape from the specimen). At low take-off angles, small changes in that angle may result in significant effects on the analysis. For analysis of chromium in a titanium matrix, Birks (1963) reported that a change in take-off angle of $\pm$ 1° from 6° resulted in a relative x-ray intensity change of 10%. If the take-off angle was 30° or higher, the change was less than 2%. In addition to depending on detector location, t depends on the specimen tilt relative to the excitation source and on the specimen topography. In a rough specimen t may be determined mainly by topography. Figure 3 graphically demonstrates possible effects of topography on the distance x-rays travel through the specimen. If x-rays are emitted at a distance d below the surface (Figure 3), the x-rays may have to travel a longer distance (Figure 3a) or a shorter distance (Figure 3b) than would be predicted for a flat specimen under the same conditions of incident beam angle and specimen tilt angle. If the concentration of an element is the same

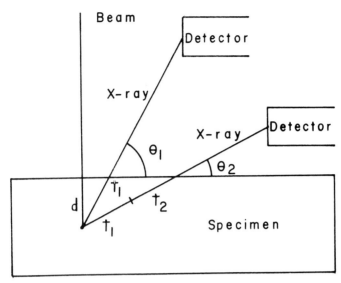

Figure 2. The effect of the take-off angle (θ) on the distance the detected x-rays travel within the specimen. Decreasing the take-off angle of a detector from θ_1 to θ_2 increases the distance the detected x-rays travel by an amount equal to t_2.

at a distance d below the surface for the flat surface (Figure 2) and the rough surfaces (Figures 3a and 3b), the detected x-ray rate from these areas may not be the same for a given set of conditions. If the differences due to topography are not recognized, the precision of the x-ray microanalysis will be adversely affected (Beaman and Isasi, 1972; Beaman and Solosky, 1972). It should be remembered that x-ray absorption through a distance t depends strongly upon sample composition (Russ, 1974). Salter and Johnson (1970) analyzed samples that had been prepared by using different final grit sizes during polishing. They concluded that rough surfaces (due to inadequate polishing) would tend to lower the measured x-ray intensity. This effect will increase as the mass absorption coefficient for the sample increases (Salter and Johnson, 1970). Even if the surface is adequately prepared for analysis, a scratch in the surface can affect the results. Sweatman and Long (1969) found that a scratch 0.5 μm deep reduced the measured magnesium concentration in a pyroxene sample by approximately 10% when the take-off angle was 20°. Hallerman and Picklesimer (1969) reported, however, that at a take-off angle of 35°, surface roughness that was smaller or equal in magnitude to the diameter of the electron beam did not cause any significant error in x-ray microanalysis.

In addition to influencing quantitative point analysis, changes in specimen topography during a line scan may have a substantial effect on

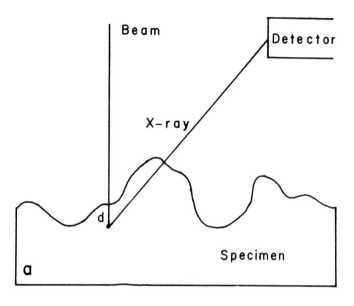

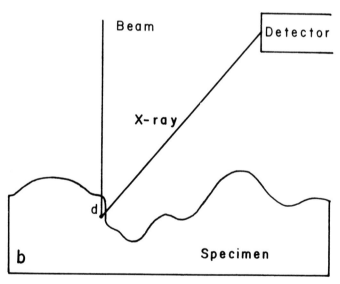

Figure 3. Influence of specimen topography on the distance that detected x-rays travel through the specimen. **a:** Specimen topography causing an increase in this distance. **b:** Specimen topography resulting in a decrease in this distance.

measured x-ray production of a specific element. The relative differences in a line scan analysis of a rough sample may result more from specimen topography than from actual specimen composition. For example, if there is an element of interest that is in an unrecognized uniform layer just below the surface of a rough specimen (Figure 4), the line scan can reflect in part the topographic variation, not the true uniformity of the layer. The degree of topographic effect will depend on the composition of the material above the layer of interest. The effect of the varied topography will be on both the incoming electrons and the outgoing x-rays. To demonstrate an actual effect of specimen topography on a line scan analysis, Hess et al. (1975) analyzed an etched 1-cm² piece of pure copper sheeting. The copper was etched by placing it in concentrated H_2SO_4. The copper specimen was then washed with distilled water, air-dried, and glued to an aluminum specimen holder with silver conductive paint. The secondary electron image (Figure 5a) revealed a very rough surface for analysis. A line scan was made across the specimen in a horizontal direction at the position marked by the straight line in Figure 5a. The copper L x-ray line scan (Figure 5b and superimposed on Figure 5a) indicated a variation in the concentration of copper; yet the sample was pure copper. Because the sample was pure copper, the variation in the line scan analysis was attributed to the topographic relief of the specimen. Specimen surface topography can change the dot density obtained during the recording of an element map. This change does not reflect any actual change in elemental concentration (Russ, 1972; Hess et al., 1975).

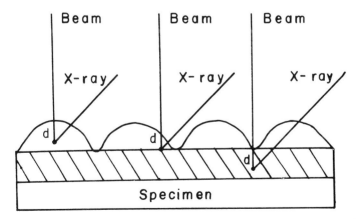

Figure 4.   A limitation in the accurate analysis of a uniform layer of an element of interest due to a rough specimen surface.

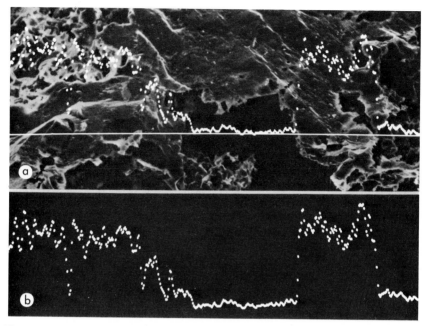

Figure 5. Line scan analysis of an etched copper plate. **a:** Secondary electron image and line scan (×1,100). The straight horizontal line is the position of the line scan on the copper plate. **b:** Line scan of **a.** There was a significant degree of variability in the line scan, even though the specimen was pure copper.

X-rays are generated from essentially the entire electron capture volume. By using a Monte-Carlo method it has been shown that in an SEM system with a 15-kV accelerating potential the excited volume of calcium in a biological tissue is 7 μm deep and 3.5 μm in diameter at the widest point (Russ, 1974). In this calculation the greatest number of x-rays come from a depth of more than 3 μm. X-rays continue to be emitted until the excitation electrons lose enough energy to fall below the critical excitation energy of the element of interest. This capture volume is much larger than the beam diameter and so prevents a small but distinct feature (resolvable by secondary electron detection) from being analyzed for x-ray emission with any degree of certainty about the accuracy of the analysis. This is because the excited volume may include areas outside the target feature that may have a higher, or lower, concentration of the element than the analyzed area. The consideration of excited volume can become important when specimens with topographic relief are being studied. In 1968 Petty and Preston used the EMP to investigate copper, chromium, and arsenic in sitka spruce "sapwood" treated with Tanalith C, a wood preservative. They reported that there was an increased accumulation of the metals near the middle lamella as compared to the lumen edges. This result may be due to an effect of topography on the excited volume. Figure 6 is a drawing that depicts a situation similar to that described in the Petty and Preston (1968) paper. In the drawing it was assumed that the cell wall thickness being analyzed was on the order of 4 to 5 μm (calculated from Petty and Preston's micrographs) and the diameter of the excited volume was 3.5 μm. Under these conditions the volume of x-ray emission is smaller at the edges of the wall than at the middle. This would result in a reduced x-ray yield if the edge of the specimen is not facing the detector. If the specimen edge is facing the detector, the x-ray yield may be enhanced at the edge of the sample because of decreased distances the x-rays must travel through the specimen. This would be similar to the situation illustrated in Figure 3b. These factors must be considered when x-ray information from different parts of a cut cell wall is interpreted. In addition, electrons and x-rays escaping from the side of the wall may impinge on other areas of the specimen and cause the excitation of x-rays that are not part of the area being analyzed. This problem is discussed in the next section.

Resch and Arganbright (1971), using an EMP, reported that "latewood" cell walls had greater amounts of the preservative pentachlorophenol than did the "earlywood" when the preservative was commercially applied with liquified petroleum gas as the carrier. The difference in thickness between the earlywood and latewood walls may have influenced the x-ray analysis. The latewood wall (right wall in Figure 7) would be thick enough to contain the entire excited volume for this matrix, whereas the earlywood wall (left wall in Figure 7) might not. If the diameter of the potential excited volume is larger than the wall thickness in the earlywood, electrons will

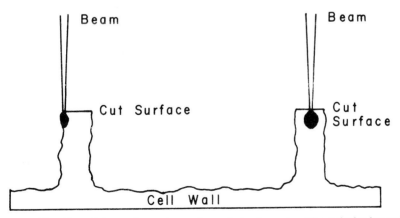

Figure 6.   The effect of the specimen edge on the excited volume size. The excited volume will be decreased at the specimen edge (left wall) when compared to the analysis of the center of the wall (right wall).

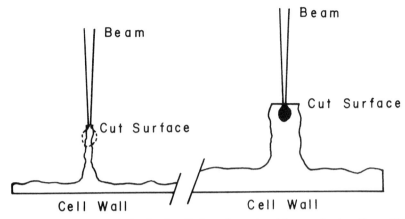

Figure 7. X-ray microanalysis of cell walls in earlywood and latewood xylem tissue. If the potential excited volume is larger than the wall thickness in the earlywood, a decrease in x-ray detection rate may occur during analysis of the earlywood.

escape from the matrix before decreasing below the critical excitation level of the element of interest. This will result in an x-ray yield below that expected for the concentration of the element in the wall. As was the case in Figure 6, the effect of a decreased pathlength of the x-rays within the specimen must be taken into consideration during interpretation of the analysis. Unless a suitable standard is constructed, the area of interest should have a physical size that is larger than the excited volume if the concentration representative of that finite area is to be accurately determined.

Specimen Topography Blocking X-Ray Detection

Whenever the emitted x-rays are absorbed by an adjacent segment of the specimen before arriving at the detector, a "shadow" will occur on the specimen (Leamy and Ferris, 1972; Bomback, 1973; Hess et al., 1975). Areas on a rough surface apparently not emitting x-rays may result from specimen topography blocking x-ray detection and not any inherent differences in the elemental distribution.

In studies using the EMP, with its large beam diameter and consequential low resolution, it is difficult to determine the specimen's topography. Because of this, unrecognized specimen topography may cause misinterpretation of element map data. In a study of the silicon distribution in rice inflorescences, Soni and Parry (1973) reported a higher silicon content associated with the "apices of the papillae." These authors recognized that the shape of the papillae may have had an influence on their element maps and line scans. To study the influence of specimen topography on the EMP element map analysis reported by Soni and Parry (1973), Hess et al. (1975) analyzed the silicon distribution pattern of the fused lemma and palea of rice inflorescences using three different specimen orientations. The EMP operating parameters were as follows: electron beam diameter, 0.5 μm; accelerating potential, 15 kV; beam current at the sample, 0.5 μA; emission current, 250 μA; x-ray detector, λ-dispersive, ADP crystal; x-ray detected, Si Kα. The brightness of a cathode ray tube in the EMP was modulated at pulse rates that were proportional to the number of x-rays produced at that area of the specimen during a scan. The analysis indicated that silicon was not distributed evenly at all locations on the rice hull (Figure 8b). Areas at the base of some papillae (arrows in Figure 8a and b) indicated no silicon deposition. Because the EMP x-ray detector was located behind and above the specimen (52.5 degrees with respect to the plane of the specimen), the uneven silicon deposition may have been caused by unrecognized topography. Careful examination of Figure 8a and b revealed that the areas indicating no silicon were on the side of the papillae (arrows in Figure 8a and b) opposite the detector. Therefore, the topography of the papillae increased the pathlength (t) within the specimen or acted as a barrier to x-rays already emitted from the specimen. If the rice hull shown in Figure 8a was rotated to change the papillae orientation relative to the x-ray detector (Figure 8c), the Si Kα x-ray image obtained (Figure 8d) was significantly different from that shown in Figure 8b. Again, the apparent silicon-free regions were regions at the bases of the papillae (arrow in Figure 8c and d). If the rice hull was rotated so that the depressions between the papillae were directly in line with the x-ray detector (Figure 8e), the silicon element map (Figure 8f) was different from either of the previous two EMP element maps. The depressions between the papillae are marked by broken lines on the element map (Figure 8f). These areas appear to have a silicon distribution similar to the papillae.

The changing patterns of the element maps recorded by EMP analysis indicated that specimen topography blocked x-ray detection at some locations on the specimen surface. To further evaluate this effect the analysis was repeated using a SEM equipped with an energy-dispersive x-ray analyzer (Hess et al., 1975). The SEM secondary electron mode image allowed a better evaluation of the specimen topography; therefore, the element maps could be critically evaluated with respect to the topography of the specimen. The SEM and x-ray detector were as follows: beam diameter, 0.12 μm; accelerating potential, 10 kV; emission current, 200 μA; x-ray detector, 10 mm^2 Si(Li) solid state-detector, 3 mil Be window; x-rays detected, Si Kα and Kβ. The fused lemma and palea of a rice inflorescence were analyzed at different rotational positions about an axis. When the long axis of the inflorescence was placed in a vertical position (Figure 9a), the Si

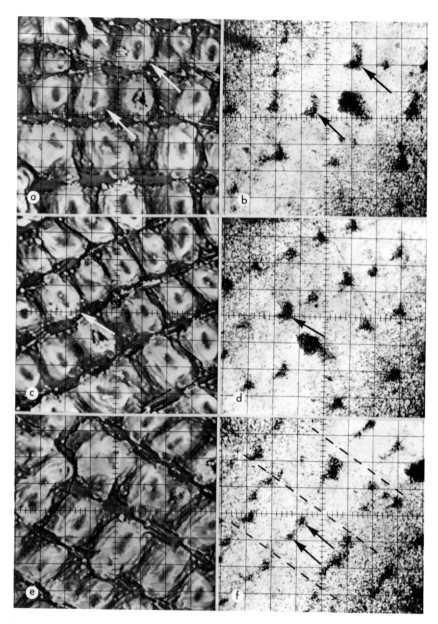

Figure 8.   EMP x-ray microanalysis of silicon deposition in a rice inflorescence. **a, c, and e:** EMP backscattered electron image (×300). **b, d, and f:** Si Kα element maps of corresponding areas of **a, c,** and **e.** The detector was above the specimen and in the direction of the upper left corner of the micrographs. Note the changing patterns in the areas where silicon was not detected.

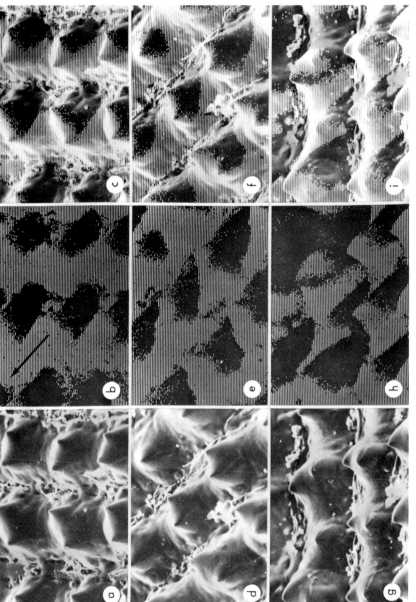

Figure 9.   SEM energy-dispersive x-ray microanalysis of silicon deposition in a rice inflorescence. **a**, **d**, and **g**: SEM secondary electron images of rice inflorescence (×300). **b**, **e**, and **h**: Si Kα and Kβ x-ray element map of the area shown in the secondary image. The x-ray detector was above the specimen and in the direction of the arrow in **b**. **c**, **f**, and **i**: Composite micrographs of secondary electron and x-ray element map images. Areas where silicon x-rays were not detected correspond to where the papillae interfered with x-ray detection.

Kα + Kβ x-ray element map (Figure 9b) indicated definite silicone-free areas. The detector was located above the specimen and in the direction of the arrow in Figure 9b. The composite micrograph (Figure 9c) revealed that the silicon-free regions were on the side of the papillae opposite the position of the x-ray detector. The shadows show that topography influenced the silicon element map by providing a physical barrier between the emission point and the detector.

With all other parameters remaining constant, the specimen was rotated counterclockwise 45° so the papillae faced the detector (Figure 9d). The x-ray element map (Figure 9e) shows a different patterning than that observed in Figure 9b. The composite micrograph (Figure 9f) implies that detection of Si Kα + Kβ x-rays had been blocked by the topography of the papillae. Rotating the sample an additional 45° (Figure 9g) resulted in a significantly different element map (Figure 9h). The composite of these latter two micrographs (Figure 9i) again shows that variance in silicon distribution was due to topographical variations of the specimen. When silicon was analyzed in a small region at the base of a papilla for 300 s at 45° increments of rotation (keeping the same region under the electron beam), the total number of Si Kα + Kβ x-ray counts recorded at each increment ranged from a low of 105,767 to a high of 975,357 (average = 534,106, standard deviation = 364,743). This variation can be explained by the papillae acting as a physical barrier to Si Kα and Kβ x-ray detection at certain specimen orientations, by the topography of the specimen changing the angle of incidence of the electron beam and take-off angle of the emitted x-rays, and by changes in the distance the x-rays must travel in the specimen before escaping on a path within the solid angle of the detector. As shown in the SEM secondary electron images (Figure 9a, d, and g), two pointed projections occur at the top of each papilla. In Figure 8f the two small shadow areas (arrows) correspond to the place where the two points of a papilla have blocked the detection of silicon x-rays.

Excitation and Detection of X-Rays from Remote Areas

It is important to remember that when the electron beam strikes a sample, emissions from the sample emanate in all directions (Hess et al., 1975; Marshall, 1975). Backscattered electrons have sufficient energy to elicit x-rays (if the electron energy is above the critical excitation energy of the element) from any area of the sample that they may encounter. Because energy-dispersive spectrometers are insensitive to source position, any x-rays reaching the detector are counted and may cause an inaccurate analysis (Russ, 1973; Yakowitz, 1974; Hess et al., 1975). In addition, x-rays emitted from the excited volume of the primary electron beam can strike the specimen at other locations and cause an element having a critical excitation energy below that of the x-ray energy to emit x-rays (secondary fluorescence). In addition to striking remote locations on the sample, backscattered electrons and x-rays can strike portions of the specimen chamber and stage, which may result in the emission of detectable x-rays. Beaman and Isasi (1972) described an experiment, presented in a paper by Bomback (1970) at the ASTM conference in Toronto, Canada, that demonstrated the effect of remote objects on x-ray microanalysis. Bomback constructed a test specimen by placing a 0.02 inch diameter nickel wire on a thorium substrate. A concentration of 12% nickel was detected when a static electron beam was positioned 10 μm from the wire. When the electron beam was placed 1,000 μm away from the nickel wire, the concentration of nickel detected was 2%.

To illustrate the extent of possible problems of x-ray excitation and detection from remote areas of rough specimens and from non-specimen areas inside the microscope chamber, a specimen was constructed as shown in Figure 10a (Hess et al., 1975). A static electron beam of the SEM was positioned to strike only on the copper. The apparent elemental composition (Figure 10b) indicated the presence of aluminum, silicon, silver, iron, and potassium but no copper. The electron beam striking the copper elicits, among other emissions, copper x-rays and backscattered electrons. The copper x-rays were unable to reach the detector because those in a direct path to the detector were absorbed by the tin and lead (Figure 10a). Backscattered electrons from the copper, striking the sand, had sufficient energy to excite the silicon and potassium to x-ray emission. The specimen-detector geometry was such that these silicon and potassium x-rays were able to reach the detector. In addition, K x-rays emitted from the copper, and striking the sand, also caused silicon and potassium x-ray emission (secondary fluorescence). Backscattered electrons and copper K x-rays also struck the aluminum specimen holder, creating aluminum x-rays. The specimen was constructed with a silver adhesive that emitted x-rays when struck by backscattered electrons from the copper. Backscattered electrons had sufficient energy to cause the emission of iron x-rays from the iron in the pole piece of the final microscope lens, which explains the presence of the iron in the spectrum shown in Figure 10b.

There is no reliable way of subtracting x-rays detected from remote areas of the specimen when using an energy-dispersive detector (Bomback, 1973). This error can be minimized by good collimation, which will restrict the "detector view" to a small area of the sample (Russ, 1970, 1973). In EMPs equipped for wavelength-dispersive analysis the problem of detection of x-rays emitted from remote areas is reduced, because, as discussed earlier, the x-rays must be from an area on the focusing circle of the EMP in order to be detected (Marshall, 1975).

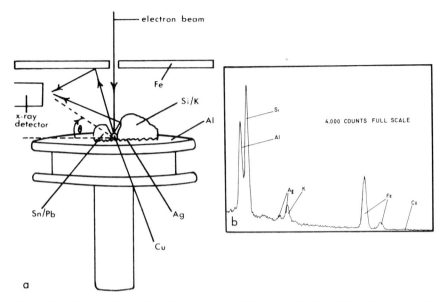

Figure 10. X-ray analysis of specimen constructed for demonstration of the effects of remote topographic points on the analysis. **a:** Specimen constructed for topography analysis. Solder (Sn/Pb), copper wire (Cu), and sand (Si, K) were adhered to an aluminum stub with silver adhesive. During analysis, the static electron beam impinged only on the copper. Copper x-rays were prevented from reaching the x-ray detector by the Sn/Pb. **b:** Spectra from specimen described in **a.**

RECOGNITION OF TOPOGRAPHIC EFFECTS

Biological x-ray microanalysis of bulk specimens requires recognition that these specimens have three dimensions. To minimize the misinterpretation of element map and line scan data, the specimen topography should be determined, whenever possible, with a SEM prior to x-ray analysis with an EMP or SEM system. Two simple methods of determining if specimen topography is interfering with line scan and element map x-ray micro-analysis are: 1) to conduct a background analysis, and 2) to conduct analyses at different specimen orientations.

When analyzing a specimen in the line scan or element map mode, it is important to record a background (Bremsstrahlung) map or line scan in addition to the map or line scan of the elements of interest. A background analysis is obtained by setting the window (x-ray energy level range) of the analyzer to an energy level adjacent to that of the analyzed element. A "background" line scan or element map is then recorded for the same area of the specimen in which the elemental analysis was conducted. It is important to make sure that the energy level range used for the background analysis has only continuous radiation contributing to the analysis and that no element with an energy level the same as the background scan is present in the sample (Beaman and Isasi, 1972). If topography is interfering with the x-ray analysis, the background element map or line scan will show variations in intensity that correspond to those variations observed in the elemental analysis (Anderson and Leitner, 1969; Beaman and Isasi, 1972; Russ, 1972). If a remote obstacle is blocking x-ray detection, a large drop in the background intensity will be observed in the background analysis. One problem with this method is that these differences in background intensity due to specimen topography effects may not be recognized in element map analyses. Point dwell times are often short and background radiation levels are relatively low; therefore, the dot density on the element map may be so low that any variation may be unrecognizable. When specimen topography is thought to be interfering with the element map analysis, the specimens can be analyzed at several different orientations (see Figures 9a, d, and g) to determine if the x-ray patterns change (see Figures 9b, e, and h). Changes in the patterns of the element map suggest that topography is influencing the analysis. When this occurs, the analysis should be interpreted with caution.

ELIMINATION OF OR
COMPENSATION FOR TOPOGRAPHIC EFFECTS

Because the SEM was designed primarily for purposes other than x-ray microanalysis, x-ray take-off angles are often not optimal. To minimize topographic effects, the x-ray detector should be installed in a way that achieves a large specimen-to-detector take-off angle. This can be accomplished by moving the detector horizontally toward the incident electron beam (Figure 11). If the specimen chamber is constructed so that the detector assembly cannot be brought close to the incident beam, another method of increasing the take-off angle can be used. An increase in the distance between the specimen and the final lens (working distance) will increase the take-off angle of the x-rays (Figure 12). As this distance is increased, it must be remembered that the solid angle of collection is being reduced, thus reducing the number of x-rays detected per unit of time. This will result in an increased analysis time to achieve an adequate number of x-ray counts. The specimen can be tilted toward the detector to increase the take-off angle; however, this will change the angle of electron beam incidence.

For reliable quantitative analysis the specimen and the standard should have a flat, well-polished surface (Sweatman and Long, 1969). The perfection of polish is of greatest importance for low atomic number elements and for low take-off angle analysis (Sweatman and Long, 1969). Biological x-ray microanalysis in a SEM system is often conducted on low atomic number elements and at low take-off angles on specimens that do not have

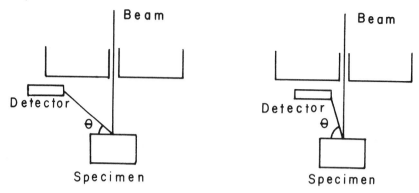

Figure 11.   Increasing the take-off angle (θ) by moving the detector closer to the electron beam.

polished surfaces. Therefore, for quantitative analysis local specimen orientation must be known in order to obtain a value for the take-off angle used in the mathematical correction calculations (Russ, 1973; Yakowitz, 1974).

Russ (1973) described a mathematical method whereby the local specimen orientation for the angle of horizontal specimen tilt and the angle of lateral specimen tilt (at right angles to the principal tilt direction) can be obtained. This is accomplished by measuring the distance between two points at two different, but known, tilt angles. These measurements can be done directly on the SEM cathode ray tube. Once the measurements are completed, the numerical values are used to solve a mathematical formula (Russ, 1973) that will yield values for the local horizontal and lateral tilt

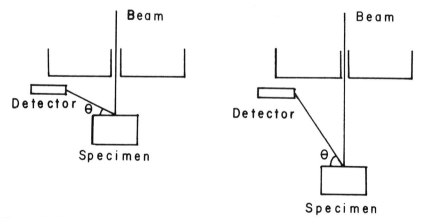

Figure 12.   Increasing the take-off angle (θ) by increasing the specimen-to-final lens distance.

angles of the specimen. A stereoscopic/computer technique of adjusting for topographic effects in quantitative analysis has been published by Bomback (1973). This technique adjusts for changes in the local angle of electron beam incidence and for changes in the local take-off angle caused by surface topography of the specimen. The procedure calculates how the specimen can be reoriented so that the local surface geometry of the specimen becomes coincident with that of the polished standard. The local region to be adjusted on the rough sample must be at least 5 to 10 μm in each dimension. After the orientation of this local plane is determined, the specimen is reoriented until a position the same as the standard is achieved. The incident electron beam current must remain constant during quantitative analysis; therefore, after the reorientation process has been completed, the refocusing procedure should be done by adjusting the distance between the final lens and the specimen. If focusing is achieved by adjusting the current in the final lens, the incident electron beam current will be changed.

Where practical, an alternative to the analysis of rough samples is the use of thin-sectioned biological samples. In this instance the analysis is conducted with the transmission mode of the electron microscope. Information about application of x-ray microanalysis to thin-sectioned biological material is currently available (see Echlin and Galle, 1975; Lechene and Warner, 1977; see also Marshall and Roomans in this volume).

If the tissue preparation of choice results in the need for analysis of bulk samples, accurate and meaningful elemental analyses are possible. Of importance is proper tissue preparation (Lechene and Warner, 1977) and an awareness of, and correction for, errors caused by specimen roughness (described in this chapter), as well as errors attributable to other causes (Marshall, 1975).

The author would like to thank Dr. R. H. Falk and Dr. D. E. Bayer for their cooperation in the SEM x-ray microanalysis projects. Acknowledgment is given to Dr. C. E. Bracker and Dr. R. H. Falk for reviewing this chapter, Mr. R. W. Whittkopp for the EMP analysis of rice inflorescences, and Ms. B. M. Hess for the drawings used to illustrate this chapter.

LITERATURE CITED

Anderson, C. H., and Leitner, J. W. 1969. Applications of a combination scanning electron microscope and electron probe. In: *Scanning Electron Microscopy* (Johari, O., and Corvin, I., eds.), p. 73. IIT Research Institute, Chicago.

Beaman, D. R., and Isasi, J. A. 1972. Electron beam microanalysis. American Society for Testing and Materials. Special Technical Publication 506. 80 pp.

Beaman, D. R., and Solosky, L. F. 1972. Accuracy of quantitative electron probe microanalysis with energy dispersive spectrometers. Analyt. Chem. 44:1598.

Berkey, E., and Whitlow, G. A. 1971. Quantitative microanalysis in Fe-Cr-Ni and Fe-Cr-Ni-Mn alloys by scanning electron microscopy using an empirical approach. In: *Scanning Electron Microscopy* (Johari, O., and Corvin, I., eds.), p. 73. IIT Research Institute, Chicago.

Birks, L. S. 1963. Electron probe microanalysis. Chemical Analysis, Volume 17. Interscience Publishers, New York. 253 pp.

Bomback, J. L. 1970. Practical limitations of x-ray analysis in the scanning electron microscope. From a talk presented at the 73rd Annual Meeting, American Society for Testing and Materials, Toronto, Canada.

Bomback, J. L. 1973. Stereoscopic techniques for improved x-ray analysis of rough SEM surfaces. In: *Scanning Electron Microscopy* (Johari, O., and Corvin, I., eds.), p. 97. IIT Research Institute, Chicago.

Echlin, P., and Galle, P., eds. 1975. Techniques et applications de la microanalysis en biologie. J. Microscopie Biol. Cell 22:125.

Hall, T. A. 1968. Some aspects of the microprobe analysis of biological specimens. In: *Quantitative Electron Probe Microanalysis* (Heinrich, K. F. J., ed.), p. 269. National Bureau of Standards, Special Publication 298, Department of Commerce, Washington.

Hallerman, G., and Picklesimer, M. L. 1969. The influence of the preparation of metal specimens on the precision of electron probe microanalysis. In: *Electron Probe Microanalysis* (Tousimis, A. J., and Manton, L., eds.), p. 197. Academic Press, New York.

Hess, F. D., Falk, R. H., and Bayer, D. E. 1975. The influence of specimen topography on x-ray microanalysis element mapping. Amer. J. Bot. 62: 246.

Läuchli, A. 1972. Electron probe analysis. In: *Microautoradiography and Electron Probe Analysis: Their Application to Plant Physiology* (Luttge, U., ed.), p. 191. Springer-Verlag, New York.

Leamy, H. J., and Ferris, S. D. 1972. Scanning x-ray emission microscopy. In: *Scanning Electron Microscopy* (Johari, O., and Corvin, I., eds.), p. 81. IIT Research Institute, Chicago.

Lechene, C. P., and Warner, R. R. 1977. Ultramicroanalysis: X-ray spectrometry by electron probe excitation. Ann. Rev. Biophys. Bioeng. 6:57.

Marshall, A. T. 1975. Electron probe x-ray microanalysis. In: *Principles and Techniques of Scanning Electron Microscopy: Biological Applications* (Hayat, M. A., ed.), Vol. 4, p. 103. Van Nostrand Reinhold Company, New York.

Petty, J. A., and Preston, R. D. 1968. Electron probe microanalysis of metals in cell walls of conifer wood treated with preservatives. Holzforschung 22:174.

Resch, H., and Arganbright, D. G. 1971. Location of pentachlorophenol by electron microprobe and other techniques in Cellon treated Douglas-fir. Forest Prod. J. 21:38.

Russ, J. C. 1970. Energy dispersion x-ray analysis on the scanning electron microscope. In: *Energy Dispersion X-ray Analysis: X-ray and Electron Probe Analysis* (Russ, J. C., ed.), p. 154. ASTM—Special Technical Publication 485.

Russ, J. C. 1972. Modern x-ray analysis II. Energy dispersive analysis of x-rays. *Methods and The Scanning Electron Microscope.* Edax International Inc., Prairie View, Illinois.

Russ, J. C. 1973. Microanalysis of thin sections, coatings and rough surfaces. In: *Scanning Electron Microscopy* (Johari, O., and Corvin, I., eds.), p. 114. IIT Research Institute, Chicago.

Russ, J. C. 1974. X-ray microanalysis in the biological sciences. J. Submicrosc. Cytol. 6:55.

Salter, W. J. M., and Johnson, W. 1970. Sources of errors in electron probe microanalysis—a report of co-operative experiments carried out by the Northern Microanalyser Group. Micron 2:157.

Soni, S. L., and Parry, D. W. 1973. Electron probe microanalysis of silicon deposition in the inflorescence bracts of the rice plant (*Oryza sativa*). Amer. J. Bot. 60:111.

Sweatman, T. R., and Long, J. V. P. 1969. Quantitative electron probe microanalysis of rock-forming minerals. J. Petrol. 10:332.

Walinga, J. 1977. X-ray microanalysis in the electron microscope. In: *Bulletin*, Scientific and Analytical Division, Philips. EM 108/1.

Woldseth, R. 1973. X-ray energy spectrometry. Kevex Corporation, Burlingame, California.

Yakowitz, H. 1974. X-ray microanalysis in scanning electron microscopy. In: *Scanning Electron Microscopy* (Johari, O., and Corvin, I., eds.), p. 1029. IIT Research Institute, Chicago.

Chapter 7

THE SKELETAL MUSCLE

Michael Sjöström

Departments of Anatomy and Neurology,
University of Umeå,
Umeå, Sweden

INTRODUCTION

Each element in the periodic table has a specific configuration of electron energy levels in its atomic structure. X-ray spectrometry makes use of the fact that atoms, when struck by electrons from an external source, yield x-rays that are characteristic of those atoms. The precise energy of the x-rays is then used to identify the element.

When the electron microscope was introduced four decades ago, its initial promise was the ability to extend magnification beyond that of the light microscope to show finer details of morphology. Efforts have then continually been made to use it to obtain a better understanding of the ultrastructure, and especially of the relationship between structure and function in the biological tissue. Toward this goal, x-ray spectrometry and electron optics have been put together. More recent developments in combining and extending these methods—x-ray microanalysis—have met with very rapid acceptance.

The analysis is performed with a wavelength-dispersive crystal spectrometer and/or an energy-dispersive x-ray analyzer (solid-state detector). These x-ray detection systems are either built into an instrument with a light optical viewing system (the electron probe microanalyzer, EMPA) or combined with an electron microscope. This microscope can be either a conventional transmission electron microscope (TEM), a scanning transmission electron microscope (STEM), or a scanning electron microscope (SEM). A TEM designed especially for x-ray microanalysis, the electron microscope microanalyzer (EMMA-4), is also used. In practice, the limits of detection lie at present in the region of 10^{-18} g, well within the physiological range of many ions in the biological tissue. Future technical developments are almost certain to increase the sensitivity (see Marshall's chapter in this volume). However, the extension of x-ray microanalysis to the study of biological specimens has been slow. This is mainly due to limitations in specimen preparation.

Applications of X-Ray Microanalysis to Muscle

There are several specific examples of applications of x-ray microanalysis to muscle research, some of which are discussed in later sections of this chapter. But, to help place the importance of the microanalytical technique in perspective, the types of problems that may occur are classified and listed below. Two main types of x-ray microanalysis are easily distinguished, according to how the element of interest is identified and localized: 1) the element itself produces the x-rays to be detected—*direct detection*, and 2) the element is identified by x-rays from a characteristic precipitate or another type of labeling—*indirect detection*. It should be emphasized that these general categories clearly and frequently overlap.

Direct Detection   Direct detection yields fundamental information about the distribution of insoluble or soluble naturally occurring elements, diffusible ions included. This is possible both at the cellular level, i.e., inside or outside the muscle fiber, and at the subcellular level, i.e., in the nucleus, the mitochondria, the sarcotubular system, etc. The application of x-ray microanalysis may also contribute to the understanding of normal physiological processes, e.g., the technique of cryo-ultramicrotomy includes possibilities for instantaneous interruption (freezing) of these processes.

In muscle pathology a large number of different structural abnormalities occur, such as dense cytoplasmic bodies, inclusions, autophagic vacuoles, etc. A better knowledge of the elemental composition of these structures will obviously be of great value to the understanding of the nature of many pathological conditions. Further, accidentally introduced foreign material can be determined.

Indirect Detection   Through indirect detection, the localization of specifically labeled drugs and experimental substances, administered by diffusion or circulation, can be traced throughout the muscle tissue so that their effects can be better related to morphological changes that occur therein.

Cytochemical reactions that result in a precipitated reaction product can be used to determine the chemical nature of a specific tissue component by subjecting the reaction product to x-ray analysis. The results of such analysis can also be used as a control, ensuring reaction specificity.

STRUCTURE OF SKELETAL MUSCLE

The smallest unit of structure in a striated skeletal muscle that can give a normal physiological response is the muscle fiber. Each fiber contains an assembly of contractile material enclosed within an electrically polarized membrane. Muscle fibers characteristically have diameters between 50 and 100 μm. The lengths of the fibers depend on the lengths and the construction of the muscles from which they are derived; commonly, as in muscles of frog sartorius and toe extensor, both of which are often used as examples in this presentation, the fibers are several centimeters long. In humans, there are fibers that are $\sim$0.3–0.5 m long.

The contractile material is contained in structures known as myofibrils (Figure 1). These are usually 1–2 μm in diameter. Each fibril extends for the whole length of the fiber in which it is contained. The myofibrils have a banded appearance along their length, due to the occurrence of interdigitating arrays of thick (myosin-) and thin (actin-containing) myofilaments, and they are usually arranged in the fiber with their band patterns in register. It is this feature that gives rise to the characteristic cross-striated appearance of the muscle fiber as a whole, readily visible in the light microscope.

In addition to the contractile material, muscle fibers contain numerous nuclei and the normal components (mitochondria, ribosomes, storage granules, and glycogen particles) of many other cell types. Glycogen often occurs in accumulations in the sarcoplasm. The muscle fibers also contain an elaborate internal membrane system, the sarcoplasmic reticulum, which

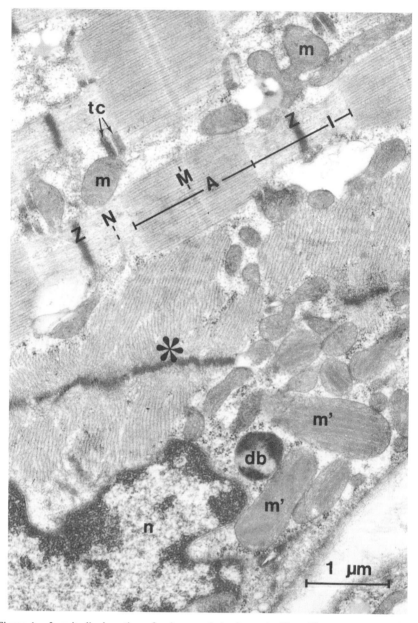

Figure 1.   Longitudinal section of a human skeletal muscle fiber. The arrangement of the normal constituents is seen as well as structural abnormalities. The figure illustrates the usefulness of a combined fine structural examination and x-ray microanalysis. The contractile material is contained within myofibrils. These have a banded appearance along their length, due to the occurrence of interdigitating arrays of thick and thin myofilaments. The band containing the thick filaments is termed the A-band (A), while the band where there is no

is concerned with switching on and off the contractile mechanism, probably by controlling the level of free calcium in the muscle fiber.

Ions in Skeletal Muscle

One characteristic of living biological cells is motility, i.e., the ability to move. Motility and contraction seem to have a common molecular basis in all biological systems. Striated skeletal muscle is the classic and most accessible contractile apparatus. Its excitation-contraction process has been thoroughly analyzed by morphological, biochemical, and physiological techniques. (For reviews see Peachey, 1968; Close, 1972; Huxley, 1972. "The Mechanism of Muscle Contraction," Cold Spring Harbor Symposia on Quantitative Biology, Vol. 37, 1972, is highly recommended.) Typical for this course of events is the redistribution of ions, such as sodium, potassium, and calcium. It has been estimated that the free calcium ion concentration in muscle in the resting state is around 3×10^{-4} mM, while the concentration in the lumen of the sarcoplasmic reticulum (Figure 1) is roughly 25 mM, of which rapidly exchangeable calcium has been estimated to be 0.2–0.5 mM (Ebashi and Endo, 1968). When events during a physiological process are reconstructed from these and other fragmentary biochemical data and the model is then compared with living muscles, discrepancies are noted. These are easily explained in view of the drastic procedures employed in obtaining the biochemical preparations, but they cannot be overlooked.

Is electron microscopy a possible approach when a better understanding of ionic events like those described above is to be obtained? For the moment, a complete answer cannot be given. However, the technique offers certain promise at its present state of development. X-ray microanalysis has been added, which allows detection of elements within a very small and well-defined volume of a biological specimen. This chapter describes and discusses different procedures of preparing the muscle tissue to be examined in the electron microscope and to be analyzed by this method.

SPECIMEN PREPARATION

Ideal tissue preparation for electron microscopy involves instantaneous cessation of all cell processes, the cell components remaining in their

overlap and where the thin filaments occur is termed the I-band (I). M-line (M), N-line (N) and Z-line (Z). Mitochondria (m) and a nucleus (n) are also seen. Components of the sarcoplasmic reticulum, especially the terminal cisternae (tc), are seen interspersed between the myofibrils. Myofibrils running at an angle of 90° to the long axis of the fiber, "ring-fibrils" (*), are shown. These or other disturbances of the regular myofibrillar organization often occur in diseased muscle fibers. A dense cytoplasmic body (db) and intramitochondrial crystalline inclusions (m') are also seen. Obviously, a better knowledge of the elemental composition of these structures will be of value for the understanding of the pathological condition in the fiber. × 22,000 natural configurations, and all elements remaining in their in vivo positions. Subsequently, in conjunction with the ultrastructural investigation, it should be possible to identify individual elements in situ and determine their in vivo concentrations by means of microanalytical techniques. Preparation of specimens by the plastic-embedding methods usually employed for transmission electron microscopy causes certain substances to leach out and to be lost or to migrate from their in vivo sites (Hayat, 1980, 1981). Other methods of preparation are being developed. The techniques of cryo-ultramicrotomy and thin sectioning of frozen tissue, in particular the "dry"-cutting variant (Appleton, 1974; Hodson and Marshall, 1970; Christensen, 1971), seem especially promising. This technique has been supposed to eliminate the risk for extraction of tissue components. It involves instantaneous freezing of chemically untreated tissue, and during subsequent sectioning and handling the specimen never comes in contact with liquid (see the chapter by Marshall in this volume).

Due to its unique morphology, skeletal muscle fiber presents quite different conditions in processing than do cells of many other tissues. This is especially true when soluble substances in the fiber are to be detected. Furthermore, because diffusible ions are of particular interest in muscle research, it is obvious that the chemical and physical phenomena that occur during different phases of various procedures of specimen preparation must be understood. Results reported from a comparative morphological examination and x-ray microanalysis of thin sectioned skeletal muscle fibers exemplify this necessity (Sjöström and Thornell, 1975b). On the other hand, if elements strongly bound to a certain intracellular structure are to be analyzed, the conventional procedure of specimen preparation (resin embedding) does not necessarily pose problems.

How to Obtain Muscle Specimens

Muscle specimens must be obtained in different ways, depending on the nature of the problem to be solved by applying x-ray microanalysis. For simplicity, let us consider two main types of problems: 1) those concerned with soluble, diffusible elements, such as naturally occurring ions, and 2) those concerned with insoluble elements (e.g., minerals) bound to subcellular components. In the former case, the muscle fibers must remain living structural entities that can give normal physiological responses until the very moment of instantaneous interruption of the physiological processes. Any chemical treatment must be avoided. Great care must be taken not to damage the fibers mechanically during dissection and removing the muscle from its in vivo position. Later in this chapter, a method is described that makes possible freezing of a living muscle in a defined functional state.

In the latter case, fixatives can normally be used and dehydration and plastic embedding can be performed (Figure 2). However, the procedure of

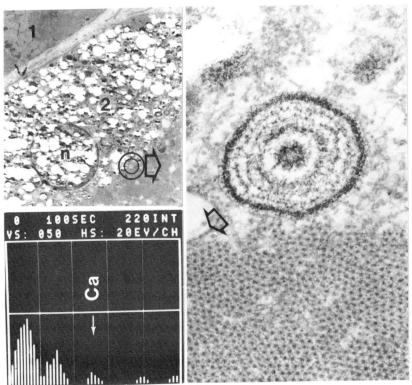

Figure 2.   Human skeletal muscle fibers (1 and 2, upper left) from a patient suffering from hyperkalemic periodic paralysis. One of the fibers (2) shows numerous vacuoles, both in the sarcoplasm as well as in the nucleus (n). A vacuole is seen at higher magnification to the right. By applying x-ray microanalysis on such a vacuole, calcium is demonstrated (the spectrum shows elemental peaks after background subtraction). (From Thornell and Sjöström, 1975.) × 1,800 and 54,000, respectively

muscle biopsy is still a critical phase of the processing and certain guidelines are worth following to ensure a mechanically undamaged and correctly stretched muscle specimen. The procedure of muscle biopsy is therefore described and commented upon in the following section.

PREPARATION OF SKELETAL MUSCLE FOR X-RAY MICROANALYSIS

Procedure of Muscle Biopsy

Muscle biopsy is a relatively simple procedure. However, it is frequently poorly done and it is therefore often unlikely to supply any meaningful information, ultrastructural as well as microanalytical, no matter how careful the processing. Based on the experience of the author, the surgical method (open biopsy) is by far the most satisfying alternative when compared with the use of percutaneous needle biopsy or other related instruments. In the latter cases the sample obtained is too small, is difficult to orient, and is impossible to stretch correctly. Open biopsy is done under local anesthesia (the skin is infiltrated with 1% Xylocain® without adrenaline). An incision about 2–3 cm long is made over the belly of the muscle in the direction of the fibers. The fascia is incised to expose the muscle fascicles. A segment of muscle can be removed by grasping one end with forceps and cutting a cylinder with a pair of sharp scissors. Alternatively, a strip of muscle can be isolated by placing a suture at either end and then removing the cylinder of muscle together with the sutures. The muscle can then be mounted, using pins, in a slightly stretched position on a plate of cork or wood. Maintaining the muscle at a slightly stretched position throughout fixation is important for preservation of the subcellular organization.

The muscle specimen is then either prepared for thin sectioning according to the routine method of plastic embedding (Table 1) or according to the methods of cryo-ultramicrotomy (Tables 2 and 3). If plastic embedding is chosen, which normally will be the case in muscle pathology on biopsied human muscle, the following routine is recommended: After plastic embedding, semi-thick survey sections, ~1 μm, are cut and stained with toluidine-blue for examination in a light microscope. Selected areas are then trimmed, and thin sections (50–150 nm) are cut, examined in the electron microscope, and analyzed.

Specimen Thickness During Analysis

The muscle specimens prepared and subjected to x-ray microanalysis can be of four main types, depending on the thickness of the specimen. These are: 1) *thin sections* (less than 200 nm), 2) *thick sections* (0.2–2.0 μm), 3) *single*

Table 1.   A method commonly used when preparing skeletal muscle for plastic embedding, ultra-thin sectioning, and transmission electron microscopy

1. Prefixation with glutaraldehyde, 1%–5% in a Ringer or a buffer solution, for 30 min to 24 h
2. Rinsing for 5–15 min in the corresponding buffer
3. Postfixation with osmium tetroxide, 1% for 2 h (Uranyl acetate staining *en bloc*)
4. Dehydration 15 min in each of 45%, 70%, 95%, and 100% acetone or ethanol
5. Plastic infiltration using Vestopal, Epon, etc.
6. Polymerization at 60°C
7. Trimming, cutting of semithick (1 μm) survey sections, contrasting with toluidine blue and examination in the light microscope
8. Trimming and cutting of ultra-thin (50–150 nm) sections of selected areas
9. Contrasting and examination in the electron microscope

Table 2.   Cryo-ultramicrotomy of skeletal muscle fibers: the wet-cutting variant

1. Stabilization with glutaraldehyde, 1%–5%, in a Ringer or a buffer solution for 5–15 min
2. Glycerol treatment, 30%, for 30 min to 1 h at room temperature
3. Rapid freezing in liquid nitrogen chilled Freon-12 or 22
4. Sectioning with a cryo-ultramicrotome at −70°C (specimen) and −50°C (knife) using trough liquid (e.g., DMSO or glycerol, 50%–60%, in water)
5. Staining
6. Air drying, examination in the electron microscope

fibers (or isolated organelles), and 4) *bulk samples*. Each of the four types presents its own problems, from the point of view of both specimen preparation and the information that it can provide. On the other hand, some problems concerned with the preparative procedure are common to all of them. This presentation deals, as pointed out earlier, mainly with the preparation of thin sections (type 1) for x-ray microanalysis. Thin sections are suitable for analysis in the TEM; the other three types must be analyzed in other modes of analytical microscopes. However, for comparison, at the end of this section examples of x-ray microanalysis of thick sections of muscle tissue are given.

Content of a Thin Section

The structural components that occur in a thin section may include, as is shown in Figure 3, a few layers of myofilaments, a thin-sectioned nucleus, a part of a mitochondrion, a terminal cistern (lateral sac) belonging to the sarcoplasmic reticulum, etc. Normally, no overlapping of cellular organelles is present. In a thick section, however, there is such an overlapping (Figure 3). This circumstance drastically decreases the resolution of the x-ray microanalysis.

ULTRAMICROTOMY OF PLASTIC-EMBEDDED TISSUE

The most widely used and thereby the most easily accessible procedure for preparing muscle tissue for thin sectioning and examination in the transmission electron microscope is the method of plastic embedding. This has

Table 3.   Cryo-ultramicrotomy of skeletal muscle fibers: the dry-cutting variant

1. Rapid freezing of living skeletal muscle fibers
2. Sectioning at −95° to −105°C with a cryo-ultramicrotome using a dry knife
3. Freeze-drying at −95° to −105°C in nitrogen atmosphere for 1–2 h
4. Warming up and storage in a desiccator
5. Rapid transfer to and examination in the electron microscope

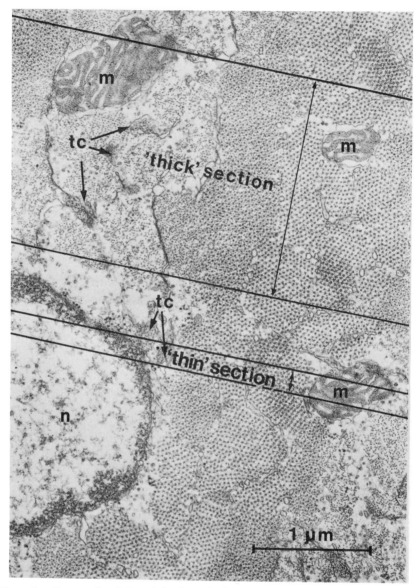

Figure 3.   Cross-sectioned frog skeletal muscle fiber. A number of cross-cut myofibrils are seen as well as a nucleus (n), mitochondria (m), and components of the sarcoplasmic reticulum (terminal cisternae = tc). Two (longitudinal) hypothetical sections of different thicknesses are indicated. The lower of these represents a thin section (less than 200 nm), the upper a thick section (up to 2 μm thick). The amount of material in each of these two types of sections is demonstrated. There is practically no overlapping of structures in the thin section. In the thick section structural overlapping frequently does occur, however. Therefore, x-ray microanalysis of subcellular components is possible when using thin sections, while such a good resolution is practically unattainable when thick sections are analyzed. × 31,500

the advantages that the step of cutting is technically relatively easy to carry out and that the sections are easy to handle. Figure 2 gives an example that shows the type of information that can be obtained when a thin section of plastic-embedded muscle fibers is subjected to x-ray microanalysis.

However, since an embedding medium is used, the mass in the section is large compared with, for example, freeze-dried sections of unembedded tissue. During x-ray microanalysis this circumstance results in a higher background, due to increased lateral electron scattering. This in turn makes detection of elements occurring in lower amounts more difficult. Further, the fact remains that the steps of chemical treatment (fixation) and dehydration, which are necessary for plastic embedding, may considerably alter the intracellular conditions.

The steps of the conventional procedure of preparing muscle tissue for plastic embedding and ultrathin sectioning are listed in Table 1. In the following discussion these steps are commented upon, especially with regard to how they may alter the elemental content of the muscle fiber. For description of the practical procedures, see Hayat (1972, 1980, 1981).

Stabilization with Glutaraldehyde

Since glutaraldehyde was introduced by Sabatini et al. in 1963, it has been the most widely employed cross-linking agent in fixation procedures for muscle tissue; nevertheless, the mechanism of its activity is still poorly understood. It was first suggested that glutaraldehyde would react with the tissue in a monomeric form, by cross-linking amino and imino groups in proteins and hydroxyl groups in polyalcohols (Sabatini et al., 1963). Others have pointed out that polymeric, rather than monomeric, forms of glutaraldehyde are likely to be most responsible for the cross-linking (Bowes and Cater, 1968; Richards and Knowles, 1968; Robertson and Schultz, 1970). Good results in muscle tissue have been achieved by Peracchia and Mittler (1972) using glutaraldehyde in its monomeric form, which could easily penetrate the fibers. They raised the pH of the solution of fixative in steps from 7 to 8 and then warmed the solution to 45°C. Their hypothesis was that the improvement in fixation they achieved was due to a better cross-linking of the polymerized forms of glutaraldehyde thus obtained. Hayat (1980) has discussed in detail the role of monomeric and polymeric species of glutaraldehyde in the cross-linking of proteins.

The aims of fixation are to preserve the muscle fiber fine structure as close to the living state as possible and to protect it against disruption during dehydration and embedding. Glutaraldehyde seems to be well suited for the first of these purposes. In fact, comparative x-ray diffraction studies of muscle tissue in vivo and of glutaraldehyde-fixed tissue indicate that glutaraldehyde stabilizes but does not denature or reorganize the myofibrillar proteins (Reedy et al., 1965; Huxley, 1969; Reedy and Barkas, 1974).

However, it must be emphasized that there is a loss of cellular material, especially of diffusible elements, during the glutaraldehyde treatment (see Figures 15 and 16). The selective permeability of cellular membranes is soon partially or totally lost during aldehyde fixation. This has been shown in red blood cells from measurements of the dielectric constant (Carstensen et al., 1971) and osmotic properties (Morel et al., 1971) as well as of a number of other physiochemical properties (Vassar et al., 1972). Concerning muscle tissue, changes in permeability have also been found, e.g., in frog skeletal muscle and cardiac muscle fibers, where electrophysiological properties have been measured (Wood and Luft, 1965; Fozzard and Dominguez, 1969) and in cardiac muscle where ion distribution has been measured before and after fixation (Krames and Page, 1968).

The ability of glutaraldehyde to protect the subcellular structures within the muscle fibers against damage during dehydration and embedding is poor, leading to extraction of many of these components. Therefore, there is normally a need for the use of post-fixation of the fibers, for example, with OsO_4.

Osmium Tetroxide Fixation

Osmium tetroxide seems to have multiple effects as a fixative (Hayat, 1980, 1981). Above all, it acts as a strong denaturing agent, leading to considerable disordering of myofibrillar structure, as determined from low-angle x-ray diffraction studies (Reedy and Barkas, 1974). Further, fixation by osmium tetroxide produces a rapid and complete loss of selective membrane permeability (Wood and Luft, 1965; Krames and Page, 1968; Bone and Denton, 1971). The changed permeability results in the flow of water and an alteration in the volume and shape of the muscle fiber (Davey, 1973; Eisenberg and Mobley, 1975). If x-ray analysis is used with the energy-dispersive detector, it must be kept in mind that osmium, which is a heavy metal, may cause confusion in identification of peaks arising from elements (e.g., silicon and phosphorus) present in the sections analyzed (Figure 4). The use of osmium tetroxide also results in the accumulation of large amounts of calcium, as has been shown in heart muscle by Krames and Page (1968).

Dehydration, Embedding, Sectioning, and Staining

The use of acetone, alcohol, and other dehydration agents causes disordering of the protein-rich myofibrillar components (Reedy and Barkas, 1974). Because the dehydration agents are strong organic solvents, there is a loss of material during dehydration. This phenomenon has been documented in many ways (Hayat, 1980). The embedding procedure, using polyesters (e.g., Vestopal W), epoxy resins (e.g., Epon), or methacrylates, includes possibilities that can make analysis difficult, since elements like silicon, sulfur, chlorine, and calcium are often contained within the medium. Further, as

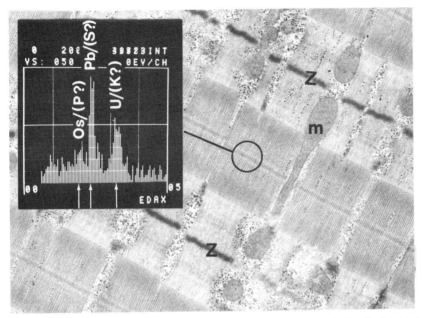

Figure 4.[1]   Elemental analysis performed over the middle of an A-band of a fixed (OsO_4), dehydrated, plastic-embedded, and thin-sectioned muscle fiber. The section has been contrasted using uranyl acetate and lead citrate. The heavy metals used for fixation and contrasting cause confusion in the identification of peaks possibly arising from elements present in the tissue as such. However, if fixation with osmium tetroxide is omitted and no contrasting is performed, no reproducible peak is obtained. $\times$ 17,500

mentioned earlier, the embedding medium makes a major contribution to the mass of the section and therefore renders the detection of elements occurring at lower concentrations difficult. It is thus largely responsible for the high general x-ray background, which reduces sensitivity. During sectioning with distilled or deionized water as trough liquid, it cannot be ruled out that errors may be introduced through contamination or loss, although this step is relatively harmless compared to fixation and dehydration. Finally, staining of tissue blocks or of thin sections using salts of heavy metals (such as lead and/or uranium) interferes with the x-ray microanalysis, since the heavy metal reagents produce x-rays that may cause confusion in the interpretation of spectra obtained when using the energy-dispersive detector (Figure 4).

---

[1] Figures 4, 13, 15 and 16 are x-ray microanalyses carried out on thin sections of longitudinally cut skeletal muscle fibers. Different specimen preparation procedures have been used, however. The analyses have been performed over the middle of an A-band. The beam spot diameter was 0.5 μm, i.e., about one-third of the length of the A-band (cf. Figure 4). Below the spectra, arrows indicate the expected sites of counts arising from a specific element. Mitochondria = m; Z-line = Z.

Grid and Support Film

Various grid materials are available (copper, nickel, gold, aluminum, chromium, platinum, titanium, carbon, nylon, stainless steel, etc.). Each of these produces its own characteristic spectrum of x-rays, which can cause difficulties in the detection of certain elements in the sectioned muscle fiber; for example, copper grids must be avoided not only if occurrence of copper is to be studied but also if elements such as sodium and zinc are of interest. This is due to overlapping of the peaks of sodium and zinc with copper L and K_β peaks, respectively. It should also be mentioned that a difficult problem to overcome is how to avoid certain elements in the specimen holder (often copper, zinc, iron, and chromium) and other surrounding metal parts within the microscope, which all can contribute strong, sometimes enormous, peaks to the spectrum. It is therefore advised that the area to be analyzed should be sought or mounted near the center of the grid to reduce this effect.

Support films are generally of polyvinyl formal (Formvar), nitrocellulose (collodion), or evaporated carbon. The plastic films sometimes contain significant amounts of silicon and sulfur.

Electron Microscopic Cytochemistry

The localization of inorganic soluble ions at subcellular level in plastic-infiltrated and thin-sectioned cells is possible by means of specific precipitation techniques. These are based on the reaction of an ion with a heavy metal, producing an electron-dense deposit that is visible in the electron microscope as a dark-contrasted spot. For example, in several such studies potassium pyroantimonate has been used (Komnick, 1962; Legato and Langer, 1969; Yarom and Meiri, 1971, 1972; Yarom and Chandler, 1974). When introduced together with osmium tetroxide during tissue fixation, it penetrates the cellular membranes and forms electron-dense precipitates with Na^+, Ca^{2+}, Mg^{2+}, Zn^{2+}, and K^+ (Lane and Martin, 1969; Legato and Langer, 1969; Klein et al., 1972). In this way, aggregates are formed in the terminal cisternae (lateral sacs), along the sarcolemma and in the I-bands of the sarcomeres in frog skeletal muscle (Legato and Langer, 1969; Podolsky et al., 1970). It has been suggested that calcium is the main cation precipitating the pyroantimonate in muscle (Yarom and Meiri, 1971; Garfield et al., 1972; Klein et al., 1972). Yarom and Chandler (1974) presented a study in which x-ray microanalysis was undertaken to investigate further the nature of this precipitate. Specimens from *m. sartorius* were fixed in a cold solution of 2% potassium pyroantimonate ($K_2H_2Sb_2O_7$) to which was added 1% osmium tetroxide. After fixation for 1 h the specimens were rapidly dehydrated with alcohol and embedded in Epon. Sections 100–150 nm thick were cut and mounted on Formvar and carbon-coated copper grids. They were left unstained for the x-ray microanalysis. Other sections were conventionally stained with uranium and lead for a better ultrastructural presentation of the morphology. The precipitates produced contained significant amounts of calcium (Figure 5) and were aggregated in the terminal cisternae of the triads, in the N-lines of the I-bands, along the sarcolemma, and in the nucleus (Figure 5). In specimens treated with osmium tetroxide alone as fixative there was no significant localization of calcium. Thus, calcium was precipitated by the antimonate inside the cell. The precise calcium localization, correlating with the good preservation of ultrastructure, in comparison with the random distribution in the specimens fixed in osmium alone, suggested that the degree of artifactual precipitation was not large. However, it must be emphasized that the analysis was of elements trapped in the muscle by the fixation procedure as well as the materials used in specimen embedding. The normal intracellular calcium concentrations and binding sites are undoubtedly disturbed by fixation and dehydration. The study of Yarom and Chandler (1974), as well as others done by Yarom and co-workers (1971, 1972, 1974, 1975) and Saetersdahl et al. (1974), illustrate the use of x-ray microanalysis of visualized intracellular biological material in conjunction with histochemical techniques. X-ray microanalysis allows the validity of histochemical interpretations be tested directly, and the best application of a method can be determined. Much useful information can be obtained by such combined studies.

There is a wide range of applications of x-ray microanalysis in electron microscopic cytochemistry, as well as in enzyme and immuno-cytochemistry on other cells and tissues. For those interested in the field, the review written by Chandler (1975) is highly recommended.

ULTRAMICROTOMY OF FROZEN TISSUE

The use of thin sections of frozen tissue in x-ray microanalysis offers other, and perhaps more promising, possibilities than does the use of sections of plastic-embedded tissue. The specimens can be treated according to two procedures which are different in principle. When processed according to the first procedure, the specimens are chemically treated to a certain degree both before freezing and also during sectioning (Table 2; Sjöström et al., 1973). When processed according to the second, the specimens or the sections are never brought into contact with any chemical solution (Table 3; Sjöström and Thornell, 1975b). For simplicity, these two procedures are here termed "wet" and "dry" cryo-ultramicrotomy, respectively. Both variants of the technique include freezing as well as sectioning at low temperatures. The major features of the technique are therefore common to both of the two procedures. In the following, first the principles of freezing are described. (For comments on the effects of chemical fixation, see the previous section.) Then the freezing of muscle fibers and the procedures of sectioning as well as section collection and handling are described in detail.

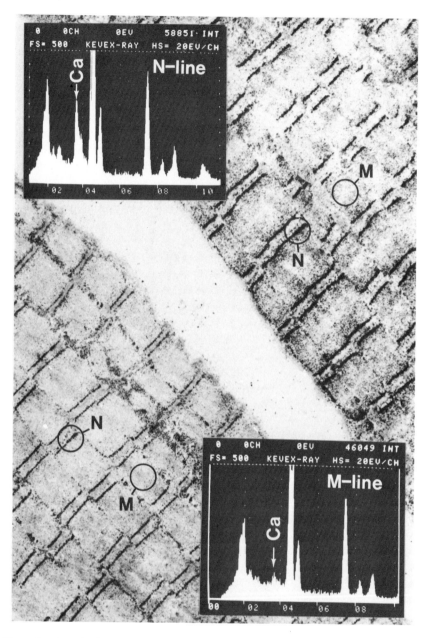

Figure 5.   Unstained section of human skeletal muscle fibers. The specimen has been fixed with osmium tetroxide and pyroantimonate. Accumulation of precipitates occurs along the N-lines (N) (cf. Figure 1). On the other hand, no precipitate is seen in the M-line region (M). X-ray microanalysis performed on the different subcellular components showed that the precipitate contained significant amounts of calcium. (Courtesy of Rena Yarom.) × 4,500

The descriptions are directly applicable to the use of an LKB Ultrotome III equipped with a CryoKit and related instruments; some of the steps must be slightly modified for work with other types of cryo-ultramicrotomes.

It should be mentioned that there are probably differences between the nature of sectioning in cutting material that has been embedded in plastic and sectioning material that has been frozen. In the latter case, sectioning is probably combined with fracturing where fracture lines follow boundaries of the cell organelles. This assumption is partly based on the fact that super-imposed structures, such as mitochondria over myofibrils, are often seen in negatively stained thin cryosections (Sjöström and Thornell, 1974, 1975a; Thornell and Sjöström, 1975). Fracturing can explain this appearance. Further, comparisons can be drawn with structural characteristics in replicas obtained by the freeze-etching method (Figure 6). This also means

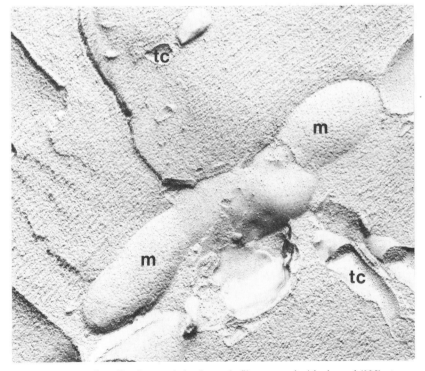

Figure 6.   Freeze-etch replica from a skeletal muscle fiber, treated with glycerol (30% at room temperature for 30 min) before freezing. No sign of ice crystals is seen. During fracturing, the fracturing lines have followed boundaries of the cell organelles. Mitochondria = m, terminal cisternae = tc. Thus, "sections" obtained when frozen muscle fibers are thin-sectioned may be of uneven thickness due to fracturing combined with sectioning instead of a pure sectioning. × 40,500

that the exact determination of section thickness is very difficult, which in turn makes quantitation of elements still more difficult.

Freezing

Principles   The principal phenomena, as defined by Luyet (1966), involved in the process of freezing of biological material may be classified into two major groups: 1) a phase transition of water into ice and 2) changes in the biological material resulting from that phase transition. The problems encountered in the first group are physical; those encountered in the second are both physical and biological.

Freezing as a physical phenomenon consists basically of two steps: (1) the initiation of freezing (nucleation), and (2) the growth of the ice phase. The main variables involved in these processes are temperature, time, rates of temperature changes, and nature and concentration of solutes. If an aqueous solution is cooled slowly and there are suitable nucleation centers available, all of the water turns into crystalline ice at 0°C. The faster the cooling, the smaller the ice crystals formed. Once formed, ice is not a static structure, however. At all temperatures between 0°C and −130°C it is subject to migratory recrystallization, with large ice crystals growing at the expense of small ones because of surface-energy considerations. If salt is added to the water and the solution is frozen, water initially crystallizes out to give pure ice. This increases the concentration of the remaining salt solution until the eutectic temperature (i.e., the lowest temperature at which a solution of the salt can remain in equilibrium with ice) and the eutectic concentration (i.e., the concentration of the still-dissolved salt at this point) are reached. The importance of eutectic formation is twofold: it leads to changes in concentrations of electrolytes and, because the different components of buffer systems have different eutectic temperatures, there may also be pH changes on freezing.

Freezing of Biological Material   Freezing of biological material differs from that of physical systems in that biological systems have a cellular structure. The development of the ice phase is hindered by the cellular components (e.g., membranes, filaments, etc.). Fundamentally, the problems to consider are those of 1) the effects exerted by the structure of the biological material on the development of the ice phase, 2) the physical (mechanical) action of the ice on the cells of tissues, and 3) the reaction of the cells or tissues when the ice invades them. In the following paragraph the basic phenomena that occur when cells are frozen are summarized. Furthermore, problems related to the first and second of the above-mentioned categories are discussed, especially those that are pertinent to the preparation of skeletal muscle fibers for electron microscopical x-ray microanalysis. The reader interested in other aspects of freezing of biological systems should consult the book *Cryobiology* (edited by Meryman,

1966a), *The Frozen Cell* (edited by Wolstenhome and O'Connor, 1970), and a review by Mazur (1970). Also, the works of Moor (1964), Bullivant (1970, 1973), and Robards (1974) are recommended.

Cooling rates . Events during cooling depend chiefly on the cooling velocity. When an aqueous solution containing cells is exposed to slow cooling (e.g., 10°C/min), large extracellular ice crystals are formed, while the cells initially supercool. Due to the increase in the concentration of the remaining extracellular salt solution that follows, the cells equilibrate osmotically by transfer of intracellular water to the external ice (dehydration). No, or very few, intracellular ice crystals are formed in the shrunken cells, and these cells often show high survival on thawing. If the cells are cooled rapidly (e.g., 1000°C/s) there is no time for such an equilibration, however. Instead, the cells will equilibrate by intracellular freezing. The higher the freezing rate, the smaller the crystal sizes. Ultrarapid freezing rates should result in extremely small ice crystals or vitrification, i.e., the production of ice crystals too small to interfere with the resolution in electron microscope specimens. Cells frozen in this way are often suitable for ultrastructural studies, but they do not normally retain viability. Usually, vitrification takes place in small, spherical cells, such as yeast or red blood cells. Vitrification is not always attainable, however. For example, it is not possible to freeze skeletal muscle fibers (long cylinders, diameter 50 to 100 μm) at such rapid rates, because of the low heat transfer resulting from the low thermal conductivity of the biological systems.

Cryoprotection (Anti-Freeze Treatment)

As discussed above, it is not possible to increase the rate of heat loss from large biological cells, e.g., muscle fibers, beyond a certain point because of the low thermal conductivity in the specimen. Therefore, methods of reducing the critical freezing rate required for vitrification, i.e., the use of cryoprotective or anti-freeze agents, must be considered. Several theories regarding the mode of action of glycerol, DMSO, and other penetrating cryoprotective agents have been advanced. They may lower the freezing point of water, increase the supercooling capacity, raise the ice recrystallization point, and/or bind or substitute for water and thereby reduce the amount of water available for freezing (Luyet and Rapatz, 1958; Doebbler, 1966; Mazur, 1970; Meryman, 1971; Diller et al., 1972). Glycerol has been widely used since its introduction by Polge et al. in 1949. Skeletal muscle fibers are known to possess contractility when immersed in solutions containing up to a few percent of glycerol (Caputo, 1968; Howell, 1969; Sjöström, unpublished observations). But such a low concentration (2%) cannot prevent crystal growth during freezing (Figures 7 and 8). For this purpose treatment in at least 30% glycerol (for at least 30 min at room temperature) must be used. However, in unfixed glycerinated fibers it is

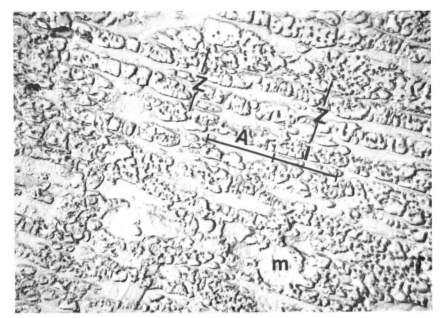

Figure 7.   Freeze-etch replica from a deep portion of a skeletal muscle fiber that was rapidly frozen without any previous anti-freeze treatment. Longitudinally oriented ice crystals are seen. They are located mainly between the myofibrils. A-band = A, I-band = I, Z-lines = Z. The mitochondria (m) are relatively unaffected. × 12,000

only the filament structure that is preserved intact; other cell components of metabolic importance in the living muscle are destroyed or removed by the treatment. The effects of glycerol treatment, as well as of other cryoprotective additives, are therefore extensive and to be avoided if accurate elemental distribution of mobile elements is desired.

Ice Recrystallization

Recrystallization is defined as a migratory growth of large ice crystals at the expense of small ones because of high surface free energies. This phenomenon takes place in frozen aqueous cellular material at temperatures higher (warmer) than $-130°C$ to $-160°C$ (for pure water) (Dowell and Rinfret, 1960; Moor, 1964). A temperature commonly used for cutting in cryo-ultramicrotomy is $-70°C$ to $-85°C$. The freeze-fracturing and etching technique has been used to study the degree of ice recrystallization at those temperatures (Sjöström, 1975). The results showed that a slight crystal growth occurred in specimens (muscle fibers) that had been kept at $-70°C$ for half an hour. This growth was interpreted as being of minor importance relative to the drastic events during freezing. Further, with the x-ray microanalytical technique at its present state of sophistication, the effects of recrystallization are of more academic than practical importance.

Freezing of Skeletal Muscle Fibers

Skeletal muscle has been the subject of extensive low-temperature research (see reviews by Love, 1966; Karow, 1969; Menz, 1971). However, the investigations have so far mainly used muscle fibers, which initially have been in rigor mortis and which have been frozen at slow rates. Furthermore, the effects of freezing on the muscles have often been studied by the use of various techniques after the muscles have passed through a state of thawing. Little information has therefore been available about the effects of rapid freezing as such on the material within the muscle fibers. The technique of freeze-cleaving and etching can be used, however, in order to obtain this knowledge and thus also give a better background for interpreting the results obtained by x-ray microanalysis. Since ice sublimes at a higher rate than biological material during etching, the etched regions represent the locations of ice. Thus, this technique is the ideal method for visualizing the

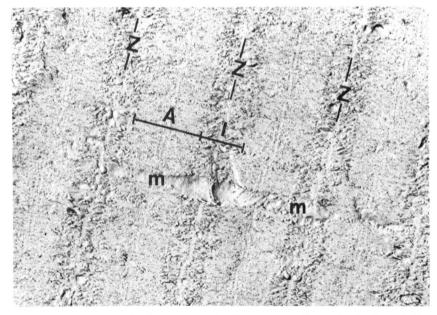

Figure 8.   Freeze-etch replica from an area of a fiber corresponding to that in Figure 7. The fiber was treated with 2% glycerol before freezing. Numerous small and scattered ice crystals are seen, especially in the I-band (I). (Cf. also Figure 7 above.) The low concentration could not prevent ice-crystal growth during freezing. For this purpose treatment in at least 30% glycerol (for at least 30 min at room temperature) must be used (cf. Figure 6). × 12,000

fine structure of freezing damage to the muscle fibers. Some details from such a study (Sjöström, 1975) are given below. They exemplify 1) the effects exerted by the organized subcellular structure of the muscle fiber on the development of the ice phase, and 2) the mechanical action of the ice on the fine structural organization within the fiber.

Ice Crystal Occurrence in Frozen Muscle Fibers   Numerous ice crystals are visualized in all muscle fibers in specimens that have been rapidly frozen (for description of the techniques of freezing, see below) and freeze-etched (Figures 7, 9, and 10). The ice crystals occupy about 55% of the fiber volume regardless of their size or location within the fiber. In general, they are smaller in the fibers in the superficial parts of the specimen and larger in more deeply situated fibers. However, the size of an intracellular crystal depends primarily on its distance from the surface of the fiber. Close to the surface (less than 1–2 μm) the crystals are numerous and scattered (Figure 9). They have a diameter of less than 50 nm and are round in profile. Deeper within the fiber, the crystals are larger. They are located mainly between the myofibrils and are longitudinally oriented (Figures 7 and 10). Approximately 10 μm from the surface, the crystals might have the dimensions 0.5–1.0 $\times$ 0.1–0.2 μm. At a depth of more than 25 μm, crystal sizes of 1.0–2.0 $\times$ 0.3 μm are found. However, the pattern of crystal sizes can vary from one fiber to another and fibers may be found with irregularly distributed islets of large crystals in an area with small crystals or vice versa. The myofilament material is laterally compressed between the crystals (Figure 10). This compression is pronounced in deeper parts of the fibers. The large crystals are normally located between the I-bands (containing thin filaments); smaller crystals are often present in the I-bands and cause splitting of the bundles of thin filaments. The Z-bands appear to be resistant to crystal expansion. The mitochondria are relatively unaffected; only a few very small crystals are found in them (Figure 7). The components of the sarcoplasmic reticulum are difficult to identify, however.

Why Ice Crystals Grow Within the Fiber   The high cooling rate normally used for rapid freezing precludes the transfer of cell water to extracellular ice, which is what occurs during slow cooling (see above; Mazur, 1970; Bank and Mazur, 1973). During rapid cooling, when the retained intracellular water changes phase and becomes ice, a critical temperature interval is traversed (i.e., down to the point of recrystallization, devitrification, or glass transition) during which the ice crystals grow. The briefer the associated time interval, the shorter is the time for crystal growth (Moor, 1964). As discussed earlier, thermal conductivity is low in biological systems. Freezing is thus slower in the depth of the muscle fibers and the crystal complexes formed there become larger.

Ice-Crystal Growth and X-Ray Microanalysis   The findings described above may explain why a reproducible specific ionic localization is not

Figure 9. Replica of a freeze-fractured and etched skeletal muscle fiber, rapidly frozen without any previous anti-freeze treatment. From close to the surface of the fiber, numerous small ice crystals are visible. They are located between the A-bands (A) and in the I-bands. Mitochondria = m, Z-line = Z. × 35,000

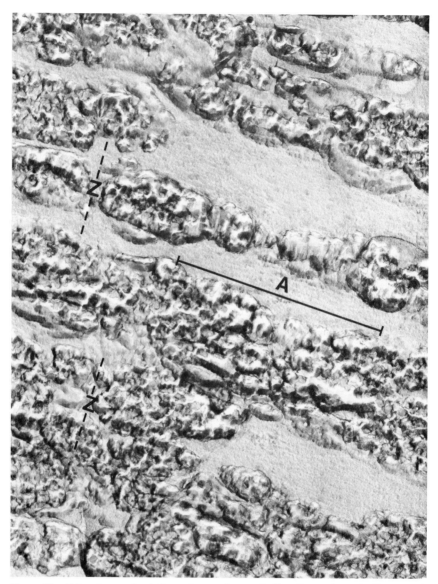

Figure 10.   Taken from a deeply located part of a muscle fiber. Processing as in Figure 9. The large ice crystals compress the myofibrils. A-band = A; Z-line = Z. Further, when the ice crystals form during freezing, the elements are concentrated around them and probably displace when the crystals expand. × 35,000

always found in deeper parts of the muscle fibers (Sjöström and Thornell, 1975b). When ice crystals form, the elements are concentrated around them and are displaced when the crystals expand (Figure 10). Nearer to the fiber surface this displacement is less (Figure 9) and is probably below the resolving capacity (Russ, 1974) of present analytical electron microscopical equipment (see also discussion in Appleton, 1974).

Conclusion   A freeze-etching study like that described above should be performed for each tissue to be analyzed. In this way a better understanding of the freezing process can be achieved in the individual case, and it is possible to make a more meaningful interpretation of data obtained regarding diffusible elements identified "in situ."

Processing

Various methods are available for freezing muscles. The choice depends on the nature of the problems to be solved. When a muscle has been stabilized with a fixative it is possible to cut it into small pieces, which can be frozen according to a relatively simple procedure (see below). On the other hand, if data on diffusible elements are of interest, a more sophisticated method must be used. Such a method is also described.

Freezing of Fixed Muscle   The pieces of muscle, 0.2–0.5 mm^3 in size, are placed on the top of clean metal specimen pins. Excess solution is removed with the aid of a piece of filter paper. The specimen and its holder are quenched either directly into liquid nitrogen ($-196°C$) or into liquid nitrogen–chilled liquid Freon 12®/Frigen 12® (difluoro-dichloromethane; boiling point $-29.8°C$, freezing point $-158°C$) and then rapidly transferred into liquid nitrogen for storage. The frozen tissue can be stored in liquid nitrogen indefinitely without suffering damage. Because frozen muscle tissue has very favorable properties for easy sectioning, there is no need for the use of tissue-supporting or encapsulation procedures.

Specimens can also be frozen by being quickly but gently brought against a liquid nitrogen-cooled metal surface (the specimen holder) for a few seconds and then both specimen and holder are lowered into liquid nitrogen (Christensen, 1971).

That one of the freezing methods described above is more advantageous than the other is doubtful. The use of a variety of coolants, except liquid nitrogen, has been tried, such as melting nitrogen, liquid helium, propane, isopentane, liquid Freons, etc. However, the size of a muscle fiber (50–100 μm in diameter) and the low thermal conductivity inherent within the biological material are probably overwhelming limitations when ultrarapid freezing rates are to be achieved. Therefore, the differences between all these methods are likely to be of no practical importance when working with muscle fibers and also with the x-ray analytical technique at its present state of development.

Freezing of Fresh Muscle   The following describes in brief a method using an apparatus that permits momentary freezing of chemically untreated skeletal muscle fibers of physiologically defined length and functional state. (For a more detailed description, see Sjöström et al., 1974.) There have been two main reasons for developing this method. The first reason was purely procedural, due to the difficulties in freezing an unfixed muscle at its in vivo length and attached to specimen holders. If not mounted, the muscle shortens maximally when cut free from its attachments, which is morphologically unacceptable. The other reason was an attempt to make microanalysis more meaningful. In that the method here described eliminates some of the effects of normally uncontrolled postmortem events (e.g., diffusion) occurring during specimen handling, it will be of value in the study of the localization of administered drugs, experimental substances, and, not the least, diffusible ions.

Description of the method.   The fresh muscles (e.g., frog toe extensor, single muscle fibers, or heart papillary muscles) are carefully dissected out, and immediately after removal are suspended in an oxygenated bath (4–10°C) containing Ringer's solution. The muscles are mounted horizontally between a force transducer and a micrometer screw (Figures 11 and 12). Electrical stimulation with alternating positive and negative square pulses of 2-ms duration is achieved by two parallel plates of platinum mounted so that they cover the two sides of the bath. If desirable, the muscle can be stimulated instead through its nerve supply. Tension recording is performed either by paper recording or by photographing the oscilloscope trace. A length-tension diagram is obtained for each muscle by measuring total tension during smooth tetanus and passive tension at different muscle lengths. The physiological parameters can be calculated from these diagrams.

The freezing apparatus includes two pneumatically controlled chilled hammers with five copper specimen holders mounted on one of them. With the muscle either stimulated or not and its length physiologically defined, the chilled hammers are instantaneously brought together and at the same time the bath is drawn away. The muscle is frozen between the hammers and attaches simultaneously to the specimen holders. When the bath is drawn away and the muscle passes the surface of the bathing solution, a small tension peak is detectable, measuring about 10%–20% of the total tension developed during tetanus. The time the muscle is exposed to air and thus without possibility for stimulation is shown to be between 5 and 10 ms on photographs of the oscilloscope trace (Figure 12). This is a shorter time than the interval between electrical stimuli leading to tetanus. The freezing takes place within a few ms (Glover and Garvitch, 1974).

The specimen holders are lowered within the hammer assembly to avoid breakage of the muscle by excessive pressure. The temperature of the

Figure 11.   A freezing apparatus. For legends see Figure 12. The apparatus makes possible instantaneous interruption (within a few ms) of defined physiological processes within the muscle fibers. The entire muscle is frozen between two liquid nitrogen–chilled hammers (h). In one of these are mounted specimen holders to which the muscle attaches. The holders are then transferred to a cryo-ultramicrotome and thin sections are cut, freeze-dried, and examined/analyzed in the electron microscope.

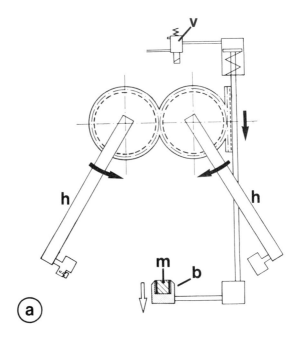

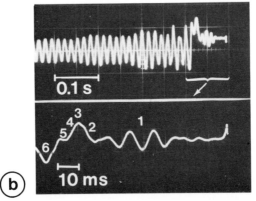

Figure 12. **a**: Schematic drawing of the freezing apparatus (cf. Figure 11). With the muscle (m) either stimulated or not and its length physiologically defined, the (chilled) hammers (h) are instantaneously brought together and at the same time the bath (b) is drawn away. The muscle is frozen between the two hammers. **b**: Oscilloscope traces demonstrating events (i.e., tension changes) occurring before and during the very moment of freezing. When the muscle passes the surface of the bathing solution, a small tension peak (see 3, lower trace) is detectable. The time the muscle is exposed to air (without the possibility for stimulation) is about 5 ms, as is revealed on the photographs of the oscilloscope trace. 1 = opening of the magnetic valve (see *v* in the drawing), hammer movement initiates; 2 = movement of the bath; 3 = the muscle passes the surface of the solution; 4 = the time the muscle is exposed to air; 5 = the hammers are brought together; 6 = post-oscillations of the system.

holders is measured continuously with a nickel resistance thermometer. The best attachment of the muscle to the specimen holders is obtained when the holders have a temperature of $-70°$ to $-75°C$. The other hammer is kept at liquid nitrogen temperature. The rapid movement of the hammers arises from their mechanical connection with a pneumatic cylinder electronically operated by a magnetic valve. The movement of the bath is driven by the same pneumatic cylinder as the hammers, but delayed to minimize the time of air exposure for the muscle. Therefore, the bath does not begin to move downward until the hammers have accelerated and reached the edge of the bath. Adhesion between muscle and bathing solution is minimized by mounting the muscle just below the surface of the solution. After freezing, the hammer with the muscle attached to the holders is immediately immersed into liquid nitrogen, where the muscle is cut off between each holder. The specimens are then stored in liquid nitrogen for subsequent cryosectioning.

Sectioning and Handling

(It is pointed out that the following procedures are applicable to the use of an LKB Ultrotome III equipped with a Cryokit; some of the steps must be slightly modified for work with other types of cryoultramicrotomes.) While still in liquid nitrogen, the specimen holder is mounted on a special holder suited for the cryospecimen head of the microtome. As quickly as possible this holder is then transferred from liquid nitrogen and mounted on the head. Time should be allowed (a few minutes) to enable thermal stabilization to occur within the cryochamber. If mounting medium has been used, trimming should now be performed. Sectioning is carried out using glass knives (diamond knives can also be used), which may have a knife angle (bevel angle) of $45°$ and are mounted $5°$ from the vertical (clearance angle). When "wet" cutting is performed the knife is equipped with a trough. "Dry" cutting is carried out with a knife without trough ("dry" knife). An anti-roll plate can be mounted on the upper face of the dry knife. Dry cutting is chosen when the aim of the study is to retain diffusible substances in situ, i.e., when chemically untreated and frozen muscle fibers are sectioned. Otherwise, wet cutting is normally preferred, since it is more easily carried out.

   Wet Cutting   During sectioning the sample holder maintains the desired temperature, e.g., $-70°C$, while the knife and the liquid in the trough ought to be kept at a temperature $10°-15°C$ higher (i.e., warmer).

   Cutting speed recommended is 2 or 5 mm/s. As trough liquid, dimethylsulfoxide (DMSO), 50% to 60% in water, or glycerol, 60% in water, can be used. The ribbons of the thin sections are removed from the trough with the aid of plastic rings and transferred to a bath of distilled water to remove the trough solution. The sections are then picked up on grids. Contrasting of the sections is in most cases not necessary, since subcellular details are identified due to contrast differences inherent within the cytoplasm (Figure 13). When the sections of the muscle fibers float on the surface of the trough solution, an extraction of elements does occur. This loss is not extensive, however, relative to that occurring during tissue fixation, for example (Sjöström and Thornell, 1975b).

Dry Cutting   This variant offers problems different from those connected with wet cutting. It is necessary that the temperature be stable, in the range of $-95°$ to $-105°C$ or lower (colder). Higher temperatures cause too rapid drying (sublimation of ice), which results in severe disruption of the sections. The author has used the following method: The temperature in the cryochamber at specimen and section level is kept at $-95°$ to $-105°C$ by the presence of liquid nitrogen in an open container on the bottom of the chamber. The lid of the CryoKit is opened only when absolutely essential. The walls and the lids are insulated by foam isolation tape. To decrease condensation of moisture in the chamber, on the knife, and on the sections, the interior of the chamber is lined with aluminum foil, which is in contact with the naked metal tubing containing liquid nitrogen.

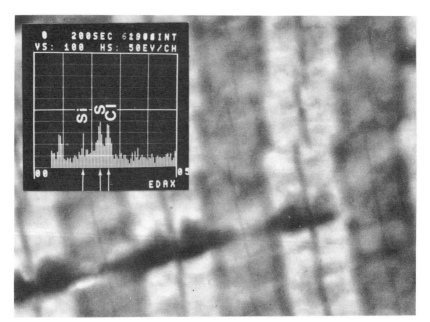

Figure 13.   A thin cryosection from a muscle fiber fixed in glutaraldehyde before the freezing. After cutting, using 50% DMSO in water as trough liquid (i.e., "wet cutting"), the section has been rinsed in distilled water. Very few elements are detected. (For further details, see Figure 4.) × 10,800

The very act of sectioning is expected to increase the possibility of redistribution of elements (Thornburg and Mengers, 1957; Hodson and Marshall, 1972). However, Appleton (1974), who performed microanalysis of ultra-thin sections of frozen carboxymethyl cellulose in which sodium chloride had been dissolved, found that the freeze-dried sections showed no sodium outside the structures over which sodium was concentrated; any diffusion that may have occurred was so slight as to be undetectable by the x-ray spectrometer used.

Section collection . A practical problem that has been difficult to overcome is how to collect the dry-cut sections of the frozen tissue, since the sections tend to curl into rolls or they may curl over the edge of the knife and be lost. Attempts to deal with this problem have included drawing the sections horizontally away from the edge of the knife by applying a suction to the leading section (Appleton, 1974), or flattening the sections by firmly pressing them onto the grid with a chilled polished copper rod (Christensen, 1971). However, the suction device is not appropriate for use with friable tissue sections that readily separate or tear, such as sections containing muscle fibers. To flatten the sections on the grids is also an uncertain procedure and apt to cause mechanical damage. A simple anti-roll plate has therefore been designed for use in preparing "dry" ultra-thin frozen sections (Hellström and Sjöström, 1974). The plate, which is made of a strip of a glass cover slip and attached close to the edge of a glass knife, prevents curling of the sections. The methodology is described in brief below.

The anti-roll plate. (See Figure 14.) A glass-knife is made. It is important that the edge of the knife be perfectly straight, otherwise it will be impossible to attach the anti-roll plate to the knife. A narrow strip, ~0–2 mm wide, is cut by hand along a ruler from a glass cover slip (e.g., Chance No. 0, thickness: 0–10 mm) with a diamond. The strip is trimmed to the length of the edge of the glass knife. It is attached along the edge of the knife with rapid-hardening Super Epoxy-glue® (Plastic Padding). Before the attachment, the glue is placed as droplets on both ends of the knife-edge as lateral as possible. The space between the surface of the strip, or anti-roll plate, and the trough surface of the knife should be quite narrow. The initial gap is ~20 μm, achieved by placing a piece of thin paper (Ross lens tissue) between the place and the knife before gluing. The place must be advanced toward the edge of the knife, to a position just short of contact with the specimen during cutting. Before the glue is completely hardened, the piece of paper is removed, and by gently pressing the plate, the gap between the plate and the knife is further decreased to ~5 μm. The ideal gap depends upon the thickness of the sections desired. The anti-roll plate appears to function better if it is slightly angled so that the gap is wider posteriorly.

Cutting starts with a regular knife without anti-roll plate, until a smooth block surface is obtained. The knife with the anti-roll plate is then

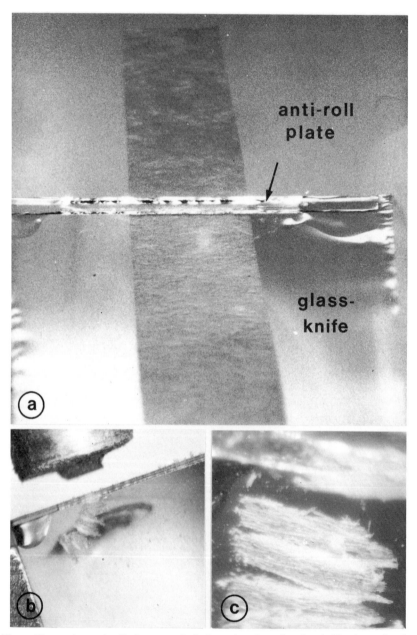

Figure 14. a: An anti-roll plate, attached along the edge of a glass-knife. A gap has been achieved by placing a piece of lens tissue paper between the knife and the plate. (For further details, see text.) b and c: During sectioning, flattened sections move in a ribbon from under the anti-roll plate.

attached to the cryo-ultramicrotome after the space between the plate and the surface of the knife has been freed from dust by a blast of compressed air. The knife is allowed to cool before cutting starts. One starts to feed manually, aiming at sections about 1 μm thick. It is important to feed very carefully, otherwise the first sections will be too thick and the gap under the plate will be plugged. The automatic feed for ultra-thin sections is then switched on and flattened sections move in a ribbon from under the anti-roll plate. After cutting, the knife with the anti-roll plate is changed.

After the very moment of cutting, parts of the sections touch the surfaces of the knife and the anti-roll plate. The friction thus generated may cause a rise of temperature in certain parts of the section and thus increase the risk of thawing and consequent diffusion or displacement of tissue components. However, the risks due to use of the plate seem to be relatively small in relation to the risks involved in the procedure as a whole.

Handling of dry sections. The ribbons of the sections are transferred with the aid of an eyelash probe from the knife to cooled grids, which have been coated with, for example, Formvar and/or carbon. The grid can be placed on the top face of the knife ~1 mm from its cutting edge. By pressing a piece of insulating tape at a distance of 4–5 mm behind the cutting edge, a shelf is provided on which the grid rests (Appleton, 1974). A ribbon of sections is transferred onto the grid with the use of the eyelash-probe. The corners of the sections are gently pressed, using the eyelash, into contact with the bars of the grid.

In addition to the suction device and polished copper rod mentioned earlier, a set of equipment that facilitates section collection, handling, and freeze-drying has been presented by Sevéus and Kindel (1974), but the author has not so far tried them. However, compared with the suction device and the metal rod, the eyelash probe normally gives the same, i.e., sufficient, amount of successfully collected and undamaged sections of muscle fibers and is also most easily made.

Freeze-drying of dry-cut thin sections. Freeze-drying of thin sections (50–150 nm) is a process quite different from conventional freeze-drying of tissue blocks. Therefore, a method for freeze-drying thin sections is described (Sjöström and Thornell, 1975b). Other methods have been presented elsewhere (see Appleton, 1974; Sevéus and Kindel, 1974). The grids, with the dry-cut thin sections attached, are placed in small, open, gelatin capsules mounted on a shelf at the bottom of the cryochamber. They are left there for 2 h in the cold (−95° to −105°C) nitrogen atmosphere. The capsules are then covered and the shelf is placed in an insulated beaker with liquid nitrogen at its bottom. The beaker is transferred to a desiccator, which then is evacuated (10^{-2} torr) and in which the sections are allowed to warm up to room temperature. The sections are stored in the desiccator until examination. Immediately before opening, the desiccator is filled with dry nitrogen. A grid is quickly removed, placed in the specimen holder, and put into the electron microscope. Thus, the freeze-dried sections are exposed to room air for only a few seconds before examination. When examined, the sections show, if properly freeze-dried, a number of longitudinal holes between the myofibrils. This is due to the occurrence of ice crystals in the frozen specimen (Figures 15 and 16; also Figures 7, 9, and 10).

As has been mentioned earlier, temperatures above (i.e., warmer than) $-90°C$ cause too rapid drying, which results in severe disruption of the sections. The drying time used at atmospheric pressure and at $-95°C$ is about 1 h, which is convenient compared with times otherwise used (days or weeks) in freeze-drying of pieces of tissue (see Meryman, 1966b). In fact, when dealing with (ultra-)thin sections the problem is how to slow down the rate of drying.

After freeze-drying it is necessary to permit the sections to warm up in vacuo, since warming of the dried, hygroscopic sections in contact with the humidity of the room air results in changes in properties of the sections and their morphology due to reabsorption of water (Figure 14). These changes are similar to those occurring when frozen-hydrated or incompletely dried sections are warmed up to $\sim -30°C$ or higher temperatures. Also, to prevent reabsorption of water, the sections must be stored in a dry atmosphere or in a desiccator until examination and x-ray analysis in the electron microscope. It may be that some distortion occurs even during the rapid transfer of the dry sections from the desiccator to the microscope. Ideally, this transfer should be carried out in a dry system or a vacuum system. Such systems have already been developed for microanalysis of sections in the scanning electron microscope (Bacaner et al., 1973; Moreton et al., 1974; Saubermann and Echlin, 1976).

Examination and Analysis in the TEM

No elements are found extracellularly in freeze-dried sections of untreated muscle (Figure 16). At present, microanalysis of frozen-hydrated tissue sections seems to be the only possible approach to investigate extracellular elements, e.g., elements in the lumen of a gland (Moreton et al., 1974). It may be that the drying process causes elements to migrate to nearby structures. Or, more probably, elements may be lost during electron microscopy. It has been shown that sodium can be lost from tissue due to electron beam damage (Spurr, 1972), and sodium loss from thin specimens of glass has been observed (Chandler, 1973) as well as migration of potassium out of electron microscope specimens (Hodson and Marshall, 1971). This volatilization of elements from the specimen due to the energy from the electron beam may prove to be a serious problem in the analysis of unbound elements. This phenomenon may partly explain the relatively low peak of potassium present in the spectra on Figure 16. The loss during the very first

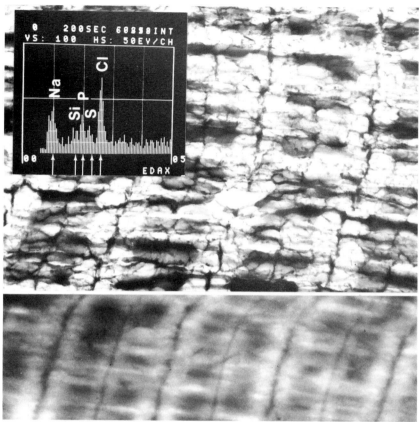

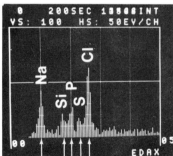

After storage
in humid air:

elements on the
supporting film

Figure 15.   Thin cryosection of a skeletal muscle fiber. The fiber has initially been fixed in glutaraldehyde and then rapidly frozen. However, in comparison with Figure 13, "dry" cutting has not been performed and the sections have been freeze-dried. (For further details, see Figure 4.) A number of peaks occur in the spectrum, of which sodium and chlorine are prominent. As may be expected, "wet" cutting of frozen material, combined with subsequent rinsing, seriously reduces the amount of diffusible elements. In fact, the spectrum reflects the composition of the Ringer's solution used. If the freeze-dried sections are kept in room air, reabsorption of water (from the air humidity) occurs. This in turn causes both changes in section morphology (b) and diffusion of elements over the immediate area around the section (c). × 10,800

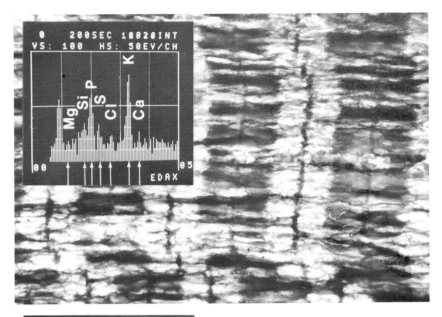

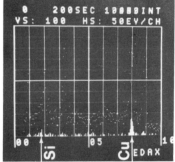

The supporting film
close to the
freeze-dried section

Figure 16.   Thin cryosection of an unfixed skeletal muscle fiber, frozen at rest length without any previous chemical treatment (cf. Figures 11 and 12), sectioned "dry," freeze-dried, and microanalyzed. (For further details, see Figure 4.) Numerous peaks are seen in the spectrum, of which the potassium peak is prominent. The spectrum is qualitatively and (semi-)quantitatively generally in accordance with the intracellular concentrations. Just outside the section, none of the elements otherwise present in the section are found (**b**). × 10,800

seconds of beam exposure and during the long counting time thus makes the determination of in vivo concentrations extremely difficult. However, rapid beam-induced loss of organic mass from cryosections under electron microprobe conditions was considerably reduced by cooling the specimens to −160°C or lower (Hall and Gupta, 1974; Moreton et al., 1974).

    The x-ray analysis of the freeze-dried cryosections results in a low background, due to reduced lateral electron scattering. This circumstance permits detection of elements that probably would be difficult to find in sections of embedded tissue. Low mass in sections also improves the spatial resolution (Russ, 1972, 1974; Chandler, 1973), but this is at present of less value for these types of investigations, in which larger volumes must be analyzed in order to pass the detection limits.

ADDITIONAL METHODS

Analyses of Deep-Frozen Unfixed Muscle in the STEM

Frozen thin sections of deep-frozen unfixed muscle have been examined and analyzed by Bacaner et al. (1973) using a method that is similar to that described and discussed in detail earlier in this chapter. However, the authors suggest that the sections they are studying still are in the frozen-hydrated state. Therefore, the interpretations diverge concerning the effects of certain phases of the preparation procedure. In brief, the method of Bacaner et al. (1973) is as follows: Small bundles of muscle fibers, ~0.5 mm in diameter, are dissected free from rabbit *m. psoas*. Two fine threads are tied about 3 mm apart and the muscle fibers are cut free just distal to the threads. The muscle bundle is mounted on a specially designed holder that permits the muscle to remain either at rest length or under some degree of stretch. The specimen is rapidly frozen by a gentle touch against a copper block cooled to liquid nitrogen temperature. After tissue trimming, thin sections, 100–140 nm, are cut at −85°C with a cryo-ultramicrotome equipped with dry glass-knives. The sections are then placed on copper grids. A second copper grid is placed on top of the first. The two grids are finally transferred in a precooled cartridge to the cryostage of the SEM, modified for transmission imaging. Attempts are made to maintain the frozen, thin sections in a dry atmosphere below −85°C at all times, inclusive of their transport off the cartridge to the microscope and when they are inserted into the cryostage.

In the microscope the sections are examined and x-ray microanalysis of the elemental composition of certain regions is performed. However, the fine morphology of the thin sections shows similarities with the freeze-dried sections demonstrated in this chapter in Figures 15 and 16. The temperature used during sectioning, around −85°C, might have been too high (warm), which has resulted in an early and rapid sublimation of ice (freeze-drying). There was an unexpectedly large amount of chlorine inside the muscle cell. In vivo, the relative concentration of intracellular chlorine is less than in serum. However, when the fine threads were tied and the muscle fibers were cut off, an immediate redistribution of large quantities of mobile elements might have occurred. In other words, there was an equilibration between the inside and the outside of the fibers. This illustrates the difficulty in the handling of skeletal muscle fibers; they are very easily damaged by mechanical influence.

Microprobe Analysis of Thick Sections

It is doubtful whether the conventional transmission electron microscope (TEM) would provide any useful data on thick (0.2–2.0 μm) sections. The image quality falls off in sections more than 0.2 μm thick unless the accelerating voltage is substantially increased above 100 keV. When thick sections are to be analyzed, the electron probe microanalyzer or scanning beam electron microscopes are more useful. In the following, two examples are given that illustrate this aspect of x-ray microanalysis.

Ingram et al. (1968, 1972, 1974, 1975) have performed quantitative analysis of thick sections (2–3 μm) of frog skeletal muscle. This procedure does not involve any use of aqueous solutions (Table 4a). The muscle is quickly frozen in liquid propane cooled by liquid nitrogen, freeze-dried in a vacuum for 2–4 weeks at between $-60°$C and $-85°$C, fixed with osmium tetroxide vapor, embedded in Epon, sectioned on a steel-knife microtome, mounted on a carbon-cast quartz slide, and then coated with about a 20-nm carbon layer. The sections are then analyzed by means of an electron probe microanalyzer. An example of results obtained from their studies of electrolyte concentrations is given in Table 5.

A different approach has been proposed by Nichols et al. (1974) in a study of the electrolyte composition of rat muscle during maturation (Table 4b). Initially, a block of muscle is frozen in liquid propane or Freon. The block is then mounted in a cryostat stage and cut in 8-μm thick sections. The sections are picked up with a silicon disc and adhere to the polished surface of the disc. The disc is then transferred to a pre-cooled ($-67°$C) flask and the sections are allowed to freeze-dry at 0.025 mm Hg pressure. Finally, the disc is stored in a desiccator until analyzed in an electron-probe

Table 4.   Preparing thick sections of skeletal muscle tissue to be analyzed using the electron probe microanalyzer

a. According to Ingram and Ingram (1975):
    1. Rapid freezing of a fresh tissue block in liquid-nitrogen chilled propane
    2. Freeze-drying in a vacuum for 2–4 weeks at $-60°$ to $-85°$C
    3. Fixation with osmium tetroxide vapor
    4. Embedding in Epon
    5. Sectioning with a microtome into 2–3 μm thick sections
    6. Mounting on a quartz slide and coated with carbon
    7. Analysis in the electron probe microanalyzer

b. According to Nichols et al. (1974):
    1. Rapid freezing of a fresh tissue block
    2. Sectioning of the frozen tissue into 8-μm-thick sections using a cryostat
    3. Mounting on a silicon disc
    4. Freeze-drying in vacuum (0.025 mm Hg) at $-67°$C
    5. Storage in vacuum until examination in the electron probe microanalyzer

Table 5.   Electron-probe measurements of electrolyte concentrations in frog skeletal muscle*

Fiber number	Concentration (mEq/kg)		
	K	Cl	Na
1	108.4 ± 4.6 SD	9.7 ± 1.0	5.1 ± 1.2
2	120.3 ± 5.0	8.4 ± 0.9	9.6 ± 1.2
3	106.4 ± 4.6	7.2 ± 0.9	3.3 ± 1.2
4	126.1 ± 5.3	7.4 ± 0.9	6.6 ± 1.2
5	104.9 ± 4.4	8.4 ± 0.9	6.8 ± 1.2
6	113.5 ± 4.9	11.5 ± 1.1	11.7 ± 1.5
7	105.9 ± 4.7	7.0 ± 1.0	7.3 ± 1.4
8	115.9 ± 4.8	6.4 ± 0.8	8.0 ± 1.2
9	104.6 ± 4.4	8.0 ± 0.9	7.6 ± 1.2
Average	111.8 ± 7.7	8.2 ± 1.6	7.3 ± 2.4

Courtesy of Duane Ingram.

* $E_0 = 10$ kV. $I_{sp} = 50$ na.

microanalyzer. Their results on intracellular electrolyte concentrations (K, 121 ± 10; Na, 15 ± 3), for example, are not directly comparable with those of Ingram and Ingram (1975; also Table 5). Above all, they are studying different species (rat vs. frog).

The major difference between the two preparation procedures now described is that the former includes embedding of the freeze-dried tissue in a plastic matrix before cutting, while in the latter the specimen block is cut while still frozen and hydrated. The sections are then subjected to freeze-drying. Principally, the latter method seems to be more attractive. However, according to Ingram and Ingram (1975), the plastic matrix is useful, since 1) it simplifies manipulation and preparation of samples; 2) sections can be stained and studied by normal techniques with light and electron microscopy; 3) the increased density of the embedded specimen compared with a similar unembedded one decreases beam penetration and electron scatter in thick specimens, resulting in improved x-ray spatial resolution; 4) the embedded specimen, being about 80% plastic, results in a more homogeneous specimen, with a flat surface to present to the electron beam, which simplifies quantitation and the interpretation of data. However, there are also complications that arise from the use of plastic-embedding materials: 1) morphologic changes may result from surface tension forces of the advancing surface front of the embedding material; 2) the plastic will redistribute plastic-soluble constituents; 3) the embedding material dilutes the cellular constituents and attenuates x-rays, adversely affecting the limits of detection; and 4) many otherwise satisfactory embedding materials contain chlorine, which adds to the uncertainty of measurements of chlorine.

In spite of the differences between the two methods and the complications listed above that arise from the use of plastic-embedding material, the results obtained by both were similar. A reasonable explanation seems to be that the degree of specimen damage causes changes that are near to or below the x-ray spatial resolution (more than 1 μm) of the electron probe.

X-Ray Analysis of Thick and Thin Sections: A Comparison

What then is the difference between x-ray analyses of thick and thin sections? As mentioned above, the x-ray spatial resolution was more than 1 μm for thick sections, which obviously limits the application of the technique. When analyzing thin sections, x-ray spatial resolution is 0.1 μm, i.e., one-tenth (or even more) of that of thick sections. This also means that subcellular components, such as mitochondria, areas of the sarcomere, and vacuoles, can be selectively analyzed. However, this high-resolution microanalysis offers, as we have seen earlier in this chapter, problems during the specimen preparation procedure that need more sophisticated methods before they can be overcome. The method of x-ray microanalysis has such high potential that, hopefully, these problems will be pursued further and solved.

I am most grateful to Drs. Sten Hellström and Lars-Eric Thornell of my own department, Dr. Rena Yarom, Pediatric Department, Hammersmith Hospital, London, and Dr. Duane Ingram, Department of Physiology and Biophysics, University of Iowa, for allowing me to use their micrographs and associated data.

LITERATURE CITED

Appleton, T. C. 1974. A cryostate approach to ultrathin "dry" frozen sections for electron microscopy: a morphological and x-ray analytical study. J. Microscopy 100:49.

Bacaner, M., Broadhurst, J., Hutchinson, T., and Lilley, J. 1973. Scanning transmission electron microscope studies of deep-frozen unfixed muscle correlated with spatial localization of intracellular elements by fluorescent x-ray analysis. Proc. Nat. Acad. Sci. 70:3423.

Bank, H., and Mazur, P. 1973. Visualization of freezing damage. J. Cell Biol. 57:729.

Bone, Q., and Denton, E. J. 1971. The osmotic effects of electron microscope fixatives. J. Cell Biol. 49:571.

Bowes, J. H., and Cater, C. W. 1968. The interaction of aldehydes with collagen. Biochem. Biophys. Acta 168:341.

Bullivant, S. 1970. Present status of freezing techniques. In: *Some Biological Techniques in Electron Microscopy* (Parsons, D. F., ed.), p. 101. Academic Press, New York.

Bullivant, S. 1973. Freeze-etching and freeze-fracturing. In: *Advanced Techniques in Biological Electron Microscopy* (Koehler, J. K., ed.), p. 67. Springer Verlag, Berlin.

Caputo, C. 1968. Volume and twitch changes in single muscle fibers in hypertonic solutions. J. Gen. Physiol. 52:793.

Carstensen, E. L., Aldridge, W. G., Child, S. Z., Sullivan, P., and Brown, H. H. 1971. Stability of cells fixed with glutaraldehyde and acrolein. J. Cell Biol. 50:529.

Chandler, J. A. 1973. Recent developments in analytical electron microscopy. J. Microscopy 98:359.

Chandler, J. 1975. Electron probe x-ray microanalysis in cytochemistry. In: *Techniques of biochemical and biophysical morphology*, Vol. 2 (Glick, D., and Rosenbaum, R., eds.), p. 307. John Wiley and Sons, Inc., New York.

Christensen, A. K. 1971. Frozen thin sections of fresh tissue for electron microscopy, with a description of pancreas and liver. J. Cell Biol. 51:772.

Close, R. I. 1972. Dynamic properties of mammalian skeletal muscles. Physiol. Rev. 52:129.

Davey, D. F. 1973. The effect of fixative tonicity on the myosin filament lattice volume of frog muscle fixed following exposure to normal or hypertonic Ringer. Histochem. J. 5:87.

Diller, K. R., Cravalho, E. G., and Huggins, C. E. 1972. Intracellular freezing in biomaterials. Cryobiology 9:429.

Doebbler, G. F. 1966. Cryoprotective compounds: Review and discussion of structure and function. Cryobiology 3:2.

Dowell, L. G., and Rinfret, A. P. 1960. Low temperature forms of ice studied by x-ray diffraction. Nature 188:1144.

Ebashi, S., and Endo, M. 1968. Calcium ion and muscle contraction. Prog. Biophys. Mol. Biol. 18:123.

Echlin, P. 1974. The application of scanning electron microscopy and x-ray microanalysis in the plant sciences. In: *Proc. 7th Ann SEM Symp.* (Johari, O., ed.), p. 477. IITRI, Chicago.

Eisenberg, B. R., and Mobley, B. A. 1975. Size changes in single muscle fibres during fixation and embedding. Tiss. Cell 7:383.

Fozzard, H. A., and Dominquez, G. 1969. Effect of formaldehyde and glutaraldehyde on electrical properties of cardiac Purkinje fibers. J. Gen. Physiol. 53:530.

Garfield, R. E., Henderson, R. M., and Daniel, E. E. 1972. Evaluation of the pyroantimonate technique for localization of tissue sodium. Tiss. Cell 4:575.

Glover, A. J., and Garvitch, Z. S. 1974. The freezing rate of freeze-etched specimens for electron microscopy. Cryobiology 11:248.

Hall, T. A. 1971. The microprobe essay of chemical elements. In: *Physical Techniques in Biological Research*, Vol. 1A, 2nd Ed., p. 157. Academic Press, London.

Hall, T. A., Echlin, P., and Kaufman, R. (eds.) 1974. *Microprobe Analysis as Applied to Cells and Tissues*. Academic Press, London.

Hall, T. A., and Gupta, B. L. 1974. Beam-induced loss of organic mass under electron microprobe conditions. J. Microscopy 100:177.

Hall, T. A., Röckert, H. O. E., and Saunders, R. L. de C. H. 1972. In: *X-ray Microscopy in Clinical and Experimental Medicine*. Charles C Thomas, Springfield, Illinois.

Hayat, M. A. 1972. *Basic Electron Microscopy Techniques*. Van Nostrand Reinhold, New York.

Hayat, M. A. 1980. *Principles and Techniques of Electron Microscopy: Biological Applications*, Vol. 1, 2nd Ed. University Park Press, Baltimore, Maryland.

Hayat, M. A. 1981. *Fixation for Electron Microscopy.* Academic Press, New York.

Hellström, S., and Sjöström, M. 1974. An anti-roll plate for flattening of "dry" ultrathin frozen sections. J. Microscopy 101:197.

Hodson, S., and Marshall, J. 1970. Ultracryotomy: a technique for cutting ultrathin sections of unfixed frozen biological tissues for electron microscopy. J. Microscopy 91:105.

Hodson, S., and Marshall, J. 1971. Migration of potassium out of electron microscope specimens. J. Microscopy 93:49.

Hodson, S., and Marshall, J. 1972. Evidence against through section thawing whilst cutting on the ultra-cryotome. J. Microscopy 95:459.

Howell, J. N. 1969. A lesion of the transverse tubules of skeletal muscle. *J. Physiol.* 201:33.

Huxley, H. E. 1969. The mechanism of muscular contraction. Science 164:1356.

Huxley, H. E. 1972. Molecular basis of contraction in cross-striated muscles. In: *The Structure and Function of Muscle,* Vol. 1 Structure/part 1 (Bourne, G. H., ed.), 2nd Ed., p. 301. Academic Press, New York.

Ingram, M. J., and Hogben, C. A. M. 1968. Procedures for the study of biological soft tissue with the electron microprobe. In: *Developments in Applied Spectroscopy,* Vol. 6 (Baer, W. K., Perkins, A. J., and Grove, E. L., eds.), p. 43. Plenum Press, New York.

Ingram, F. D. and Ingram, M. J. 1975. Quantitative analysis with the freeze-dried, plastic embedded tissue specimen. J. Microscopie 22:193.

Ingram, F. D., Ingram, M. J., and Hogben, C. A. M. 1972. Quantitative electron probe analysis of biologic tissue for electrolytes. J. Histochem. Cytochem. 20: 716.

Ingram, F. D., Ingram, M. J., and Hogben, C. A. M. 1974. An analysis of the freeze-dried, plastic embedded electron probe specimen preparation. In: *Microprobe Analysis as Applied to Cells and Tissues* (Hall, T., Echlin, P., and Kaufman, P., eds.), p. 119. Academic Press, London.

Karow, A. M. 1969. Biological effects of cryopreservation as related to cardiac cryopreservation. Cryobiology 5:429.

Klein, R. L., Yen, S. S., and Thuresson-Klein, A. 1972. Critique on the K-pryoantimonate method for semiquantitative estimation of cations in conjunction with electron microscopy. J. Histochem. Cytochem. 20:65.

Komnick, H. 1962. Electronmikroscopische Lokalisation von Na^+ and Cl^- in Zellen und Geweben. Protoplasma 55:414.

Krames, B., and Page, E. 1968. Effects of electron-microscopic fixatives on cell membranes of the perfused rat heart. Biochim. Biophys. Acta 150:24.

Lane, B. P., and Martin, E. 1969. Electron probe analysis of cationic species in pyroantimonate precipitates in Epon-embedded tissue. J. Histochem. Cytochem. 17:102.

Läuchli, A. 1972. Electron probe analysis. In: *Microautoradiography and Electron Probe Analysis* (Lüttge, U., ed.), p. 191. Springer-Verlag, Berlin.

Legato, M. J., and Langer, G. A. 1969. The subcellular localization of calcium ion in mammalian myocardium. J. Cell Biol. 41:401.

Love, R. M. 1966. The freezing of animal tissue. In: *Cryobiology* (Meryman, H. T., ed.), p. 317. Academic Press, New York.

Luyet, B. J. 1966. Anatomy of the freezing process in physical systems. In: *Cryobiology* (Meryman, H. T., ed.), p. 115. Academic Press, New York.

Luyet, B. J., and Rapatz, G. 1958. Patterns of ice formation in some aqueous solutions. Biodynamica 8:1.

Mazur, P. 1970. Cryobiology: the freezing of biological systems. Science 168:939.

Menz, L. 1971. Structural changes and impairment of function associated with freezing and thawing in muscle, nerve and leucocytes. Cryobiology 8:1.

Meryman, H. T. (ed.). 1966a. *Cryobiology*. Academic Press, New York.

Meryman, H. T. 1966b. Freeze-drying. In: *Cryobiology* (Meryman, H. T., ed.), p. 609. Academic Press. New York.

Meryman, H. T. 1971. Cryoprotective agents. Cryobiology 8:173.

Moor, H. 1964. Die Gefrier-Fixation lebender Zellen und ihre Anwendung in der Elektronenmikroskopie. Z. Zellforsch. 62:546.

Morel, F. M. M., Barker, R. F., and Wayland, H. 1971. Quantitation of human red blood cell fixation by glutaraldehyde. J. Cell Biol. 48:91.

Moreton, R. B., Echlin, P., Gupta, B. L., Hall, T. A., and Weis-Fogh, T. 1974. Preparation of frozen hydrated tissue sections for x-ray microanalysis in the scanning electron microscope. Nature 247:113.

Nichols, B. L., Soriano, H. A., Sachen, D. J., Burns, L., Hazlewood, C. F., and Kimzey, S. L. 1974. Electron probe localization of electrolytes in immature muscle. Johns Hopkins Med. J. 135:322.

Peachey, L. D. 1968. Muscle. Annu. Rev. Physiol. 30:401.

Peracchia, C., and Mittler, B. S. 1972. New glutaraldehyde fixation procedures. J. Ultrastruct. Res. 39:57.

Podolsky, P. J., Hall, T. A., and Hatchett, S. L. 1970. Identification of oxalate precipitates in striated muscle fibres. J. Cell Biol. 44:699.

Polge, C., Smith, A. U., and Parkes, A. S. 1949. Revival of spermatozoa after vitrification and dehydration at low temperature. Nature 164:666.

Reedy, M. K., and Barkas, A. E. 1974. Disordering of myofibril structure due to fixation, dehydration and embedding. J. Cell Biol. 63:282a.

Reedy, M. K., Holmes, K. C., and Tregear, R. T. 1965. Induced changes in orientation of the cross-bridges of glycerinated insect flight muscle.. Nature 207:1276.

Richards, F. M., and Knowles, J. R. 1968. Glutaraldehyde as a cross-linking agent. J. Mol. Biol. 37:231.

Robards, A. W. 1974. Ultrastructural methods for looking at frozen cells. Sci. Prog. Oxf. 61:1.

Robertson, E. A., and Schultz, R. L. 1970. The impurities in commercial glutaraldehyde and their effect on the fixation of brain. J. Ultrastruct. Res. 30:275.

Russ, J. C. 1972. Resolution and sensitivity of x-ray microanalysis in biological sections by scanning and conventional transmission electron microscopy. In: *Proc. 5th Ann. SEM Symp.* (Johari, O., ed.), p. 73, IITRI, Chicago.

Russ, J. C. 1973. Microanalysis of thin sections in the TEM and STEM using energy-dispersive x-ray analysis. In: *Electron Microscopy and Cytochemistry* (Wisse, E., Daems, W. Th., Molenaar, J., and van Duijn, P., eds.), p. 223. North-Holland Publ. Co., Amsterdam.

Russ, J. C. 1974. X-ray microanalysis in the biological sciences. J. Submicr. Cytol. 6:55.

Sabatini, D. D., Bensch, K., and Barrnett, R. J. 1963. Cytochemistry and electron microscopy. The preservation of cellular ultrastructure and enzymatic activity by aldehyde fixation. J. Cell. Biol. 17:19.

Saetersdal, T. S., Myklebust, R., Berg-Justesen, N. P., and Olsen, W. C. 1974. Ultrastructural localization of calcium in the pigeon papillary muscle as demonstrated by cytochemical studies and x-ray microanalysis. Cell Tiss. Res. 155:57.

Saubermann, A. J., and Echlin, P. 1976. The preparation, examination, and analysis of frozen hydrated tissue sections by scanning transmission electron microscopy and x-ray microanalysis. J. Microscopy 105:155.

Sevéus, L., and Kindel, L. 1974. Dry cryo-sectioning of human and animal tissue at a very low temperature (−140°C). In: Proc. 8th Int. Congr. Electron Microsc., Canberra, 1974, Vol. II, p. 52.

Sjöström, M. 1975. Ice crystal growth in skeletal muscle fibres. J. Microscopy 105 (in press).

Sjöström, M., Johansson, R., and Thornell, L. E. 1974. Cryo-ultramicrotomy of muscles in defined state. Methodological aspects. In: Electron Microscopy and Cytochemistry (Wisse, E., Daems, W. Th., Molenaar, J., and van Duijn, P., eds.), p. 387. North-Holland Publ. Co., Amsterdam.

Sjöström, M., and Thornell, L. E. 1974. Cryo-ultramicrotomy in the study of myofibrillar fine structure. I. The preparation procedure. Sci. Tools 21:26.

Sjöström. M., and Thornell, L. E. 1975a. Cryo-ultramicrotomy in the study of myofibrillar fine structure. II. The A-band. Sci. Tools 22:7.

Sjöström, M., and Thornell, L. E. 1975b. Preparing sections of skeletal muscle for transmission electron analytical microscopy (TEAM) of diffusible elements. J. Microscopy 103:101.

Sjöström, M., Thornell, L. E., and Cedergren, E. 1973. The application of cryo-ultramicrotomy in the study of the fine structure of myofilaments. J. Microscopy 99:193.

Spurr, A. R. 1972. Freeze-substitution systems in the retention of elements in tissues studied by x-ray analytical electron microscopy. In: Proc. Symp. on Thin Section Microanalysis, St. Louis, Missouri (Russ, J. C., and Panessa, B. J., eds.), p. 49.

Thornburg, W., and Mengers, P. E. 1957. An analysis of frozen section techniques. I. Sectioning of fresh-frozen tissue. J. Histochem. Cytochem. 5:47.

Thornell, L. E., and Sjöström, M. 1975. The myofibrillar M-band in the cryo-section. Analysis of section thickness. J. Microscopy 104:263.

Vassar, P. S., Hards, J. M., Brooks, D. E., Hagenberger, B., and Seaman, G. V. F. 1972. Physiochemical effects of aldehydes on the human erythrocyte. J. Cell Biol. 53:809.

Wolstenholme, G. E. W., and O'Connor, M. (eds.) 1970. The Frozen Cell. A Ciba Foundation Symposium. J & A Churchill, London.

Wood, R. L., and Luft, J. H. 1965. The influence of buffer systems on fixation with osmium tetroxide. J. Ultrastruct. Res. 12:22.

Yarom, R., and Chandler, J. A. 1974. Electron probe microanalysis of skeletal muscle. J. Histochem. Cytochem. 22:147.

Yarom, R., Hall, T. A., and Peters, P. D. 1975. Calcium in myonuclei: electron microprobe x-ray analysis. Experientia 31:154.

Yarom, R., and Meiri, U. 1971. N line of striated muscle; a site of intracellular Ca^{2+}. Nature 234:254.

Yarom, R., and Meiri, U. 1972. Ultrastructural cation precipitation in frog's skeletal muscle. J. Ultrastruct. Res. 39:430.

Yarom, R., Peters, P. D., Scripps, M., and Rogel, S. 1974. Effect of specimen preparation on intracellular myocardial calcium. Histochemistry 38:143.

Chapter 8

LIQUID DROPLETS AND ISOLATED CELLS

Joseph V. Bonventre, Kristina Blouch, and Claude Lechene

National Biotechnology Resource in Electron Probe Microanalysis, Harvard Medical School

INTRODUCTION

The electron microprobe, an instrument widely used for elemental analysis of non-biological specimens, is a much-needed addition to the armamentarium of the biological scientist. With the ultimate possibility to analyze attoliter (1×10^{-18} l) volumes of biological samples for all the elements of the periodic table with atomic number greater than 5 (boron), electron probe microanalysis is certain to open entirely new areas and push back frontiers of biological research.

Electron probe microanalysis is a spectrometric technique that takes advantage of the unique response of each element in the periodic table to emit x-rays of characteristic energies when bombarded with high-energy electrons. The characteristic x-rays are produced by a two-step process. First, the high-energy electrons knock an electron out of the inner shell of the atom. The vacancy produced is then quickly filled by an outer shell electron. The outer shell electron thus loses energy, which leaves the atom as an x-ray photon. Since each element of the periodic table has electron shells with unique energy characteristics, the properties of the emitted x-ray photons will be distinctive for each different element. If the incident electron does not displace an inner shell electron, the incident electron will lose energy by interacting with the nucleus. The loss of energy results in the release of x-ray photons of varying energies. This process leads to the production of a continuous x-ray energy spectrum, referred to as the Bremsstrahlung, or "white radiation."

Electron probe microanalysis can provide both qualitative and quantitative analysis of the elements present in the sample. Through examination of the x-ray spectrum emitted from the sample in response to electron beam bombardment, characteristic x-ray frequencies are identified that indicate the presence of particular elements in the sample. The intensity of the x-rays emitted at a particular characteristic frequency is correlated with the amount of the corresponding element present in the sample. The intensity of the Bremsstrahlung is correlated with the amount of mass of sample excited by the incident electron beam.

Moseley (1913) introduced the concept of using characteristic x-rays for chemical analysis. Castaing (1951) fitted an x-ray spectrometer to a modified electron microscope and invented the electron microprobe. He formulated the theoretical and practical aspects of microanalysis using x-ray spectrometry with electron beam excitation. There are excellent reviews (see Birks, 1969; Anderson, 1973; Goldstein and Yakowitz, 1975; Reed, 1975a; Siegel and Beaman, 1975) that discuss at length the principles of electron probe microanalysis. We present only a few salient aspects of the technique to clarify the subsequent discussion.

The electron microprobe consists of a source of electrons that are accelerated and focused as they move through a vacuum column. A condenser lens controls the amount of electron beam passing through an intermediate diaphragm. An objective lens defocuses the beam to a size slightly larger than the sample, so that the entire sample will be bombarded. A sample bombarded by these electrons emits x-rays. The x-rays are then collected and analyzed. From the wavelength and intensity of the x-rays, the elements and their amounts present in the sample are determined. Classically, x-rays emitted from the sample have been analyzed by wavelength-dispersive spectrometry. Alternatively, with the introduction of detectors that have the capability of distinguishing the different characteristic radiations by their energies, x-ray analysis may be performed by energy-dispersive spectrometry.

In wavelength-dispersive spectrometry the x-rays emitted from the sample are collimated and directed onto a crystal. The angle of the crystal is adjusted according to Bragg's law so that x-rays of a wavelength characteristic of the element of interest are reflected toward the proportional counter. Actually, the crystal is curved, to maintain a constant Bragg angle over a larger solid angle in order to increase efficiency. The selectively diffracted x-ray photons strike a proportional counter, resulting in a voltage signal proportional to the number of incident x-ray photons. This signal is then passed through a pre-amplifier, an amplifier, and a pulse-height analyzer. For x-ray photons of a particular energy, the proportional counter and amplifier will produce a voltage with a Gaussian distribution. The pulse-height analyzer is set to select the peak signal corresponding to the appropriate x-ray energy. Wavelength-dispersive spectrometry takes full advantage of the unique x-ray emission characteristics of each element.

The sensitivity of electron probe microanalysis is generally considered to be on the order of 100 ppm or 10^{-16} g in 1 μm^3 (10^{-15} l). Since the technique is nondestructive, it may be possible to achieve even higher sensitivities with longer counting times and highly controlled analytical conditions. Shuman and Somlyo (1976) were able to detect 2.3 $\times$ 10^{-19} g of iron in a single ferritin molecule.

An additional major advantage of electron probe microanalysis over other analytical techniques in biology lies in the ability to measure more than one element simultaneously. The nondestructive nature of the technique makes it possible to determine the concentration of many elements of interest in the same sample.

The application of electron probe microanalysis to biology has been limited, not by the physical method or the instrumentation but by the techniques of sample preparation. To fully exploit the capability of electron probe microanalysis, samples must be prepared for analysis so that the in vivo distribution of the elements of interest is not altered. This is particularly important in biology, since the systems studied are extremely heterogeneous, with large concentration differences of diffusible elements over very short distances.

Analysis with the electron probe is primarily performed on three types of biological samples: 1) liquid droplets, 2) isolated cells, and 3) tissue. This chapter deals with techniques involved in the manipulation and analysis of liquid droplet and isolated cell samples. These techniques have been applied to the study of many problems in physiology.

LIQUID DROPLET PREPARATION TECHNIQUES

The process of biological sample preparation is least susceptible to systematic error if a homogeneous sample can be collected directly from the living system under well-controlled conditions. In liquid droplet analysis, liquid samples as small as 5 pl are collected by microtechniques and prepared for electron probe analysis by a series of controlled manipulations suited to the nature of the sample. The goal of the preparative techniques is to dry the samples in such a way that small, homogeneous crystals are obtained. These crystals then present an isomorphic geometry to the incident electron beam during the analysis.

The use of the electron probe to analyze the content of dried liquid droplets was first described by Ingram and Hogben (1967). Samples were placed on a quartz slide and allowed to air dry. Cortney (1969) found that small samples, air dried to small spots, resulted in falsely low readings of elemental contents by electron probe microanalysis. He placed his sample droplets on dried spots of urea to spread out the area over which the droplets would dry and to make all the spots relatively uniform in size. Morel and his collaborators (Morel and Roinel, 1969; Morel et al., 1969), using calibrated micropipettes, placed samples on the polished surface of a cooled beryllium block. The samples were then frozen with dry ice and lyophilized under vacuum. However, the preparative technique did not consistently lead to the formation of small, homogeneous, uniformly distributed crystals. Crystal heterogeneity limited the reproducibility of this technique.

Lechene (1970) introduced the use of oil to cover the sample droplets on the beryllium support. The oil was washed with m-xylene without disruption of the droplets. The samples were then frozen in isopentane cooled to $-160°C$ and freeze-dried under vacuum. This resulted in the routine preparation of sample drops with a homogeneous spread of small crystals. This preparation eliminated the problems caused by poorly controlled drying and resulted in excellent x-ray count linearity with element concentration over wide concentration ranges. Le Grimellec, Roinel, and colleagues (Le Grimellec, 1975; Le Grimellec et al., 1973a,b, 1974a,b, 1975) subsequently also used oil to cover their samples on the beryllium support and washed the oil with xylene. They followed the xylene wash with a chloroform wash. Roinel (1975) reported washing oil with chloroform alone.

Quinton (1975, 1978a,b) and Rick et al. (1977) have modified these techniques for use with energy-dispersive spectrometry. Because of the low signal-to-background ratios that result when this analysis technique is used with samples prepared on a beryllium support, these investigators have used thin film (parloidin or collodion) supports mounted on metal grids.

We describe techniques developed primarily in our laboratory that make the analysis of liquid droplets reproducible and accurate. Some of these techniques have been described in other publications (Lechene, 1970,

1974; Bonventre and Lechene, 1974; Lechene and Warner, 1977; Gregor et al., 1978). We have attempted to provide the reader with enough information to enable him to reproduce any of the techniques we describe. These methods are generally applicable to the determination of any elemental component above atomic number 5 in a biological sample. They are particularly useful when the biological sample obtainable is too small to analyze by more classic techniques. At times small aliquots of larger samples are analyzed by the liquid droplet techniques because of improved reliability of electron probe microanalysis relative to alternative methods of analysis. In both large and small samples, multiple element analysis of the same sample, by the same technique, expedites multiple correlation studies and makes the comparison and correlation of elemental handling by the biological system much easier than would be possible by using standard techniques (e.g., flame photometry, coulometry, absorption spectrophotometry) on different sample aliquots.

General Techniques of
Microanalysis as Applied to Microdroplets

Biological samples, ordinarily in the volume range of 10 pl to 10 nl, are collected with micropipettes from a biological compartment during in vivo experiments. The samples are isolated by saline-saturated paraffin oil in the micropipette to prevent evaporation of water. They are then introduced onto a siliconized glass slide under a cover of saline-saturated oil to avoid evaporation. All manipulations are carried out under stereomicroscopic control with a Sensaur de Fonbrune micromanipulator (Curtin Matheson Scientific, Inc., Woburn, Mass.). Other micromanipulators may be used as well (e.g., Beaudoin, AZI Instrument Corp., Dedham, Mass.; Brinkman, Brinkman Inst. Inc., Westbury, N.Y.; Hacker, Hacker Inst. Inc., Fairfield, N.J.; Leitz, E. Leitz, Inc., Rockleigh, N.J.; Narishige, Labtron Scientific Corp., Farmingdale, N.Y.; Stoelting/Prior, Stoelting Co., Chicago, Ill.). Once a set of samples to be analyzed is collected on glass slides, a specially designed volumetric micropipette is used to take identical volume aliquots (usually 20–80 pl) from each sample. Fifteen to forty sample aliquots can be taken up into a single volumetric pipette, with the samples separated by an oil column. The liquid samples are then delivered onto a specially polished beryllium support covered with saline-saturated paraffin oil. Up to one thousand samples can be deposited on the same beryllium support. Once all the samples are introduced onto the support, the oil is removed and the droplets are freeze-dried under vacuum. The freeze-dried samples are then analyzed with an electron probe. A schematic representation of the liquid droplet technique is presented in Figure 1.

In the following sections of this chapter we examine in detail each of the steps involved in the liquid droplet preparation. We also describe variations

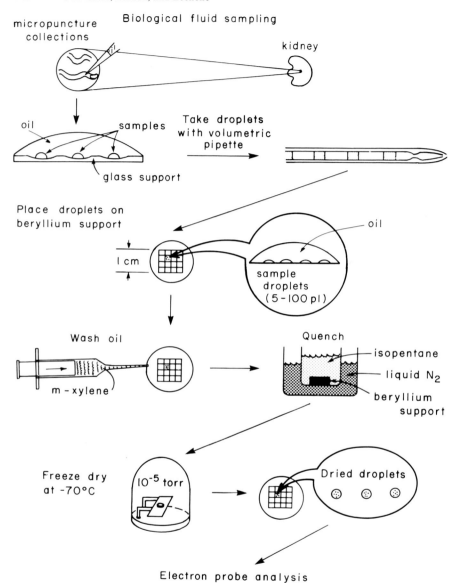

Figure 1. Schematic representation of liquid droplet technique. Biological fluid samples are collected by standard micropuncture techniques. The samples are placed on a glass support under saline-saturated oil. Aliquots are then collected with a volumetric pipette and placed under oil on a beryllium support. After all the sample and standard solution aliquots have been placed on the support, the oil is washed with m-xylene and the droplets, now covered with m-xylene, are quenched in a bath of isopentane cooled to −160°C by liquid nitrogen. The beryllium block is then transferred to the cooled stage of a freeze-drier and the droplets are freeze-dried at −70°C and below 10⁻⁵torr. The dried droplets are then analyzed with the electron microprobe.

on the use of the technique that have extended its applicability to the analysis of organic substances as well as inorganic elements.

Sampling and Transfer Pipettes

Microliter Pipettes   For routine collection or transfer of biological samples (e.g., plasma) in the microliter volume range, we use Pasteur disposable pipettes (VWR Scientific Inc., Boston, Mass.). The tapered end of the pipette is heated in a propane flame and pulled, producing a long narrowed region. The glass is then broken 3–5 cm below the upper border of this narrowed area.

Nanoliter Pipettes   Biological samples (e.g., renal tubular fluid) in the nanoliter volume range can be collected directly from the animal with micropipettes. These micropipettes can also be used as transfer pipettes for small volumes.

Pyrex capillary tubes (O.D., 1 mm; wall width, 0.20 mm; Fredrick and Dimmock, Inc., Montville, N.J.), prewashed with acetone, are placed in a vertical (Stoelting Co., Chicago, Ill.) or horizontal (Narishige, Tokyo, Japan) pipette puller. The two ends of the capillary are pulled apart by applying a force to both ends of the capillary glass while the middle is heated. Two micropipettes are thus formed, with narrowed tips formed at the site of separation. Each micropipette is then sharpened. The

Figure 2.   Technique used for sharpening micropipettes. The micropipette tip is advanced to the surface of a rotating grinding stone. The sharpening process is monitored with a stereomicroscope. The rotational velocity of the grinding stone is regulated with a variable transformer.

micropipette is mounted in a micromanipulator (Prior 22, Stoelting Co., Chicago, Ill.). Under stereoscopic control (Horizontal Spencer binocular light microscope, American Optical Co., Buffalo, N.Y.), the pipette tip is lowered onto a turning 2.5 cm × 2.5 cm Arkansas grinding stone (Chandler and Farquhar Co., Boston, Mass.) (Figure 2). The grinding stone is driven at approximately 345 rpm by an Electrohome fan motor (Kitchner, Canada) plugged into a variable transformer (W5LMT3 autotransformer, General Ruclin Co., West Concord, Mass.), which is set at 42 V. Distilled water is placed on the grinding stone just before the pipette tip is lowered onto the stone. The micropipette should make contact with the stone at a 40° angle. After sharpening, each pipette is washed in spectro-grade acetone.

The sharpened micropipettes are mounted on the stage of a microscope by pressing the untapered portion into clay. With 430× magnification, the pipette tip is examined. The tip should contain no glass or dust particles. The bevel of the pipette should be in the form of an ellipse, with one major axis approximately twice the size of the other. An eyepiece micrometer calibrated with stage micrometer disc (American Optical Co., Buffalo, N.Y.) is used to measure the tip diameter. The pipettes used most frequently in our laboratory have a tip diameter of 6–8 μm. The pipettes are then mounted in a metal sleeve and siliconized as described below.

Nanoliter and Picoliter Volumetric Pipettes   Analysis of biological samples in the nanoliter or picoliter volume range is expedited greatly by the use of calibrated volumetric micropipettes to routinely measure and transfer aliquots of these small volumes with high precision. Volumetric pipettes for these volume ranges have been constructed (Prayer et al., 1965; Quinton, 1976) by placing a small glass tube within a larger micropipette. These pipettes are self-filling but are severely limited in that they can hold only one sample at any one time. The pipettes we use can hold 30–40 samples simultaneously.

Volumetric pipettes are made from Pyrex capillary tubes with an outer diameter of 1 mm, a wall width of 0.20 mm, and a length of 150 mm (Fredrick and Dimmock, Inc., Montville, N.J.). The capillary tubes are washed with acetone (Spectro-grade, Eastman Kodak Co., Rochester, N.Y.) in the following manner. Two hundred tubes are placed in a 100 ml graduated cylinder with 30 ml of acetone. A stopper with a penetrating glass tube seals the top of the cylinder. A vacuum is then applied to the glass tube and acetone is drawn up into the capillaries for approximately 4 min. During this time the graduated cylinder is tilted at various angles by hand to ensure that all the capillaries are washed thoroughly. This process is repeated three times, using fresh acetone each time. Once the last acetone bath has been emptied, the graduated cylinder is covered with a piece of gauze, which is taped in place. The glass cylinder is then placed in an oven warmed to 100°C for about 30 min to dry the capillaries.

Once dry, a segment of each capillary is narrowed by holding the capillary at both ends and pulling the ends apart over an acetylene flame, using a welding torch with a #3 tip (issue AB-95 from Tescon Corporation, Minneapolis, Minn.). The acetylene tank valve is opened wide and oxygen is supplied at approximately 15 lb of pressure. The oxygen and acetylene are then finely adjusted through valves at the tip of the torch to obtain a small inner flame. The capillary is held in two hands and rotated in the orange part of the flame just above the inner blue core. Once the glass is soft enough (after about three quick rotations between the fingers) the capillary is given one quick pull to form a thin segment of capillary between two wide segments. The middle of the thin segment is then burned through and each of the thin ends of the two capillary pieces is melted at a point about 1 cm from the end, to form a hook (Figure 3a).

The pulled capillary is then placed in a Sensaur Microforge de Fonbrune (Curtin Matheson Scientific Co., Woburn, Mass.). The wall of the pulled section of the capillary should have an outer diameter of approximately 50–80 μm and inner diameter of 35–60 μm. A paperclip is placed on the hooked end of the capillary in order to provide weight and stability (Figure 3b).

A second pulling of the capillary is required to make volumetric pipettes for transfer of volumes in the picoliter range. A current is passed through the nichrome heating filament of the microforge, causing the wire to glow with an orange-yellow color. The filament is then brought as close as possible to a site approximately 1 cm from the upper end (Figure 4a) of the narrowed segment of capillary. Care must be taken to avoid contact between the filament and the capillary, which will result in adherence of the glass to the filament. The filament is held at one spot until the capillary has been warmed sufficiently to begin to melt. As the glass melts, the capillary will be pulled by the weight of the paper clip, forming a longer narrowed region (Figure 4b). Once this "second narrowing" appears, the capillary is moved downward past the filament to extend the area of capillary narrowing. The distance between filament and capillary must be continually adjusted in order to produce a constant diameter along the length of the capillary. The capillary should be drawn out to a length of approximately 5 mm, with a diameter of 30–40 μm along the entire length.

Once the capillary has been narrowed over a sufficient length, a tip is made by drawing the lower portion of the narrowed capillary to an outer diameter of approximately 10 μm (Figure 4c) and then breaking off the portion of the capillary below this region by touching the glass to the filament, which has been turned off, or by applying some vibration to the microforge. The filament can be used to enlarge the capillary lumen at the tip of the micropipette if necessary by hitting it against the tip of the pipette until the glass breaks. A constriction is then made in the pipette by applying heat to

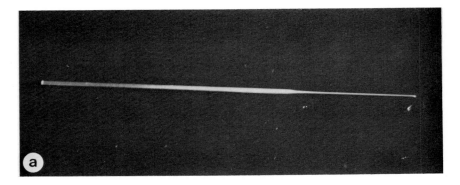

Figure 3.   **a:** Pulled segment of micropipette with hook at tip. This represents the initial step in construction of picoliter volume micropipettes. **b:** Pulled capillary tube of Figure 3a placed in a de Fonbrune microforge, with paper clip placed on the hooked end to provide weight and stability.

the capillary, with the filament placed far enough above the tip so that the heat does not cause the narrowing to extend all the way to the tip (Figure 4d). The heat may be applied alternately next to two sites separated by 180° along the circumference of the pipette or may be applied from one side of the micropipette, causing the micropipette to bend at the site of the constriction. Alternatively, the filament can be formed into a loop surrounding the micropipette and the glass can be heated evenly on all sides. We prefer to heat the glass from only one side because the angled tip facilitates sample manipulation and microscopic visualization during manipulation of

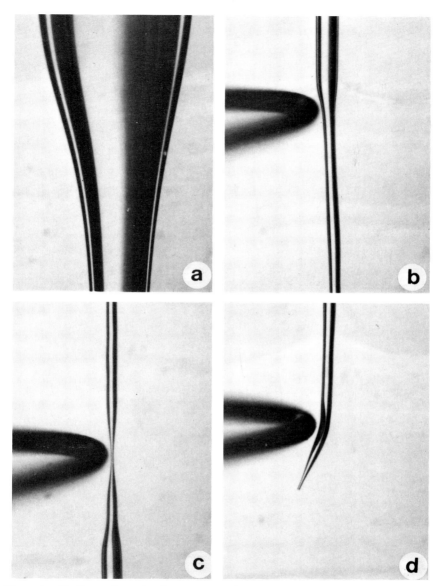

Figure 4.   Picoliter volumetric pipette construction. **a:** Upper end of narrowed portion of capillary after first pulling. **b:** As microforge filament heats capillary glass approximately 1 cm from the capillary section depicted in Figure 4a, the capillary is pulled by the weight of the paper clip, forming a longer narrowed region. **c:** The lower portion of the narrowed capillary segment is then heated to an outer diameter of approximately 10 μm. The capillary is broken at this site by touching the glass to a cooled filament or simply by vibrating the table upon which the microforge rests. **d:** A constriction is then placed in the capillary by bringing the heated filament close to the glass. The filament should be placed close to the tip of the pipette but far enough away so that the narrowing does not extend all the way to the tip.

droplets. The constriction must be small enough to slow the flow of fluid drawn up into the pipette by pneumatic aspiration, making it possible to stop the fluid at exactly the same point every time. However, if the constriction is too small, the pipette will offer high resistance to the flow of oil and sample and will not be suitable for the transfer of many droplets simultaneously.

The volume of fluid contained from the constriction to the tip of the pipette must be such that when fluid is pulled up beyond the constriction into the stem of the pipette the droplet will appear to form a square (i.e., its diameter will approximately equal its length) (Figure 5). If the length of the drop is shorter than the diameter of the pipette, the drops will tend to run together. On the other hand, if the length is very much greater than the diameter of the pipette, the stem will probably not be able to hold enough drops. A good pipette should be able to hold about 30 drops without developing a high resistance to movement through the pipette.

Mounting of Volumetric and Sampling Pipettes   The nontapered end of the glass pipette is inserted into one end of a 2″ length of #16 gauge needle tubing (Hamilton Co., Reno, Nev.) deburred at both ends. The pipette is held in place by a large drop of melted sealing wax (Dennison Mfg. Co., Framingham, Mass.) placed at the end of the tubing into which the pipette stem extends. Before the wax cools and thoroughly hardens, the pipette can be adjusted in the tubing to obtain the desired overall length. The pipette should then be held in a horizontal position as the wax on the pipette is passed through the flame once more and allowed to harden to make the seal airtight. The total length of volumetric pipettes (glass and metal tubing) should be between 10 and 12 cm.

Figure 5.   Picoliter volumetric micropipette containing fluid droplet aliquots (dark) separated by oil (light). Note that the droplets in the barrel of the micropipette have a diameter approximately equal to their length. From Lechene (1974).

Siliconization of Volumetric and Sampling Pipettes   In order to facilitate the handling of small droplets without wetting of the pipette glass, a technique of siliconization was developed with the help of Professor E. G. Rochow (Professor of Chemistry Emeritus, Harvard University). Newly mounted pipettes are connected to a 10–20 cm long segment of polyethylene tubing (PE-205: I.D., 0.062″, O.D., 0.082″; VWR Scientific, Boston, Mass.) mounted on a 20 cc syringe with a 16-gauge needle. Acetone, followed by chloroform, is aspirated into the pipette to clean it. Each solvent is taken up into the pipette and discharged approximately five times. Once clean, the pipettes are placed in a covered plastic chamber containing a small amount of water. They are left in the chamber overnight or longer to hydrate the glass surface. This hydration is required for proper siliconization. When the pipette is removed from the chamber it is rinsed again with acetone and chloroform and then siliconized with a 7% solution of Dri-Film Sc87 (now Surfasil, Pierce Chemical Co., Rockford, Ill.) in chloroform. The silicone solution is aspirated into the pipette one or two times. The pipette is then rinsed one or two times in toluene (Spectro-grade, Eastman Kodak Co.), which results in a more uniform distribution of silicone over the glass surface. The pipettes are then stored under beakers or in a desiccator to minimize exposure to dust.

The technique used for siliconization results in a thin, uniform, water-repellent film that is very durable. The procedure is convenient and does not require dry heating the glass or placing the micropipettes in boiling water (Rochow, 1951; Messing, 1976).

Cleaning of Micropipettes   After a micropipette has been used it is rinsed in acetone and stored in a covered container. Before re-use it is washed in baths of acetone, chloroform, and silicone (each solution is aspirated into the micropipette 3–4 times) followed by toluene (aspirated into micropipette 1–2 times).

If the micropipette becomes plugged or develops a high resistance to flow at any time during use, it can be cleaned with acetone. If there are sample aliquots in the pipette when it becomes plugged, it is sometimes possible to dip a piece of lens paper into acetone and run it gently over the tip of the pipette. If this is unsuccessful, or there are no droplets in the pipette, acetone can be aspirated into the pipette and then expelled four or five times. If this is unsuccessful, then holding the tip of the pipette in boiling water often unplugs it. As a last resort, the micropipette can be baked in an oven (Figure 6) at 500°C for 15 to 20 min. If the micropipette is baked, it must be resiliconized by the four-step process indicated above.

Calibration of Volumetric Micropipettes   Once the volumetric micropipettes have been constructed, mounted, and siliconized, their exact volume is determined by aspirating tritiated water of known activity into the new micropipettes and comparing the total radioactivity of these droplets

Figure 6.   Oven used to heat micropipettes to high temperatures in order to remove obstructing substances in the micropipette lumen. Temperature of oven is maintained at 450°C with a constant-temperature controller.

with that of known volumes of tritiated water. Solutions of tritiated water of different radioactivities are prepared so that it is possible to measure the volume of the smallest pipettes by comparing them with a larger one, itself calibrated by comparison to a 1-μl Lang-Levy pipette. Aliquots of tritiated water are placed under saline-saturated paraffin oil. The pipette to be calibrated is first filled to its constriction with oil, then tritiated water is taken up to the level of the constriction. The pipette is then removed from the tritiated water sample and a small amount of oil is taken into the pipette so that the sample is isolated by oil. The content of the pipette is then emptied into a scintillation vial containing 10 ml of scintillation fluid (Biofluor, New England Nuclear, Boston, Mass.). Five samples are taken for each pipette and the radioactive content determined. This is compared with samples taken with volumetric micropipettes of known volume in order to determine the precise volume of fluid that the pipette contains between tip and constriction.

Oil Saturation

All of our liquid samples are kept under saline-saturated paraffin oil to avoid concentrating of the droplets by loss of water to the oil. In order to saturate the oil, 75–100 ml of normal saline and 300–325 ml of paraffin oil

Table 1.   Increase in droplet elemental
concentrations after 3 hr under saline-saturated
oil at room temperature*

| Element | Percent increase | |
	176-pl droplet	8-nl droplet
Na	92	22
Cl	94	11
K	99	21
P	94	18
S	75	11
Ca	46	2
Mg	96	21

From Lechene and Warner (1979).

* Means of droplet x-ray counts obtained from three aliquots of each sample compared to x-ray counts obtained at time zero.

(white, light, laboratory grade; Saybolt viscosity 125/135; Fisher Co., Medford, Mass.) are placed into 500-ml wide-mouth polyethylene bottles. These bottles are then sealed with a screw top, clamped onto a wrist-action shaker (Burrell Corp., Pittsburgh, Penn.) and agitated continuously for at least 24 hr. Before the oil is used, it is centrifuged in a Sorvall GLC-2 centrifuge with an HL-4 head (Sorvall, Inc., Norwalk, Conn.) at 3,000 rpm for 15 min. The resulting supernatant is then taken for immediate use. Nanoliter or picoliter samples can be stored under saline-saturated oil before preparation for analysis. Samples will concentrate, however, if they are stored at room temperature (Table 1). By contrast, samples can be stored under oil at −80°C for at least 53 days without concentration (Tables 2 and 3).

Table 2.   Mean elemental concentrations in 21-nl drop of skate endolymph maintained at −80°C for 58 days*

Element	Day 1 (mM)	Day 58 (mM)
Na	286	273
Cl	489	488
K	115	105
Ca	5.02	5.08
S	1.95	1.67
P	0.85	0.72
Mg	0.24	0.24

From Lechene and Warner (1979).

* There was no statistical difference in any elemental concentration from day 1 to day 58.

Table 3.   Comparison of ^{3}H-inulin concentration in 23-nl droplets over 53 days

	Day												
	1	4	7	9	14	16	18	21	23	31	39	46	53
A/B*	0.98	1.04	1.04	1.08	1.11	1.19	1.28	1.31	1.48	1.57	1.69	1.81	1.98

From Lechene and Warner (1979).

* A: ^{3}H-inulin concentration in droplet that was maintained at room temperature for 6 hr before an aliquot was taken for measurement on each day indicated. B: ^{3}H-inulin concentration in droplet from which sample aliquots were taken immediately upon thawing on each day indicated. In each case the droplets were returned to freezer at $-80°C$ immediately after sample aliquots were collected. They were maintained at $-80°C$ between tests.

Polishing and Cleaning Beryllium Support Blocks

A beryllium support is used for liquid droplet analysis because of this metal's low background x-ray signal characteristics. In addition, its good electrical and thermal conductivity made it ideal as a sample support. The beryllium is 99.4–99.7% pure, with less than 28 ppm magnesium, and can be obtained from Brush Beryllium, Cleveland, Ohio, or Berylco, Hazelton, Penn. The beryllium support we use is a cylinder with a diameter of 1.9 cm and a thickness of 0.63 cm.

All machine work (i.e., polishing, engraving) and cleaning of beryllium are done under a hood with continuous air flow and a high-efficiency particulate-loading filter. Gloves and a mask are worn while polishing beryllium to avoid any risk of toxicity from beryllium dust (Reeves, 1977).

Beryllium blocks are supplied with a surface grain size of less than 15 microns. The surface is further polished in our laboratory using a Buehler Minimet polishing machine (Buehler Ltd., Evanston, Ill.) with a (Microcloth) polishing cloth, to which has been applied three drops of Buehler automet lapping oil and a small amount of Buehler Metadi 3-μm diamond-polishing compound. The surface is then polished to a mirror finish with a Buehler Metadi 6-μm diamond-polishing compound. The block surface is engraved with a grid 1 cm × 1 cm, constructed of 25 squares, each 2 mm × 2 mm. This facilitates location of samples. The block is then washed nine times in organic solvents. It is held with forceps and rinsed in a 100-ml polyethylene beaker filled to ~60 ml with trichloroethylene (Eastman Kodak). The block is then carefully placed in a second beaker that is filled to about 60 ml with trichloroethylene. The beaker is placed in the ring holder of an ultrasonicator (Bronwill Biosonic IV, VWR Scientific, Boston, Mass.). The beaker is raised so that the ultrasonic probe is ~1 cm above the surface of the block. The ultrasonicator is turned on and adjusted to a frequency producing maximal ultrasonication without producing movement of the block, which might cause the support to scrape against the ultrasonic probe. The block remains in the ultrasonicator for 2 min. After use, beakers employed in the first two washing steps are disposed of in special beryllium waste containers. The beryllium block is then ultrasonicated for 2 min in trichloroethylene, one bath of acetone, three baths of absolute alcohol, and two baths of fresh spectrograde acetone. After the last wash, the block is held with forceps and blown dry as quickly as possible with an Effa-duster (Fullam, Inc., Schenectady, N.Y.). The block is then stored in a plastic petri dish inside the laboratory hood.

To clean the beryllium surface after droplet samples have been analyzed, a polishing procedure that is much less extensive than the initial block polishing procedure is employed. The block is polished lightly for ~5 min with the 3-μm diamond-polishing compounds and rinsed by the nine-step procedure described above.

Preparation of Standard Solutions

Solutions containing standard concentrations of sodium, potassium, calcium, magnesium, chlorine, phosphorus, and sulfur are made with the following four salts: NaCl, CaCl$_2$, KH$_2$PO$_4$, and MgSO$_4 \cdot$7H$_2$O. NaCl, CaCl$_2$, and KH$_2$PO$_4$ are heated to ~125–150°C for 4–5 hr and weighed when warm. The glassware used is thoroughly cleaned and sterilized. We use ultrapure deionized, essentially organic free water with a specific resistance in excess of 18 Mohm/ml^3 (Type C Ultrapure Water Service, Hydro Service and Supplies, Inc., Durham, N.C.). Two primary standard solutions consist of: a) 400 mM NaCl and b) 20 mM CaCl, 20 mM MgSO$_4 \cdot$7H$_2$O, 20 mM KH$_2$PO$_4$. From these two solutions, 50 ml of each of five standard solutions for electron probe analysis are mixed. The compositions of these five standards are listed in Table 4. The standards are divided into 1-ml aliquots and placed in sterile ampules that are then heat sealed. When needed, one ampule of each standard is opened; approximately 20 μl is removed and placed on a siliconized microscope slide under saline-saturated oil. After an aliquot of each standard is deposited under oil, the dish is immediately frozen at −80°C. The samples are thawed to room temperature only for the minimal time necessary to deposit them on beryllium blocks. The remainder of the contents of each ampule is analyzed for sodium, potassium, calcium, and chlorine by macrotechniques immediately after opening the ampule. These macro-measurements are then entered into the

Table 4. Elemental composition (mM) of standard (ST) solutions

| Solution | Element | | | | | | |
	Na	Cl	K	Ca	Mg	S	P
ST 1	200	220	10	10	10	10	10
ST 2	140	150	5	5	5	5	5
ST 3	100	105	2.5	2.5	2.5	2.5	2.5
ST 4	40	42	1	1	1	1	1
ST 5	20	21	0.5	0.5	0.5	0.5	0.5

computer used for accumulating and calculating data from electron probe analysis and are used to create standard curves.

Preparation of Liquid Droplets on a Beryllium Block

Aliquots of biological samples and standard solutions to be analyzed are generally placed under saline-saturated oil on siliconized glass concavity slides. If the biological sample volumes are less than 20 pl they are placed directly on a beryllium surface under oil. A calibrated volumetric pipette, usually 20–100 pl, with an air syringe attached to the end of the pipette by a flexible polyethylene tube, is secured on a micromanipulator de Fonbrune. The concavity slide containing the samples and standards is placed under a dissecting microscope (Figure 7a). Then, under direct stereomicroscopic observation with a magnification of 7.8–50×, the calibrated pipette is lowered into the oil in a concavity containing a sample droplet. By gently pulling on the plunger of the air-filled syringe, oil is first pulled up into the pipette to protect the first sample droplet from evaporation and also to increase the resistance to flow in the pipette, making it easier to stop the sample aliquot at the pipette constriction. The amount of oil taken up at first depends upon the hydrodynamic resistance to uptake in the pipette; less oil is taken up into pipettes with large amounts of intrinsic resistance.

Once the appropriate amount of oil is introduced into the pipette, the pipette tip is introduced into the sample drop. The micropipette enters the drop at the point where a plane perpendicular to the vertical axis intercepts the droplet with maximal area. This permits maximal visibility of the tip and therefore maximal micropipette control. It is important to have as little contact between the pipette and the drop (whether standard or samples) as possible, otherwise the droplets may wet the outside of the pipette and make subsequent discharge of the droplet onto the beryllium block more difficult. The sample is drawn up into the pipette until its leading edge reaches the pipette constriction. With its tip maintained under oil, the pipette is removed from the sample and a small amount of the oil is drawn up into the pipette. The pipette is then reinserted into the sample solution droplet and the process is repeated. Six aliquots are taken from each of the standards and at least three drops from each of the biological samples. The standards drops are taken in order of decreasing concentration. A small amount of oil should be drawn up after the last drop to prevent evaporation from the tip of the pipette. After the last drop is taken up and while the tip of the micropipette remains under oil, the polyethylene tubing connecting the syringe to the pipette is disconnected from the syringe to allow equilibration between the pressure in the micropipette and the atmosphere. This prevents accidental aspiration of air and mixing of the droplets in the pipette when the tip is removed from the oil.

After the pipette is lifted out of the oil, the concavity slide is removed and replaced by a beryllium block (Figure 7b). The surface of the beryllium

Figure 7.   **a:** Sample and standard aliquots are collected into a volumetric pipette controlled with a de Fonbrune micromanipulator. The samples and standards are under oil on glass slides in a petri dish. Aliquots are collected by aspiration into the micropipette using a syringe attached to the micropipette with a polyethylene tube. **b:** Droplets are placed on a beryllium support covered by saline-saturated oil.

block is covered by three or four drops of saline-saturated light paraffin oil that was filtered through a 0.5-μm filter (Millipore Corp., Bedford, Mass.). The beryllium surface grid (Figure 8a) should be examined to find the region where the surface is most highly polished and clean. Once the grid square to be used has been chosen and noted, the tip of the pipette is lowered down onto the upper left-hand corner of the square. When placing drops on the block it is often better to have the pipette tilted at a somewhat greater angle (closer to the vertical) than was necessary for picking up the drops. With the tip of the pipette resting lightly on the surface of the block, gentle pressure is applied to the air syringe to force the first drop out onto the block. This gentle pressure must be released abruptly when the drop is out of the pipette to avoid introducing oil into the drop. The tip of the pipette is then withdrawn from this drop. Withdrawing the tip of the pipette must be done very carefully in order to avoid smearing the drop. Use of the fine control on the micromanipulator allows the operator to bring the pipette away from the droplet along the micropipette axis. The tip of the pipette is then set down on the block just to the right of the first drop and the process is repeated, setting the drops down in rows on the beryllium surface (Figure 8b). It is possible to place more than 100 drops in each 2 mm × 2 mm square.

A beryllium block with droplets should be kept on ice whenever it is not being worked with; for example, when other sample drops are being taken up into the pipette from concavity slides. This slows down possible concentration and prevents formation of large crystals in the drops already on the block. If storage over a longer period of time is necessary, the block is placed in a covered petri dish and kept in a freezer at −80°C. When the block is removed from the freezer, it is kept in the covered petri dish until it warms to room temperature. This is done to avoid the development of a large amount of condensation on the surface of the oil, especially when there is high humidity in the laboratory. The moisture not only distorts visibility but may settle to the surface of the block, resulting in fusion of droplets.

Once all the standards and samples have been deposited on the block and all the locations carefully noted and labeled, the block should be taken as quickly as possible through the oil washing and freeze-drying procedures described in the next section.

Oil Removal from the Beryllium Block

The paraffin oil covering the sample drops must be washed off without disrupting the samples or causing large asymmetric crystal formation. Prior to oil removal from the beryllium surface a beaker of 2-methylbutane (isopentane; Eastman Chemical Co., Rochester, N.Y.) is prepared and cooled in a styrofoam box containing liquid nitrogen. The isopentane is stirred until it reaches its freezing point (−160°C). The block is placed in a

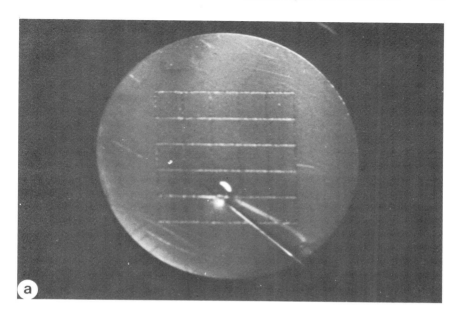

Figure 8.   **a:** Beryllium block with grid inscribed covered by saline-saturated oil. Micropipette is advanced through oil to surface of block and droplets are deposited. **b:** Higher magnification view of beryllium block with droplets placed in rows on the surface. Each of the droplets is 170 pl. From Lechene (1974).

large glass petri dish. An oil solvent, m-xylene (Spectro-grade, Eastman Chemical Co.), is drawn up into a 30-ml glass syringe, to which is attached a filter adapter (Swinny, Millipore Corp.) containing a 0.5-m filter. A #18 gauge needle is attached to the adapter with a small piece of polyethylene tubing placed over the end of the needle. With the tubing just above the edge of the beryllium block, the m-xylene is pushed through the filter and over the surface of the block. A steady, even flow of the m-xylene should be maintained and it should be directed first to one side of the block and then to the other. This results in a clockwise, and then counterclockwise, swirling of the mixture of oil and m-xylene. When the swirling is no longer apparent, and the surface appears to form waves, the oil is sufficiently washed. This washing procedure should take no more than 30 s. Alternatively, oil may be washed from the beryllium block surface by placing the block in a beaker filled with an oil solvent (Roinel, 1975).

Immediately after removal of oil, the block, with samples facing up and covered with m-xylene, is held with a pair of long forceps and carefully lowered into the beaker of cooled isopentane. The block is then kept in the cold isopentane during transport to the freeze-dryer. Using the long forceps again, the block is removed from the isopentane and is placed in the freeze-dryer chamber on a pre-cooled ($-70°C$) copper support. Roinel (1975) recommends the use of dry ice as a cooling agent rather than isopentane because of the possibility of ". . . explosive bursting of liquid bubbles with the risk of overturning the block when the support is introduced into the lyophiliser." We have never experienced this problem, which may be caused by excessive speed in drawing a vacuum in the freeze drier. We feel that the rapidity of freezing with isopentane or other rapid quenching agents is important to the ultimate formation of small, homogeneously distributed crystals.

Freeze-Drying

Freeze-drying of liquid droplets may be done with commercially available freeze-driers [Virtis Co., Gardiner, N.Y. (Agus et al., 1973) or Leybold EPA 100, Leybold-Heraeus Vacuum Products, Inc., Monroeville, Penn. (Rick et al., 1977)]. The freeze-drier used in our laboratory (Figure 9) was custom made to allow maximum flexibility for general laboratory development of liquid droplet, isolated cell, and tissue preparation techniques. The sample support, constructed of copper, is cooled by circulating liquid N_2. A heating coil is attached to the support, making possible control of support temperature. A thermocouple is placed on the support to monitor the temperature and to serve as an imput into an electronic control circuit that maintains constant temperature. The freeze-drying chamber is pumped by a mechanical pump and a diffusion pump. The diffusion pump is equipped with a cold trap that prevents vapors from recondensing on the surface of

Figure 9. Beryllium block on cooled support in freeze-drier cooled by circulating liquid nitrogen. A heating coil and thermocouple are attached to the support to maintain a constant temperature of $-70°C$ during the freeze-drying process. After the samples are freeze-dried, the support is heated to $50°C$ before removal of the beryllium block.

the sample. The chamber is maintained at a pressure below 10^{-5} torr. The total cooling period is controlled by a timing circuit. After the preset duration of freeze-drying at $-70°C$, usually 5–6 hr, the liquid N_2 circulation is stopped and the sample stage is warmed to $50°C$. Warming the block avoids condensation when it is removed from the freeze-drier. We usually run the freeze-drying procedure overnight under automatic control. It is likely, however, that the time required for each of the steps could be shortened without changing the characteristics of the dried crystals. The block is stored slightly above room temperature under vacuum in the presence of a silicon desiccant until it is analyzed with the electron probe.

Dried Sample Characteristics

The procedure used in preparing liquid droplets for electron probe analysis results in the deposition of very small salt crystals on a beryllium support. As stated previously, the purpose of the whole preparation is to obtain homogeneous dried spots of very small crystals (Figure 10a,b). The crystals form a nearly amorphous powder, homogenously spread at sites where droplets had been initially in contact with the surface of the support. All of the sample spots have a circular shape and have approximately the same

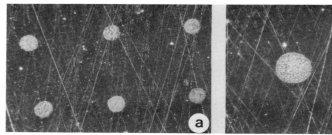

Figure 10. Dried samples as they appear on the beryllium block after the freeze-drying process. **a:** Each dried sample is 40 μm in diameter and represents the dried crystals of sample droplets of 31.1 pl containing 100 mM sodium, 105 mM chlorine, and 2.5 mM potassium, calcium, sulfur, phosphorus, and magnesium. **b:** Higher magnification appearance of two of the dried samples. Note the homogeneous distribution of small crystals.

diameter. Under these conditions, the geometry of each dried specimen is the same with respect to the exciting electron beam. The deposition of a uniform layer of thin crystals eliminates any possible difficulties in quantitative analysis due to asymmetric penetration of the electron beam or absorption of x-rays emitted from the sample. In addition, the uniformity in diameter of each of the dried specimens results in a constant relationship of sample diameter to beam diameter from sample to sample. In Figure 11 a scanning x-ray map of a dried sample spot is shown for four elements (sodium, chlorine, phosphorus, and magnesium). The scanning maps demonstrate a uniform elemental distribution throughout the dried spot.

Samples Containing Protein

Biological samples that contain large amounts of protein or other organic materials, such as plasma or epididymal fluid, are generally filtered before preparing liquid droplets for electron probe microanalysis. The presence of protein in fluid samples in the picoliter volume range can interfere with quantitative elemental analysis of these samples. The protein makes accurate delivery of picoliter aliquots more difficult due to adsorption of the biological fluid to the siliconized pipette glass and the emulsification of lipoproteins with oil in the pipette. This difficulty can be largely overcome, however, by delivering droplets of the protein-containing biological fluid onto the beryllium support one at a time. In this case, however, the pipette may have to be washed and re-siliconized between each droplet transfer.

High sample protein concentrations can result in protein coating of the salt crystals in the freeze-dried droplet. This can cause x-ray absorption and artificially lower the intensity of characteristic x-ray line signals from the protein-containing sample. This is particularly important for elements of lower atomic number, such as sodium, where the characteristic x-rays emitted are of lower energy. Finally, the presence of protein may alter the pattern of crystallization in the sample and thereby alter the conditions of analysis. In studies performed with Dr. M. Borland, bovine serum albumin was added in a concentration range of 1.0–7.0 g/100 ml to a stock solution of known elemental composition, and we found that protein concentrations above 5 g/100 ml may result in a decrease in elemental x-ray signals, resulting in falsely low determination of elemental concentrations. Biological fluids with protein concentration greater than 5 g/100 ml (e.g., mouse serum) can be diluted 1:2 or 1:3 and manipulated by the standard liquid droplet technique, which will provide accurate and reproducible elemental concentration determinations as compared with more classical macrotechniques of analysis. Nevertheless, to ease technical manipulation and to determine the amount of unbound element in the biological fluid sample, protein and other high molecular weight compounds should be removed

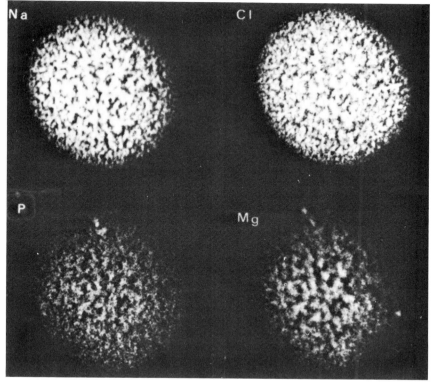

Figure 11.   Scanning x-ray map for sodium, chlorine, phosphorus, and magnesium of one of the droplets in Figure 10. Note the homogeneous distribution of each of the elements. From Lechene (1974).

from biological samples whenever possible prior to analysis. Protein removal is especially relevant for calcium and magnesium determinations, where the elemental component unbound to protein is of primary importance, and for phosphorus and sulfur, which exist in considerable amounts in protein. The presence of protein in samples analyzed for these elements will result in values that do not accurately represent free elemental concentration.

Microultrafiltration   To remove the protein and other high molecular weight substances from small biological samples we have developed a technique of microultrafiltration. Initially we developed a method whereby the biological sample was introduced into a small hollow filtration fiber supplied by Amicon (Amicon Corp., Lexington, Mass.). The fiber was then bent on itself and centrifuged in a specially constructed capillary micro-cuvette (Lechene and Warner, 1979). However, we have abandoned this method because the necessary fibers are no longer available. The technique we have developed and routinely use involves passing the sample from one glass capillary tube to another through an ultrafiltration membrane of 2.4-nm pore dialysis tubing. Samples are placed in the upper capillary glass under oil. An ultrafiltration apparatus, consisting of upper and lower capillary glass separated by ultrafiltration membrane, is spun in a microhematocrit centrifuge and ultrafiltrate is collected under oil in the tapered lower capillary glass (Figures 12 and 13a,b). This technique has proven to be very reliable and is routinely used in our laboratory. With it we can microultrafilter more than 200 samples a day. We present the technique in detail, since it has not been previously published.

Microultrafiltration centrifuge tubes.   A microhematocrit capillary tube (75 mm long × 1.3–1.5 mm in diameter; Fisher Scientific Co., Medford, Mass.) is filled approximately to one-third of its capacity with saline-saturated, light paraffin oil. The tube is then heated over flame at a site ~30 mm from the empty end. When the glass is softened, the tube is pulled out by hand so that the pulled portion is ~50–60 mm long and 0.5 mm in outer diameter. The tube is then broken through the pulled region into two segments so that the end containing oil has a narrowed taper ~30 mm long. Oil is then allowed to run down into the narrow tapered part of the tube. When the oil has completely filled the tapered end of the capillary glass, the end is sealed over a flame. A small residual air bubble remains at the tip, but it disappears during subsequent centrifugation. The exterior of the closed tapered end of the tube is then cleaned with acetone and chloroform and siliconized. The tapered capillary tube is then etched with a diamond ~1.5 cm from the beginning of the taper. The tube is broken at the etched site into two tubes, one which is tapered (lower capillary glass), and one which is nontapered (upper capillary glass). One end of the upper capillary glass is then passed through a flame to smooth the rough glass edge and to

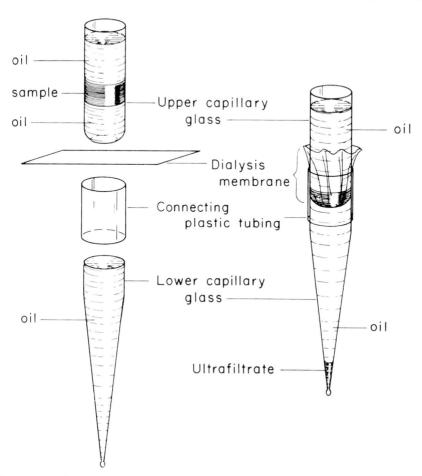

Figure 12. Schematic representation of the process used to microultrafilter samples of nanoliter and microliter volumes. Details of construction of the various parts are given in the text. The sample is placed in the upper capillary glass and is ultrafiltered through a 24Å pore dialysis membrane by centrifugation in a microhematocrit centrifuge. The tip of the lower capillary glass is then cut off and the sample is expelled onto a glass slide covered with saline-saturated oil.

decrease the area of opening to approximately one-half the inner diameter of the tube.

Ultrafiltration membrane. The microultrafiltration membrane is constructed of dialysis tubing with a pore size of 24Å (VWR Scientific Co., Boston, Mass.). The tubing is never touched with bare or gloved hands. All manipulations are done with forceps. The tubing is supplied in rolls. One and one-half to two-foot sections of tubing are slit down the edges into two

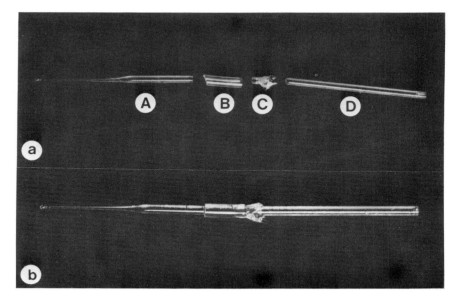

Figure 13. **a:** Photograph of disassembled microultrafiltration apparatus. A = lower capillary glass. B = polyethylene tubing. C = 24Å pore dialysis membrane. D = upper capillary glass. **b:** Assembled microultrafiltration apparatus.

halves as the tubing is unwound from the roll. To remove any contaminating substances, particularly the sulfur in which it is packed, the tubing is placed in a 2-liter beaker containing 0.6 M sodium acetate. The solution, with tubing submerged in it, is heated to 100°C and subsequently allowed to cool to room temperature. Once the tubing has cooled, the sodium acetate solution is drained off and the tubing is rinsed thoroughly by letting deionized water run continuously through the beaker for 1–2 hr. This rinsing procedure is repeated eight times over a period of 2–3 days. The tubing is then stored in a covered beaker of deionized water. Since microorganisms may eventually attack the tubing, benzoic acid or formaldehyde may be added to the water if the tubing is to be stored a long time; however, we have had no difficulty with tubing deterioration and have not used any preservative.

Approximately 1–3 weeks before use, four pieces of dialysis tubing, ~10 cm long, are cut and spaced apart in a 100-ml plastic centrifuge tube filled with saline saturated oil. One end of each piece of tubing is folded over the edge of the centrifuge tube and held in place with a lid. The tubing is then spun in a Sorvall GLC-2 centrifuge in an HL-4 head at 3,000 rpm for 30 min to remove any water present on, or in, the tubing. After centrifugation, any portion of the dialysis tubing not completely submerged in oil during the centrifugation is cut off and discarded and the remaining tubing is stored in a small covered beaker full of saline-saturated oil for 2–3 weeks. The long period of contact with oil makes the tubing more pliable, so that it will not break during subsequent manipulations. Shorter lengths of oil exposure result in leaky membranes, as do periods of oil exposure greater than 4–5 weeks.

Microultrafiltration assembly. On the day prior to ultracentrifugation, the dialysis tubing is removed from its storage bath of saline-saturated oil and mounted on the upper capillary glass tube in the following manner (Figure 14). A pinchcock, flatjawed clamp placed on the laboratory bench holds two small pieces of polyethylene tubing, one inside the other, in a vertical position. The outer tubing (I.D., 0.160″; O.D., 0.218″) is 13 mm long and the smaller piece of inside tubing (I.D., 0.1″; O.D., 0.145″) is 10 mm long. A piece of dialysis tubing ∼ 0.5 cm square is cut and laid over the opening of the polyethylene tubes. The non-tapered ultrafiltration tube is then held in such a way that oil completely fills the area of the rounded, fire-polished end. Then, with the worker's finger over one end of the tube to

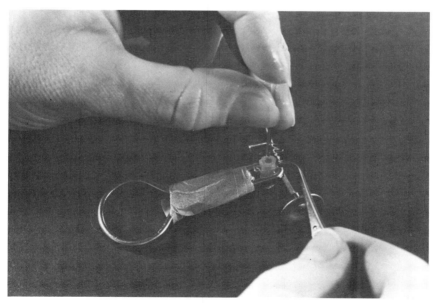

Figure 14.   Technique used to mount dialysis tubing over end of upper capillary glass. A flat-jawed clamp resting on the laboratory bench is used to hold two small pieces of polyethylene tubing, one inside the other. The dialysis membrane (held with forceps) is placed over the opening of the polyethylene tubes and the upper capillary glass (held by the hand on the left) is pushed through the opening. The dialysis membrane then folds over the end of the glass capillary.

prevent the oil from leaking out, the fire-polished end of the tube is pressed into the opening of the polyethylene tube, thereby forcing the dialysis tubing to cup around the end of the upper capillary glass. The pinchcock is then turned onto its side and the glass tube is pushed on through until just its tip, covered with the dialysis tubing, protrudes from the polyethylene tubes. In order to hold the dialysis tubing in place over the end of the ultrafiltration tube, a piece of PE-205 tubing 7–8 mm long (I.D., 0.062″; O.D., 0.082″) is pushed 3–4 mm over the dialysis tubing. It is important not to push the PE tubing sleeve too far over the dialysis tubing, otherwise the membrane will stretch and break. The tubes are placed under vacuum in a desiccator with a silica gel desiccant for 2–3 hr. Then the empty halves of the PE-205 tubing sleeves are filled with saturated oil and the tubes are placed (dialysis tubing pointed downward) into a 30-ml beaker of saturated oil and stored this way overnight in a Plexiglas high-humidity box.

On the day filtrations are to be done, the tapered lower capillary glass tubes are filled with saline-saturated oil and inserted into the polyethylene sleeves. The oil in the upper capillary glass is then exchanged with fresh saline-saturated oil by flushing this new oil into the tube with a blunt hypo-dermic needle attached to a syringe. The tapered end of the assembled ultrafiltration apparatus is then placed in a 16-gauge metal sheath to protect the tip against breakage. The tubes are spun for 5 min at 4,300 rpm in a microhematocrit centrifuge (model MB, International Equipment Co., Needham, Mass.) powered with a variable transformer with the voltage set at 50 V. After centrifugation, both upper and lower capillary glass tubes are closely examined. If the oil level in either tube has dropped, it is presumed to be leaky and the ultrafiltration apparatus is discarded.

Microultrafiltration procedure. The tips of Pasteur pipettes are heated and pulled out to a long thin taper. These tapered pipettes are then used to transfer the biological sample to be filtered into the ultrafiltration tubes. The tip of the pipette is pushed about three-fourths of the way down into the top portion of the upper capillary glass tube and 0.2–1 μl of sample is placed into the tube. The ultrafiltration apparatus, protected by the metal sheath, is placed in the hematocrit centrifuge and spun for 5 min at 3,200 rpm with the variable transformer set at 40 V. After centrifugation the tubes are closely examined again: those tubes with a small amount of fil-trate about halfway down the narrow tip of the lower capillary glass and with most of the biological sample remaining in the upper capillary glass above the filter will have good filtrates. If no sample remains in the top of the tube, the filter leaked. If no filtrate appears in the lower capillary glass and sample still remains above the filter, the tube can be centrifuged again at a higher speed. When an adequate ultrafiltrate has been obtained, the metal sheath is removed from the bottom half of the ultrafiltration apparatus and the lower capillary glass is separated from the upper. The

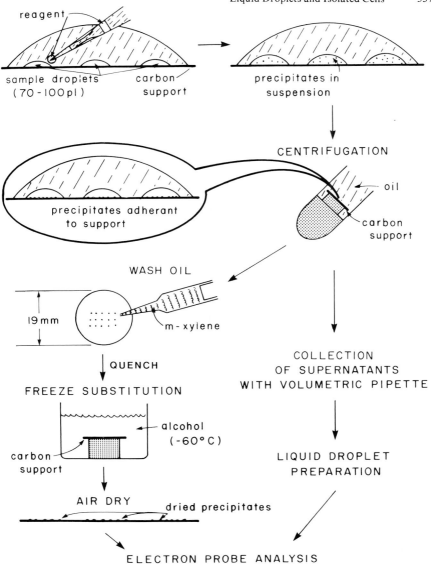

Figure 15.   Microprecipitation technique for analysis of organic components in picoliter liquid droplets. A reagent that will quantitatively precipitate the organic compound of interest in the sample is introduced into each droplet. The carbon disc, with samples containing precipitates in suspension, is then centrifuged. This causes the precipitates to adhere to the carbon support. At this point aliquots of the supernatant can be collected from each droplet and prepared by the usual liquid droplet techniques for electron probe analysis. If the precipitate is to be analyzed, the oil is washed from the support with m-xylene and the samples quenched in isopentane at $-160°C$ and then freeze-substituted in alcohol at $-60°C$. This results in removal of the supernatant from the precipitate. The precipitates are then air-dried and analyzed by electron probe microanalysis.

lower capillary glass tube is then connected to an air syringe by a polyethylene tube placed at the non-tapered end. The bead of glass at the tip is broken off with a razor blade and the filtrate is pushed out under oil. These filtered samples are then treated, as described above, like other biological samples that contain no protein.

Organic Solute Determination

Although the liquid droplet technique is conventionally used only for the measurement of inorganic elements, organic substances can be measured under certain experimental conditions. There are two techniques used for organic analysis: 1) microprecipitation, and 2) direct measurement of the carbon, nitrogen, and oxygen content of the sample.

Microprecipitation Techniques   A method (Figure 15) developed in our laboratory (Bonventre and Lechene, 1974) has provided a way to extend the liquid droplet technique to the determination of concentrations of organic components in biological samples as small as 20 pl. Using this method, a biological sample containing an unknown amount of organic component of interest is combined with a reagent that will specifically and quantitatively precipitate the organic component. In order to apply this precipitation procedure to electron probe microanalysis, the precipitate, and thus the reagent, must contain an element that can be quantitatively analyzed by the electron probe. The sample and the reagent must be combined quantitatively in a system in which no evaporation is permitted. The precipitate must then be washed of all non-reacted reagent and sample components. Finally, the precipitate must remain insoluble, and when counted by the electron probe, must produce counts directly correlated with the amount of organic component of interest in the original sample. In addition to organic compound determinations, this technique is also useful for distinguishing the amount of an element that may exist in the organic and inorganic state, and may be used to determine the oxidation state of an element. For example, sulfur and phosphorus are contained in organic substances as well as in various oxidation states in inorganic salts. Using microprecipitation techniques, electron probe analysis of droplets can distinguish between these various states. For example, barium added to the sample aliquots will react with sulfur present in the sample as $SO_4^=$ ions to form a $BaSO_4$ precipitate.

Samples are placed, by means of a calibrated volumetric pipette, under saline-saturated paraffin oil onto a pyrolytic graphite disc that has been ashed for 5 s in an oxygen plasma (Coleman Model 40 Low Temperature RF Reactor, Coleman Instruments, Maywood, Ill.) and washed with acetone. Ashing the carbon surface makes it slightly more porous and facilitates sample droplet and precipitate adhesion to the carbon. An identical volume of reagent is then introduced into each of the sample droplets, resulting in the formation of a precipitate. The carbon disc, with the droplets containing precipitates, is then placed in a centrifuge tube. The disc is supported by a fitted Lucite base at a distance of ~ 5 cm from the bottom of the tube. The centrifuge tube containing the disc is then placed in a swinging bucket HL-4 head and centrifuged at 3,000 rpm for 1 hr in a Sorvall GLC-2 centrifuge. The centrifugation results in a flattening of the precipitates against the carbon disc at the bottom of the sample drops, fixing them in position. The carbon disc is removed from the centrifuge tube and washed with m-xylene to remove the oil and then immediately quench-frozen in isopentane at −160°C. As with all liquid droplet manipulations, care is taken during the washing and freezing procedures not to expose the drops to the air at any time.

The carbon support is then removed and placed in a constant-temperature bath of absolute alcohol at −60°C for 2–5 min until the frozen droplets sublimate in the liquid alcohol and the alcohol-soluble components of the drops diffuse away. As a result, only the precipitates, washed of the reagent and of the other ionic components of the original sample droplet, remain on the carbon disc. The absolute alcohol is then removed from the bath and the carbon disc is allowed to dry. Precipitates appear as thin (1–2 μm thick) spots on the disc. They are comparable to each other in size and shape and their surface area is equal to or smaller than the surface area of contact between the original droplets and the carbon support. The washed and dried precipitates are ready to be examined by electron probe microanalysis. Figure 16 shows that this technique is quantitative over a wide range of sample component concentration. AgNO$_3$ reagent was added to standard

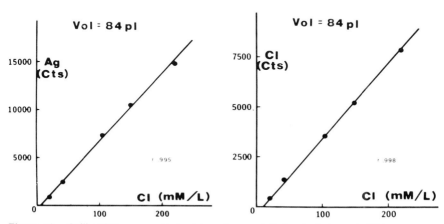

Figure 16. Calibration curves obtained by microprecipitation technique. AgCl in excess was added to 84-pl droplets containing standard amounts of NaCl. Ag (cts) and Cl (cts) represent the total silver and chlorine x-ray counts recorded in 10 s from the AgCl precipitate formed. r is the correlation coefficient. From Bonventre and Lechene (1974).

droplets containing varying amounts of chlorine. A quantitative precipitate of AgCl was formed that was analyzed by electron probe microanalysis. The relationship of silver or chlorine signal from the precipitate to the initial chlorine concentration of the sample was linear from 5 to 220 mM Cl.

A variation of this technique has been used to determine the sulfate content of picoliter micropuncture samples taken from the renal distal tubule. The total sample sulfur content was determined in aliquots of samples taken before precipitation. Barium chloride was then added to the droplets, resulting in the formation of a $BaSO_4$ precipitate. After centrifugation, aliquots were collected from the supernatant of each droplet. The amount of sulfur present in each of these supernatant aliquots was then determined by electron probe analysis. The total sulfate content of the sample was determined by the difference between total sulfur present in the initial sample and the sulfur present in the supernatant of precipitated droplets.

Beeuwkes et al. (1977) developed a method for urea analysis with the electron probe. They placed sample aliquots on silicon supports covered with oil. The oil was saturated with a reagent (thioxanthen-9-ol). The reagent diffused from the oil into the water of the sample droplets and, in the presence of a substrate that promoted crystallization (dixanthyl-urea), quantitatively precipitated urea present in the samples. The oil and reagent were then washed with m-xylene and the samples air-dried. The amount of sulfur present in the dried sample was measured by the electron microprobe and was quantitatively correlated with the amount of urea present in the sample droplet. This technique permits the analysis of inorganic elements and urea in the same droplet. It also obviates pipetting of reagent into the sample droplet. Urea determination by this technique is limited in sensitivity due to the presence of endogenous sulfur in biological samples.

Measurement of Carbon   Under certain conditions, the carbon signal measured from the sample by electron probe microanalysis is reflective of the amount of an organic compound that is present in the sample. A low atomic number element, such as carbon, has a $K\alpha$ emission wavelength of 44.5 Å (Compton and Allison, 1935) and requires the use of a diffraction crystal with lattice spacing comparable to this wavelength. In addition, the counter window must be very thin. In our laboratory carbon is measured with the electron probe using a lead stearate diffraction crystal and a collodion window. For this technique to be useful, the carbon present in the sample must be primarily in the form of the organic component of interest. For example, this method has been used to measure raffinose concentrations in 200-pl droplets obtained from rat proximal tubules (Warner and Lechene, 1979). In standard solutions there was a linear correlation of carbon x-ray counts with known concentrations of raffinose over a range of 0–300 mM.

Transportation of Samples

In collaborative research with other laboratories it is frequently necessary to transport nanoliter volumes of liquid biological samples over long distances. To ensure safe transport and handling and unchanged composition of these small samples, we have devised a specific protocol. Since these techniques may be of general use, we briefly describe them here.

Samples greater than 10 μl can be transferred in sealed plastic tubes, e.g., polyethylene centrifuge tubes (Beckman Instruments, Inc., Fullerton, Cal.), or glass capillaries (e.g., Critocaps, 1.3–1.5 mm in diameter; Sherwood Medical Industries, St. Louis, Missouri) sealed at both ends with soft beeswax. Samples in the nanoliter volume range are isolated by saline-saturated paraffin oil in individual glass capillary tubes that are approximately 1 mm in diameter and that have been pulled at one end to a taper approximately 1 cm long. The tip diameter of the tapered end is 10–20 μm. The sample should be placed in the tapered end with at least 5 mm of oil separating the sample from the pipette tip and 1 cm of oil behind the sample. The pipettes are sealed with a small amount of soft beeswax at the tip. The samples are then immediately frozen. Each pipette is secured in a metal or plastic box (a large plastic petri dish with putty is frequently used), so that the pipettes will be firmly supported and protected from damage.

Just prior to shipment, the samples, in their supporting containers, are packed in a styrofoam box with dry ice. Enough dry ice is used so that the samples will still be frozen when they arrive at the destination laboratory.

The technique we describe for transporting samples is different from that used by Quamme et al. (1978). These investigators transported their droplet samples in the dried state and then rehydrated and lyophilized them prior to electron probe analysis.

APPLICATIONS OF THE LIQUID DROPLET TECHNIQUE

The liquid droplet technique has been applied in the areas of renal physiology, reproductive physiology, auditory physiology, digestive physiology, exocrine physiology, and cytochemistry. The technique has been used most extensively in the study of fluid and electrolyte handling in the kidney. Analysis of samples collected by micropuncture from the vascular and tubular structures of the kidney with classical techniques are difficult and are usually limited to a few components because of destruction of the sample aliquots during the analysis and the lack of convenient analytical microtechniques. With the liquid droplet technique and electron probe microanalysis, small sample volumes can be collected and analyzed for many elemental components without sample destruction.

Cortney (1969) and Morel et al. (1969) were the first investigators to use liquid droplets and electron probe microanalysis to study renal function. Cortney examined renal tubular handling of electrolytes in adrenalectomized rats. Morel et al. (1969) and Le Grimellec et al. (1973a, b, 1974a, b) measured the chlorine, sodium, potassium, calcium, and phosphorus content in micropuncture samples collected from proximal and distal tubules of rats. Le Grimellec and colleagues examined the composition of the fluid in Bowman's capsule (Le Grimellec et al., 1975a) and early proximal tubule of the rat kidney (Le Grimellec, 1975) in order to study the process of glomerular ultrafiltration and the renal handling of electrolytes in the early proximal tubule. Shirley et al. (1976) measured phosphorus, calcium, and magnesium fluxes into the rat proximal convoluted tubule by using a modification of the split-droplet microinjection technique (Shipp et al., 1958). Rouffignac et al. (1973) studied the electrolyte handling in the loop of Henle by using the liquid droplet technique to analyze fluid collected directly from the loop.

Proximal tubule ion and water transport has been studied by isolating fluid volumes as small as 200 pl between oil drops in the tubule (Warner and Lechene, 1976). The droplet is then reaspirated from the tubule and movement of ions and water into and out of this droplet are studied as a function of time and the initial composition of the droplet. The use of small volumes permits functional analysis of various segments of the proximal tubule. This modification of the classical "standing droplet" technique has been possible only because of the analytical capabilities of electron probe analysis and the liquid droplet preparation.

The site of sulfate reabsorption along the nephron has been studied (Lechene, 1974; Lechene et al., 1974). It was possible to separate the sulfur present as the sulfate ion ($SO_4^=$) from the rest of the sulfur in the sample by using the microprecipitation technique, as described above.

The renal handling of calcium and phosphorus and the effect of parathyroid hormone on tubular handling of these elements has been extensively studied by the liquid droplet technique and electron probe microanalysis (Agus et al., 1973, 1977; Beck and Goldberg, 1973; Knox and Lechene, 1975; Baumann et al., 1975; Knox et al., 1976; Gregor et al., 1977; Lechene et al., 1977b; Pastoriza-Munoz et al., 1978; Bengele et al., 1979; Harris et al., 1979). The ion and water handling all along the nephron of the elasmobranch little skate, *Raja erinacea* (Stolte et al., 1977), and the mudpuppy, *Necturus maculosus* (Garland et al., 1975), have also been studied using these techniques.

The liquid droplet technique is especially useful for the study of distal tubule function in the kidney, since the tubule flow rate is slow in this part of the nephron and the large collections required for classical analytical techniques may result in retrograde flow from the collecting duct and from other nephrons (Colindres and Lechene, 1972).

In the female reproductive tract, Borland (1977), Borland et al. (1977a, b), and Biggers et al. (1978) have applied the liquid droplet technique and electron probe microanalysis to the study of ion and water transport in rabbit and mouse blastocysts. The technique permitted complete analysis of small amounts of fluid collected from one blastocele cavity containing a total volume of fluid less then 150 pl. Chong et al. (1977) studied the elemental composition of human follicular fluid and found it to be similar to the composition of blood. The oviduct environment of mouse embryos was also studied by micropuncture collection of samples followed by liquid droplet analysis (Roblero et al., 1976; Borland et al., 1977b). In the male reproductive tract, Howards et al. (1979) have studied the concentration of sodium, potassium, chlorine, magnesium, phosphorus, and sulfur in the rete testis and along the various regions of epididymis. Samples collected with micropipettes from perilymphatic and endolymphatic regions of the inner ears of anesthetized cats, lizards, and skates were analyzed with the electron probe using the liquid droplet technique (Peterson et al., 1978). Skadhauge et al. (1979) studied osmoregulation in the teleost fish, *Tilapia grahami*, by using electron microprobe analysis to analyze ionic concentrations in the intestinal contents along the length of the gut.

Quinton (1978a) has studied the characteristics of sweat and saliva using a modified liquid droplet technique with analysis by energy-dispersive x-ray spectrometry. The techniques of liquid droplet analysis have been used to identify a potent endogenous inhibitor of (Na^+, K^+) ATPase present in muscle-derived ATP. This inhibitor was found to be vanadium (Cantley et al., 1978). This work has led to the use of this transition metal as a probe to study the (Na^+, K^+) ATPase (Cantley et al., 1978; Beauge and Glynn, 1978) and to study the role of this enzyme in the renal handling of salt and water (Grantham et al., 1978; Bonventre et al., 1979).

SINGLE CELL PREPARATION TECHNIQUES

One of the ultimate goals of electron probe analysis is the measurement of cellular composition of tissues in various physiological states. The application of electron probe analysis to the study of biological tissues has been limited, not by the techniques of analysis, but by the techniques of biological sample preparation. Samples must be prepared in a way that maintains in vivo tissue distribution of biological elements. A logical system with which to approach the study of biological tissues is an isolated cell system, where cells can be manipulated and studied in an extracellular environment that can be well controlled by the investigator.

In order to study cells by electron probe analysis, they must be prepared so that: 1) the cells are separated from each other; 2) the normal biological milieu of the cells is removed, so as not to interfere with the cellular measurements; and 3) there is no change in cellular composition during the preparation.

One method of cell preparation is air-drying, which has been used to study *Amphiuma* blood cells (Anderson, 1967), human blood cells (Carrol and Tullis, 1968; Beaman et al., 1969), human and chick erythrocytes (Colvin et al., 1975), adult chicken erythrocytes and isolated intestinal cells (Barrett and Coleman, 1973), bacterial cells (Greaves, 1974), and human sperm cells (Maynard et al., 1975; Chandler and Battersby, 1976). Coleman and colleagues (Coleman et al., 1972, 1973a, b) modified the technique of air-drying in their study of large unicellular protozoa, *Tetrahymena pyriformis* and *Amoeba proteus*. These large cells were rapidly heat-dried on a silicon disc passed through the flame of a propane torch. These methods of drying in air, with or without heat, may result in elemental redistributions within cells or between intracellular and extracellular spaces. Other investigators (Hales et al., 1974; Timourian et al., 1974; Osborn and Hamilton, 1977; Davis, 1978) have fixed cells in a liquid fixation medium. This procedure may result in redistribution of diffusible ions in the cell and entry of elements from the media or loss of diffusible ions to the surrounding media.

To avoid ionic translocation, quench-freezing of the cells has been employed. After quench-freezing, investigators (Chandler and Battersby, 1976; Roomans and Seveus, 1976) have freeze-dried, under vacuum, the individual cells or sections containing single cells. Others (e.g., Ingram et al., 1974) have embedded the freeze-dried sample in plastic before sectioning and analysis. However, freeze-drying and freeze-drying embedding may result in marked disruptions of cells due to rapid expansion of trapped gas within the cells under vacuum (Nei, 1962; MacKenzie, 1965; Southworth et al., 1975).

Red blood cells, because of their easy availability and the wealth of information available regarding their physiology, are ideal single cells to use as a model system for electron probe microanalysis. Information can be obtained on single cells that can be checked by routine macrotechniques to validate the electron probe analytical techniques. The technique of preparation of red blood cells, as with other biological samples, is of critical importance. Initial electron microprobe studies of red blood cells involved preparation of the cells by simply air-drying a smear of whole blood (Anderson, 1967; Beaman et al., 1969; Colvin et al., 1975). In this type of preparation, plasma drying on top of the cells adds to the elemental x-ray signal from the cells in an unpredictable way, and thereby results in inaccurate cellular elemental determinations. Kimzey and Burns (1973) analyzed cells that had been washed in isotonic media and smeared onto silicon discs. The results obtained with this technique were only qualitative, since standards were inadequate.

Another method of cell preparation involved sandwiching cells between a collodion film and beryllium support (Roinel and Passow, 1974). The cells were then freeze-dried, at which time the collodion film was disrupted in some areas and a few red blood cells were exposed and available for analysis. The majority of cells that were covered by the collodion film were unavailable for analysis because the film absorbed the sodium-characteristic x-rays. A third method of preparation involved centrifuging cells onto beryllium discs and then coating the cells with a thin layer of dibutyl-phthalate to displace the suspending medium and to prevent loss of cellular contents (Kirk et al., 1974). This technique, however, may result in varying absorption of x-rays by the dibutylphthalate film, causing errors in the electron probe determination of cellular content.

A technique that provides for routine preparation of many single red blood cells on a pyrolytic graphite support has been developed in our laboratory (Lechene et al., 1977a). The cells maintain their ionic content and there is excellent reproducibility in cellular elemental x-ray signals by electron probe analysis.

Preparation of Cells

Fresh human blood is drawn into heparinized syringes. The blood is then centrifuged and the plasma and white cells are removed by aspiration. Approximately 2 ml of packed cells are washed twice in 50 ml of acid sucrose (0.285 M sucrose with 0.005 M $MgSO_4$, pH 6.0).

Preparation of Cell Standards

Red cells containing standard concentrations of sodium and potassium are prepared by a modification of the method of Cass and Dalmark (1973). One-tenth of a milliliter of nystatin stock solution is added to suspending media containing varying amounts of potassium and sodium, adjusted to a pH of 7.2 with CO_2, and cooled to 0°C. The nystatin stock solution of 20–50 $\mu g/ml$ is prepared in dimethyl sulfoxide from Mycostatin powder (Squibb). One-half milliliter of washed packed red cells is added to 50 ml of media and incubated 30 min at pH 7.1–7.2 and 0°C. The nystatin induces a finite but reversible increase of permeability of the red cell membrane to monovalent cations. The cellular concentration approaches that of the medium. At the end of the incubation period the suspension of cells is diluted at room temperature with an equal volume of medium identical to the suspending medium except for the absence of nystatin. The diluted cell suspension is then centrifuged and the supernatant removed from the cells.

The cells are washed twice with 30 ml of acid sucrose medium to remove all remaining extracellular substances.

Average Cell Sodium and Potassium Determination

The technique of Kirk et al. (1974) is used for determination of average sodium and potassium cellular concentration. Hematocrit, hemoglobin, and number of cells per unit volume of whole blood are determined. The hematocrit is determined by centrifugation in microhematocrit tubes (Fisher Scientific Co., Medford, Mass.). The number of red cells per unit volume is measured in quadruplicate with a hemocytometer. Hemoglobin measurements are made in duplicate in Kampden-Zijlstra reagent (Van Kampden and Zijlstra, 1961) using a Coleman spectrophotometer (Model 6/20, Perkin Elmer, Norwalk, Conn.). Sodium and potassium determinations are made in duplicate after cell lysis in 0.004 M CsCl and 0.25 ml Acatronix (Scientific Products, Evanston, Ill.) per liter of solution. The sodium and potassium concentrations are determined with a flame photometer (Corning Scientific Instruments, Medford, Mass.) and compared with electron probe measurements.

Preparation of Cells for Electron Probe Analysis

We initially prepared the red cells by smearing them onto a pyrolytic graphite support (E. Fullam, Schenectady, N.Y.). However, this technique resulted in large variations in elemental x-ray signals on different areas of the same support (Lechene et al., 1974). Low signal intensities on the cells were associated with high x-ray signal intensities in the medium beside the cells. Salt crystals containing sodium and potassium could be found outside the cells by exploring the surface of the support with a focused beam with diameter less than one micron.

After attempting many different methods of red cell preparation we found that the most reliable technique involved atomizing a cell suspension and spraying the cells onto pyrolytic graphite discs. The cells are diluted by a factor of 10 in either acid sucrose or plasma. Nitrogen gas is released at 30 psi through a 22-gauge needle that is positioned at the upper end of a 0.8 mm (I.D.) capillary tube. The lower end of the capillary tube is in the cell suspension (Figure 17). The carbon disc is held ~ 10 in from the top of the capillary tube. The suspension of cells is sprayed onto the support for 1–2 s. A special holder (Figure 18a,b) allows us to spray successively up to four preparations on the same support by shielding three-quarters of the support during each spray. Thus it is possible to analyze and to compare several batches of cells under the same electron probe microanalysis conditions. Sprayed cells, relatively uniform in appearance, are well separated on the support and dried very rapidly once on the carbon disc. The cells are maintained under vacuum until analyzed with the electron probe. The appearance of the cells on the carbon support is shown in Figure 19.

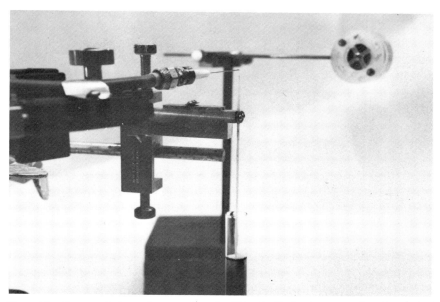

Figure 17.   Spray technique of preparation of red blood cells for electron probe microanalysis. Nitrogen gas is released at 30 psi through a 22-gauge needle that is positioned at the upper end of a 0.8-mm (I.D.) capillary tube. The lower end of the capillary tube is placed within a test tube containing a 1:10 dilution of cells in suspension. The cells are sprayed onto a carbon disc in a specially designed support.

Kirk et al. (1978) have modified the techniques described above. They use a commercially availabe EFFA spray mounter (Fullam, Inc., Schenectady, N.Y.) powered by nitrogen gas at 30 psi. They dried the sprayed cells in a nitrogen gas stream. Kirk et al. compared cellular iron and potassium distributions in cells prepared by spraying with cells prepared frozen and cut at −70°C with an ultramicrotome. The distributions of iron and potassium cellular signals were similar under both conditions of preparation.

ELECTRON PROBE MICROANALYSIS

Instrumentation and Automation

The instrument used for analysis in our laboratory is a Cameca MS46 microprobe with four wavelength-dispersive spectrometers (Cameca, Courbevoie, Paris). Our microprobe is equipped with a light microscope that allows us to: 1) align and center the electron beam and carefully set its diameter, 2) move the samples under the beam, 3) focus the sample in the Rowland circle of the spectrometers, and 4) observe directly any visual changes in the sample during electron beam bombardment that might indicate mass loss or contamination.

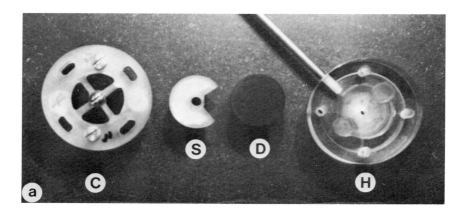

Figure 18.   Holder used for cell spray technique. **a:** Disassembled holder. Carbon disc (D) is placed into the well in the lucite holder (H). A lucite shield (S) is then placed above the carbon disc, which shields approximately three-fourths of the disc. A cover (C) is then placed over the lucite holder to stabilize the carbon support and shield. **b:** Assembled holder for carbon disc. After one group of cells is sprayed onto the disc the shield is rotated to expose another part of the carbon and another cell suspension is sprayed onto the support. It is thus possible to analyze and compare several groups of cells under the same electron probe microanalysis conditions.

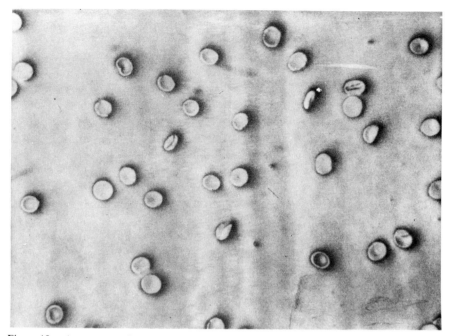

Figure 19.   Appearance of red blood cells, with their characteristic biconcave form, on a carbon support. The cells are ready for electron probe microanalysis. From Lechene et al. (1977a).

We use a specially designed specimen stage (Cameca) that is controlled by a 2100A Hewlett-Packard minicomputer to allow for automated analysis of droplet samples. Stepping motors are attached to each of the spectrometers so that spectrometer tuning is automatically controlled by the computer. A completely automated system has been described (Moher and Lechene, 1975) that includes: specimen stage control, spectrometer tuning, analysis conditions, monitoring, data acquisition, and data handling. The design facilitates easy interaction between investigator and systems. A schematic diagram of hardware and software and their interaction in the system is shown in Figure 20.

Each sample is brought under the electron beam with a joy stick, which controls step motors. The coordinates of each sample are fed into the computer. The approximate spectrometer crystal angle settings for the elements of interest are also fed into the computer. The spectrometers are automatically finely tuned by placing under the beam a standard sample that contains a very high concentration of the element of interest. The computer controls the counting of many droplet samples, usually numbering more than one hundred. Each sample is counted for a set time (10–100 s). The

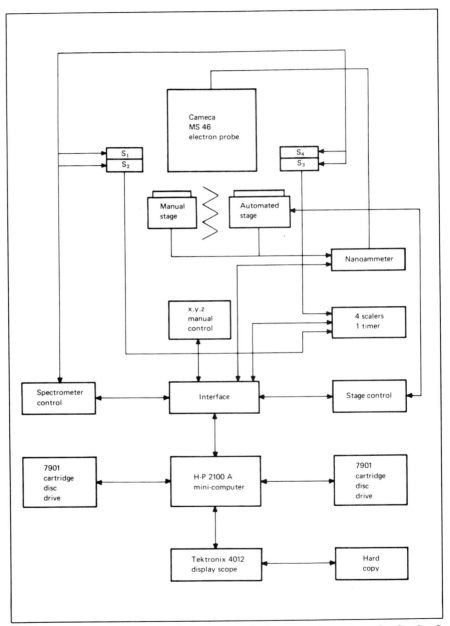

Figure 20. Schematic diagram of the automated electron probe system. S_1, S_2, S_3, S_4 represent the four spectrometers, which can be controlled automatically. The automated stage and spectrometers are interfaced with a Hewlett-Packard 2100A minicomputer. Interaction between user and system is facilitated by a Tektronix 4012 display scope. From Moher and Lechene (1975).

four spectrometers enable us to analyze four different elements simultaneously. By automatic tuning of the spectrometers to another set of four elements, multiple runs can be performed sequentially. The x-ray counts are stored in the computer; the data can be retrieved and statistical analysis performed in a simple and flexible way.

Data can be listed as raw counts or with background subtracted. By comparison with counting data obtained on aliquots of standard solutions, regression lines are established and concentrations determined. A plotting program is used extensively for the illustration of data. Frequently during the analytic run, the same standard is returned to and counted to ensure that counting conditions are not changing. This information is stored and then retrieved from the computer in the form of a set of graphs showing percent deviation of counts from first determinations as a function of time (Figure 21). In addition, the sample beam current is monitored continuously and plotted as a function of time (Figure 22). A graphic display terminal and software program permit ready illustration of data and easy user interaction with the computer.

Droplet Analysis

Method The support, containing unknown samples to be analyzed along with samples obtained from standard solutions and samples that will be used to tune the spectrometers, is placed onto a specimen stage and

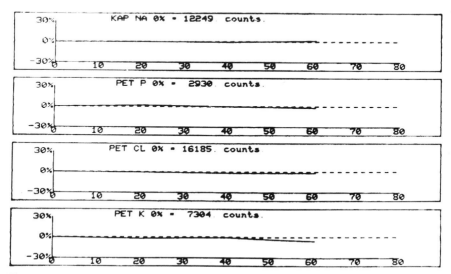

Figure 21.   Set of curves showing constancy of sodium, phosphorus, chlorine, and potassium x-ray counts with time from a sample droplet returned to at varying time intervals over a period of 60 min. Time is represented on the abscissa and percent deviation from initial counts on the ordinate. KAP = potassium acid phthalate crystal; PET = pentaerythritol crystal.

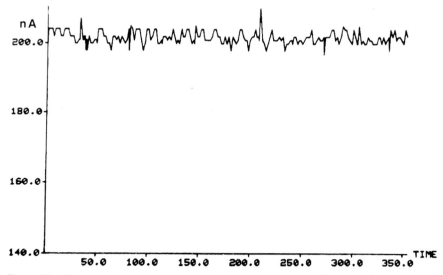

Figure 22.   Continuous recording of sample beam current over 350 min during an electron microprobe run. Note the constancy in beam current. nA = nanoamps.

introduced into the column of the electron microprobe. The column is then vacuum pumped. Current is passed through a tungsten filament, which serves as the source of electrons. The electrons are collimated by electromagnetic lenses. The accelerating voltage of the electron beam is set at 11 kV. Beam current is 200 nA. We have found that these analysis conditions provide reliably good beam and sample stability and signal-to-background ratios.

A typical sample run would include analysis of the samples for chlorine, potassium, sodium, and background. We count chlorine, potassium, and background with pentaerythritol spectrometer crystals and sodium with a potassium acid phthalate or thallium acid phthalate crystal. When all the samples have been counted the spectrometer settings are changed and the spectrometers are automatically tuned to four different elements (e.g., sulfur, calcium, phosphorus, and magnesium). Under conditions used in our analysis there is no loss of x-ray counts due to electron excitation (Table 5); therefore, the elemental determinations are independent of the order of the microprobe run in which the element is counted. During the run, frequent determinations of the background counts are made by moving the specimen stage to a region on the support where there is no sample. Background x-ray counts are determined on each spectrometer. An alternative method of background determination would involve detuning the spectrometers and measuring the x-ray signal off peak. We have compared

Table 5.   Effect of the beam bombardment on x-ray counts*

Element	Concentration (mM)	A (cts/10 s)	Time under the beam (min)	B (cts/10 s)
Mg	2.5	313 ± 4	8	321 ± 5
P	2.5	354 ± 3	10	357 ± 5
Ca	2.5	521 ± 5	10	532 ± 6
K	2.5	568 ± 6	10	566 ± 6
Na	100	2729 ± 14	7	2753 ± 14

From Lechene (1974).

* Sample droplets were 36.6 pl. Beam diameter was 50 μm. Accelerating voltage was 11 kV. Sample current was 300 nA. Each sample was counted for 3 min. A: Maintained under the beam for 7–10 min and then recounted again for 3 min. B: There was no statistically significant difference in counts obtained during the two periods. X-ray counts were expressed per 10s (mean ± S.E.).

background determined by both methods and have found that in our analytical conditions there is no significant difference between the two techniques. From this result it can be concluded that the contribution to the background of the mass of droplet crystals is negligible and that essentially all background signal is due to the beryllium substrate. The reproducibility of the liquid droplet preparation technique with electron probe microanalysis of the dried droplets is excellent (Table 6).

Quantitation

Calibration curves.   The concentration of each element in the unknown sample is determined by comparison with standard samples of known composition. Each dried standard drop contains a predetermined amount of each of the elements to be measured. Calibration curves relating

Table 6.   Reproducibility of x-ray counts from dried droplets*

Element	Concentration (mM)	Counts/10 s (mean ± SD)				
		A	B	C	. . . J	M
Na	200	1773 ± 39	1723 ± 36	1744 ± 55	. . .	1777 ± 78
Ca	10	477 ± 26	480 ± 28	462 ± 32	. . .	483 ± 27
Mg	10	216 ± 16	210 ± 16	208 ± 14	. . .	216 ± 23
P	10	173 ± 11	170 ± 9	166 ± 11	. . .	169 ± 12

From Lechene (1974).

* Ten aliquots (A through J) from the same standard solution were counted 10 times each for 10 s. Columns A, B, and C represent the mean ± SD of counts on each of 3 dried droplet samples. Column M represents the mean of the first 10-s counting periods for each of 10 (A through J) aliquots. There is no statistical difference between the mean x-ray counts in 10 measurements on any droplet and the mean x-ray counts of all the droplets. Therefore, the variation of the liquid droplet preparation technique is on the order of the variation in counting statistics with the electron microprobe. Sample droplets were 77 pl. Beam diameter was 100 μm. Beam accelerating voltage was 11 kV. Sample current was 500 nA.

characteristic x-ray counts to chemical element concentrations are easily obtained with excellent linearity and reproducibility (Figure 23). There is negligible electron beam or x-ray absorption by the dried samples. It may be possible to quantitate the amount of an element present in liquid droplets by comparison with one standard (Lechene and Warner, 1979).

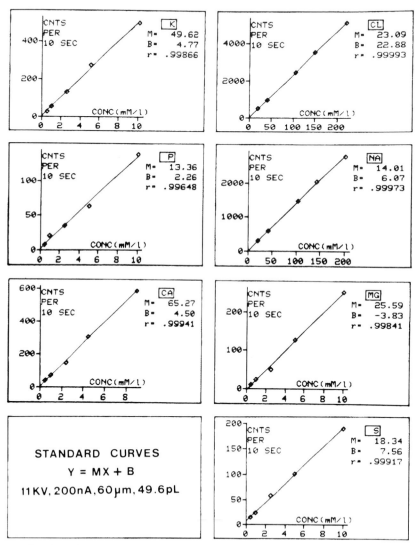

Figure 23.   Set of calibration curves for potassium, chlorine, phosphorus, sodium, calcium, magnesium, and sulfur for droplet samples of 49.6 pl. The accelerating voltage was 11 kV and sample current was 200 nA. The beam diameter was 60 μm. These curves were taken directly from the computer output. r = correlation coefficient.

Chlorine loss. Of the elements commonly studied with the liquid droplet technique, chlorine appears to be the most susceptible to loss as a consequence of electron beam bombardment of the sample. Lechene (1974) reported loss of chlorine from samples analyzed with beam currents higher than 300 nA with a beam diameter of 100 μm. Chlorine loss was avoided with lower current densities. Roinel (1975) suggested that chlorine loss could be prevented by adding urea to the samples. Neither our laboratory nor that of Quinton (1978b) has been able to confirm a protective action of urea to prevent chlorine loss. Quinton (1978b, 1979) found loss of chlorine primarily in acidic samples and suggested that it was due to the formation of HCl. He was able to prevent chlorine loss by the addition of LiOH to his acidic samples. Our laboratory (Lechene and Warner, 1979) has found a striking correlation between chlorine loss and the amount of calcium present in the sample. In summary, the cause of chlorine loss in electron microprobe analysis remains controversial.

Minimum detectability. Our laboratory has recently determined the minimum detectable elemental concentrations of sulfur, phosphorus, calcium and magnesium with the liquid droplet technique (Quinton et al., 1979). Using the expression derived by Kotrba (1977), the minimum detectable concentrations were found to be 0.08, 0.10, 0.07, and 0.09 mM, respectively. The concentrations were lowered to 25%–50% of these values when the analysis was performed on thin carbon-coated Formvar films mounted over nickel grids rather than the usual beryllium substrates.

Single Cell Analysis

Red blood cells, well isolated on the graphite support, are analyzed in our laboratory with a Cameca MS46 electron probe equipped with four wavelength-dispersive spectrometers. The cells are analyzed for sodium, potassium, chlorine, magnesium, phosphorus, sulfur, and iron. Sodium and magnesium characteristic x-ray K lines are selected with KAP (potassium acid phthalate) crystals; chlorine, potassium, phosphorus, and sulfur with PET (pentaerythritol) crystals; iron is selected with a LiF crystal. Our studies have revealed that the best elemental signal-over-background x-ray ratio is obtained with the electron beam accelerating voltage between 11 kV and 15 kV. The beam diameter is set at approximately 10 μm, so that the beam will excite the entire cell. The beam current is fixed at 200 nA. With these beam conditions the x-ray counting rate is relatively high and there is no decrease in characteristic x-ray signal counts due to volatilization or contamination of the sample, or due to migration of the beam during the x-ray analysis. Under these conditions chlorine counts [chlorine is the element most susceptible to decrease in counts with time of electron probe analysis (Lechene and Warner, 1979)] are stable over periods of 30 min of irradiation with the electron beam. Background x-ray counts are determined by moving the support so that the x-ray beam falls on an area between the cells or an area of the support not exposed to the spray.

The graphite sample supports are mounted on a special automated stage controlled by a joy stick and driven by a step motor. The stage is moved so that every cell of more than one hundred cells is sequentially localized beneath the beam. The location of each cell is stored in the memory of a Hewlett-Packard 2100A minicomputer. Recording the location of 300 cells in this manner takes approximately 90 min. The analysis is then begun for four elements simultaneously and each cell is moved beneath the beam automatically. Each cell is analyzed for 200 s. The beam current is continuously monitored and recorded during the analysis. Frequent checks on the stability of the system are made by returning to a control sample. Background is measured with the spectrometers remaining tuned. After the completion of one run the spectrometers are set to four new elements and the entire run is repeated. The second run can also be used to analyze the same elements after changing the probing conditions (e.g., accelerating voltage, beam diameter, or beam current). Up to eighteen hours of counting time may be necessary for one run. The electron probe parameters remain stable during these long periods of time. Figure 24 shows a typical set of cellular elemental x-ray count distributions from a set of cells prepared and analyzed by the above techniques. Figure 25 shows the correlation found between sulfur and iron characteristic x-ray line counts in these cells.

Cells can be analyzed by other means than spot mode analysis. Kirk et al. (1978a,b) used raster mode analysis with a 1-μm electron beam scanning an area containing a single cell. This results in uniform electron beam density over the entire cell.

Energy-Dispersive Spectrometry

In energy-dispersive analysis a Si(Li) detector collects all x-rays emitted in a large solid angle from the sample. Voltage pulses with amplitude proportional to the x-ray wavelength incident on the crystal are then emitted from the detector. These voltage signals are separated by a multichannel analyzer. One advantage of this technique is the ability to obtain an entire energy spectrum of the sample without scanning for each element, as is necessary in wavelength-dispersive systems. In addition, because the solid angle of collected x-rays is larger, energy-dispersive systems can provide high signals with less electron bombardment of the sample.

There are many disadvantages, however, of the energy-dispersive system method of elemental analysis: 1) X-ray energy resolution is poor and it is frequently difficult to determine an element corresponding to the peak on the energy spectrum or to resolve one element from another corresponding to adjacent sites on the energy spectrum. 2) In energy-dispersive analysis, peak-to-background ratios are much smaller than they are in

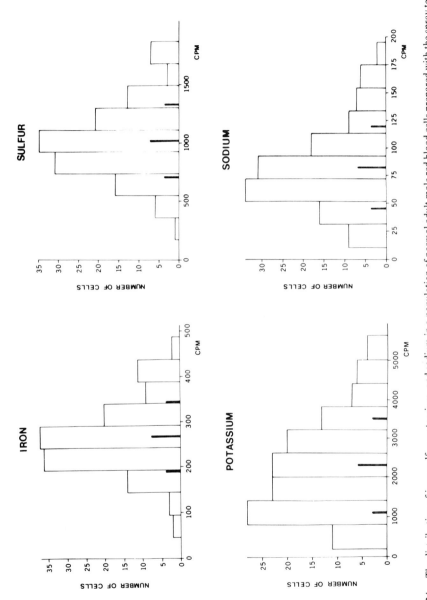

Figure 24. The distribution of iron, sulfur, potassium, and sodium in a population of normal adult male red blood cells prepared with the spray technique and analyzed by electron probe microanalysis. Large bars represent the mean values, while smaller bars represent standard deviations of the mean. Electron beam accelerating voltage was 11 kV. Beam diameter was 10 μm. Sample current was 100 nA. CPM = x-ray counts per minute. From Lechene et al. (1977a).

X-RAY INTENSITY

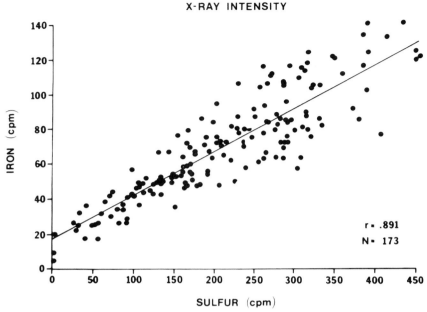

Figure 25. Correlation between x-ray count intensity of iron and sulfur in normal red blood cells. Electron beam accelerating voltage was 11 kV. Beam diameter was 12 μm. Sample current was 50 nA. cpm = x-ray counts per minute; r = correlation coefficient; N = 173 cells counted. From Lechene et al. (1977a).

wavelength-dispersive systems. Therefore, the precise determination of the background is more critical in an energy-dispersive system. However, because of the breadth of peaks on the energy-dispersive spectrum, it is frequently not possible to obtain an adequate measure of the background at a particular energy. The background is determined by estimation with complicated algorithms that rely on many parameters of the system that are difficult to predict or control (Reed, 1975; Fiori et al., 1976). 3) Energy-dispersive systems have 10 times lower detectable levels of elements than wavelength-dispersive systems (Geller, 1977). 4) The wide solid angle of detection allows for entry of spurious x-rays due to backscattered electrons from sources other than the sample in the column and secondary fluorescence of the sample. 5) The localization in the sample of the source of x-rays collected by the detector is severely limited by the extended tails on the electron beam that result from diffraction and scattering from the final column aperture. Bolon and McConnell (1976) demonstrated that a beam focused through a 20-μm molybdenum aperture resulted in x-ray emission and detection from a nickel support 1 mm away from the center of the beam.

Quinton (1975, 1978a,b) and Rick et al. (1977) have shown that energy-dispersive spectrometry can be used for analysis of liquid droplets. Because of the relatively low signal-to-background ratio inherent to energy-dispersive spectrometry, it was necessary to use a thin film support for the samples in order to reduce background. The technique used by Rick et al. involved placing the samples on collodion films mounted on nickel grids, washing the oil with isopentane and quenching in liquid propane at $-180°C$, followed by freeze-drying.

Quinton (1975, 1978a,b) prepared liquid droplets on thin parlodion films mounted on metal grids without freeze-drying. In order to achieve a thin layer of crystals with good quantitation, however, he found it necessary to add an equal volume of either manganese acetate (1975) or glycerol (1978a,b) to the sample droplets. After the hexadecane overlying the samples was washed with hexane, the samples were flash evaporated at room temperature. Quinton found that, with the exception of samples containing high NaCl concentration, the desiccated droplets appeared as spots of uniform small crystals.

FUTURE OF DROPLET AND CELL PREPARATION AND ANALYSIS

Liquid droplet analysis will likely be applied to many additional problems in biology as more investigators become comfortable with its use. In the future it may be possible to greatly increase the sensitivity of the technique by changing background or beam excitation conditions. Since most of the background derives from the droplet support, we are presently investigating the use of thin film supports, which should markedly reduce this background. Use of thin film supports decreases the minimum elemental detectability. In addition, background may be markedly reduced by using proton beam (Horowitz et al., 1976) or synchrotron radiation (Watson and Perlman, 1978).

Organic compound analysis in liquid droplets will be facilitated by modification of the general techniques of droplet manipulation to take advantage of microfluorescence techniques. Mroz and Lechene (1979) have scaled down fluorescence assays using a capillary tube as a flow-through cell and measuring fluorescence with a microscope-fluorometer. With this method 0.5 mM of urea can be measured in a sample of 20 pl.

We believe that the future of isolated cell analysis (as well as tissue analysis) lies in the preparation and analysis of cells by techniques that maintain them frozen-hydrated. This will minimize translocation of diffusible elements. Gullasch and Kaufmann (1974) demonstrated that analysis of red blood cells and *Tetrahymena pyriformis* was far superior for cells shock frozen in melting nitrogen slush and analyzed at liquid nitrogen temperatures. It is likely that the speed of freezing of the samples will be a critical parameter. Propane and Freon 22 (chlorodifluoromethane) have been shown to be good quenching agents (Rebhun, 1972); however, the former is very explosive and the latter contains fluoride and chlorine, which may interfere with the determination of these elements in the sample. Furthermore, fluoronated hydrocarbons may decompose into stable, corrosive compounds that will not be evacuated by the vacuum of the probe column. We have found that solid nitrogen, which is made by pumping a vacuum over liquid nitrogen in a dewar, is an excellent quenching agent.

After quenching, the cells can be prepared for bulk analysis or for thin section analysis by cryosectioning or cryofracturing. After preparation the cells should remain frozen throughout the analysis. Therefore, the specimen stage must be maintained at very low temperatures. The sample must be introduced into the stage using devices that avoid warming of the sample or exposure to the atmosphere. Cold stages and transfer devices have been described by Conty (1976) for the Camebax Microprobe (Cameca Co., Courbevoie, France), by Saubermann and Echlin (1975) for a modified Cambridge Stereoscan S-4 scanning electron microscope (Cambridge Scientific Instrument Ltd., Cambridge, U.K.), and by Taylor and Burgess (1977) for the J.E.O.L. JXA 50A microanalyzer (J.E.O.L. USA, Inc., Medford, Mass.).

The authors would like to acknowledge the technical assistance of Phillip Clark, Kathy Edgerly, Janet Davis Lehr, Thomas Moher, Khiem Oei, Edith Smith, and Lesley West. Bridget Fox carefully typed the manuscript.

The work from the authors' laboratory described in this chapter has been supported by National Institutes of Health grants RR 00679, AM 19449, and AM 16898. Joseph V. Bonventre is a recipient of a Fellowship from the National Kidney Foundation.

LITERATURE CITED

Agus, Z. A., Chiu, P. J. S., and Goldberg, M. 1977. Regulation of urinary calcium excretion in the rat. Am. J. Physiol. 232:F545.
Agus, Z. A., Gardner, L. B., Beck, L. H., and Goldberg, M. 1973. Effects of parathyroid hormone on renal tubular reabsorption of calcium, sodium and phosphate. Am. J. Physiol. 224:1143.
Anderson, C. A. 1967. An introduction to the electron probe x-ray microanalyzer and its application to biochemistry. In: Methods of Biochemical Analysis, XV (Glick, D., ed.), pp. 147–270.
Anderson, C. A. 1973. Microprobe Analysis. John Wiley and Sons, Inc., New York.
Barrett, E. J., and Coleman, J. R. 1973. Sodium and potassium content of single cells: effects of metabolic and structural changes. Proc. 8th Conf. Electron Probe Analysis, New Orleans, p. 60A.
Baumann, K., Rouffignac, C. de, Roinel, N., Rumrich, G., and Ullrich, K. H. 1975. Renal phosphate transport: inhomogeneity of local proximal transport rates and sodium dependence. Pfluegers Arch. 356:287.
Beaman, D. R., Nishiyama, R. H., and Penner, J. A. 1969. The analysis of blood diseases with the electron microprobe. Blood 34:401.

Beauge, L. A., and Glynn, I. M. 1978. Commercial ATP containing traces of vanadate alters the response of (Na+, K+) ATPase to external potassium. Nature 272:551–552.

Beck, L. H., and Goldberg, M. 1973. Effects of acetazolamide and parathyroidectomy on renal transport of sodium, calcium and phosphate. Am. J. Physiol. 224:1136.

Beeuwkes, R. III, Amberg, J. M., and Essandoh, L. 1977. Urea measurement by x-ray microanalysis in 50 picoliter specimens. Kidney Int. 12:438.

Bengele, H. H., Lechene, C., and Alexander, E. A. 1979. Phosphate transport along the inner medullary collecting duct of the rat. Am. J. Physiol. 237:F48–F49.

Biggers, J. D., Borland, R. M., and Lechene, C. P. 1978. Ouabain-sensitive fluid accumulation and ion transport by rabbit blastocysts. J. Physiol. 280:319.

Birks, L. S. 1969. *X-Ray Spectrochemical Analysis*. John Wiley and Sons, Inc., New York.

Bolon, R. B., and McConnell, M. D. 1976. Evaluation of electron beam tails and x-ray spatial resolution in the SEM. Scanning Electron Microscopy. IIT Research Institute, Chicago, Ill. 163.

Bonventre, J. V., and Lechene, C. 1974. A method for electron probe microanalysis of organic components in picoliter samples. Proc. 9th Microbeam Soc. Ottawa, p. 8A.

Bonventre, J. V., Roman, R., and Lechene, C. 1979. Effect of vanadate on renal sodium and water excretion. Fed. Proc. 38:1042.

Borland, R. M. 1977. Transport processes in the mammalian blastocyst. In: *Development in Mammals*, Vol. 1 (Johnson, M.H., ed.), p. 31, North-Holland Publ. Co., Amsterdam.

Borland, R. M., Biggers, J. D., and Lechene, C. P. 1977a. Fluid transport by rabbit preimplantation blastocysts in vitro. J. Reprod. Fert. 51:131.

Borland, R. M., Hazra, S., Biggers, J. D., and Lechene, C. P. 1977. The elemental composition of the environments of the gametes and preimplantation embryo during the initiation of pregnancy. Biol. Reprod. 16:147.

Cantley, L. C., Resh, M. D., and Guidotti, G. 1978. Vanadate inhibits the red cell (Na+, K+) ATPase from the cytoplasmic side. Nature 272:552.

Carrol, K. G., and Tullis, J. L. 1968. Observations on the presence of titanium and zinc in human leucocytes. Nature 217:1172.

Cass, A., and Dalmark, M. 1973. Equilibrium dialysis of ions in nystatin-treated red cells. Nature New Biol. 244:47.

Castaing, R. 1951. Application des sondes electroniques a une methode d'analyse ponetvelle chimique et crystallographique. Ph.D. thesis, University of Paris, Onera No. 55, 92 pp.

Chandler, J. A., and Battersby, S. 1976. X-ray microanalysis of ultrathin frozen and freeze-dried sections of human sperm cells. J. Microscopy 107:55.

Chong, A. P., Taymor, M. L., and Lechene, C. P. 1977. Electron probe microanalysis of the chemical elemental content of human follicular fluid. Am. J. Obstet. Gynecol. 128:209.

Coleman, J. R., Nilsson, J. R., Warner, R. R., and Batt, P. 1972. Qualitative and quantitative electron probe analysis of cytoplasmic granules in *Tetrahymena pyriformis*. Exp. Cell. Res. 74:207.

Coleman, J. R., Nilsson, J. R., Warner, R. R., and Batt, P. 1973. Electron probe analysis of refractive bodies in *Amoeba proteus*. Exp. Cell. Res. 76:31.

Coleman, J. R., Nilsson, J. R., Warner, R. R., and Batt, P. 1973. Effects of calcium and strontium on divalent ion content of refractive granules in *Tetrahymena pyriformis*. Exp. Cell. Res. 80:1.

Colindres, R. E., and Lechene, C. 1972. Technical problems associated with collection of distal tubular fluid in the rat. Yale J. Biol. Med. 45:233.

Colvin, J. R., Sowden, L. C., and Male, R. S. 1975. Variability of the iron, copper and mercury contents of individual red blood cells. J. Histochem. Cytochem. 23:329.

Compton, A. H., and Allison, S. K. 1935. *X-Rays in Theory and Experiment.* Van Nostrand Company, Inc., Princeton, N.J.

Conty, C. 1976. Mouvement-object refroidi pour applications biologiques et physiques. J. Microsc. Spectrosc. Electron 1:475.

Cortney, M. A. 1969. Renal transfer of water and electrolytes in adrenalectomized rats. Am. J. Physiol. 216:589.

Davis, D. 1978. Analysis of iron content in individual human red blood cells by electron microprobe and scanning electron microscope. Micron 9:175.

Fiori, E. C., Myklebust, R. L., Heinrich, K. F. J., and Yakowitz, H. 1976. Prediction of continuum intensity in energy-dispersive x-ray microanalysis. Anal. Chem. 48:172.

Garland, H. O., Henderson, I. W., and Brown, J. A. 1975. Micropuncture study of the renal responses of the urodele amphibian *Mecturus muculosus* to injections of arginine vasotocin and an anti-aldosterone compound. J. Exp. Biol. 63:249.

Geller, J. 1977. A comparison of minimum detection limits using energy and wavelength dispersive spectrometers. Scanning Electron Microscopy I:281. IIT Research Inst., Chicago.

Goldstein, J. I., and Yakowitz, H. 1975. Practical Scanning Electron Microscopy. Plenum Press, New York.

Grantham, J. J., Balfour, W. E., and Glynn, I. M. 1978. Vanadate: A potent inhibitor of renal (Na + K)ATPase and a powerful diuretic in unanesthetized rats. Kidney Int. 14:760 (abst.).

Greaves, H. 1974. Energy dispersion x-ray analysis by scanning electron microscopy for measuring cellular elemental composition in bacterial cells. App. Microb. 27:609.

Greger, R. F., Lang, F. C., Knox, F. G., and Lechene, C. P. 1977. Absence of significant secretory flux of phosphate in the proximal convoluted tubule. Am. J. Physiol. 232:F235.

Greger, R., Lang, F., Knox, F. G., and Lechene, C. P. 1978. Analysis of tubular fluid. In: *Methods in Pharmacology* (Martinez-Maldonado, M., ed.), p. 105. Plenum Publishing Corp., New York.

Gullasch, J., and Kaufman, R. 1974. Energy-dispersive x-ray microanalysis in soft biological tissues: reliance and reproducibility of the results as depending on specimen preparation. (Air drying, cryofixation, cool-stage techniques.) In: *Microprobe Analysis as Applied to Cells and Tissues* (Hall, T., Echlin, P., and Kaufmann, R., eds.), p. 175. Academic Press, New York.

Hales, C. N., Luzio, J. P., Chandler, J. A., and Herman, L. 1974. Localization of calcium in the smooth endoplasmic reticulum of rat isolated fat cells. J. Cell Sci. 15:1.

Harris, C. A., Burnatowska, M. A., Seely, J. F., Sutton, R. A. L., Quamme, G. A., and Dirks, J. H. 1979. Effects of parathyroid hormone on electrolyte transport in the hamster nephron. Am. J. Physiol. 236:F342.

Horowitz, P., Aronson, M., Grodzins, L., Ladd, W., Ryan, J., Merriam, G., and Lechene, C. 1976. Elemental analysis of biological specimens in air with a proton microprobe. Science 194:1162.

Howards, S., Lechene, C., and Vigersky, R. 1979. The fluid environment of the maturing spermatozoon. In: *The Spermatozoon* (Fawcett, D., and Bedford, J. M., eds.), p. 35. Urban & Schwarzenberg, Baltimore.

Ingram, M. J., and Hogben, C. A. M. 1967. Electrolyte analysis of biological fluids with the electron microprobe. Anal. Biochem. 18:54.

Ingram, F. D., Ingram, M. J., and Hogben, A. M. 1974. An analysis of the freeze-dried, plastic embedded electron probe specimen preparation. In: *Microprobe Analysis as Applied to Cells and Tissues* (Hall, T., Echlin, P., and Kaufmann, R., eds.), p. 119. Academic Press, New York.

Kimzey, S. L., and Burns, L. C. 1973. Electron probe microanalysis of cellular potassium distribution. Ann. N.Y. Acad. Sci. 204:486.

Kirk, R. G., Bonner, C., Barba, W., and Tosteson, D. C. 1978a. Electron probe microanalysis of red blood cells. I. Methods and evaluation. Am. J. Physiol. 235:C245.

Kirk, R. G., Lee, P., and Tosteson, D. C. 1978b. Electron probe microanalysis of red blood cells. II. Cation changes during maturation. Am. J. Physiol. 235:C251.

Kirk, R. G., Crenshaw, M. A., and Tosteson, D. C. 1974. Potassium content of single human red cells measured with an electron probe. J. Cell. Physiol. 84:29.

Knox, F. G., Haas, J. A., and Lechene, C. P. 1976. Effect of parathyroid hormone on phosphate reabsorption in the presence of acetazolamide. Kidney Int. 10:216.

Knox, F. G., and C. Lechene. 1975. Distal site of action of parathyroid hormone on phosphate reabsorption. Am. J. Physiol. 229:1556.

Kotrba, Z. 1977. The limit of detectability in x-ray electron probe microanalysis. Mikrochim. Acta II:97.

Lechene, C. 1970. The use of the electron microprobe to analyze very minute amounts of liquid samples. Proc. 5th Conf. Electron Probe Analysis, p. 32A. New York.

Lechene, C. 1974. Electron probe microanalysis of picoliter liquid samples. In: *Microprobe Analysis Applied to Cells and Tissues* (Hall, T., Echlin, P., and Kaufmann, R., eds.), p. 351. Academic Press, New York.

Lechene, C., Bonner, C., and Kirk, R. G. 1974. Electron probe microanalysis of isolated cells: Red blood cells. Proc. 9th Conf. of the Microbeam Soc., Ottawa, p. 9A.

Lechene, C. P., Bonner, C., and Kirk, R. G. 1977a. Electron probe microanalysis of chemical elemental content of single human red cells. J. Cell. Physiol. 90:117.

Lechene, C., Colindres, R. E., and Knox, F. G. 1977b. Electron probe microanalysis of the renal effect of parathyroid hormone. Excerpta Med. 421:230.

Lechene, C., Smith, E., and Blouch, K. 1974. Site of sulfate reabsorption along the rat nephron. Kidney Int. 6:64A. (Abst.)

Lechene, C. P., and Warner, R. R. 1977. Ultramicroanalysis: x-ray spectrometry by electron probe excitation. Ann. Rev. Biophys. Bioeng. 6:57.

Lechene, C., and Warner, R. R. 1979. Electron probe analysis of liquid droplets. In: *Microbeam Analysis in Biology* (Lechene, C., and Warner, R. R., eds.), p. 279. Academic Press, New York.

Le Grimellec, C. 1975. Micropuncture study along the proximal convoluted tubule. Electrolyte reabsorption in first convolutions. Pfluegers Arch. 354:133.

Le Grimellec, C., Poujeol, P., and Rouffignac, C. de. 1975. ^{3}H-Inulin and electrolyte concentrations in Bowman's capsule in rat kidney. Pfluegers Arch. 354:117.

Le Grimellec, C., Roinel, N., and Morel, F. 1973a. Simultaneous Mg, Ca, P, K, Na and Cl analysis in rat tubular fluid. I. During perfusion of either inulin or ferrocyanide. Pfluegers Arch. 340:181.

Le Grimellec, C., Roinel, N., and Morel, F. 1973b. Simultaneous Mg, Ca, P, Na and Cl analysis in rat tubular fluid. II. During acute Mg plasma loading, Pfluegers Arch. 340:197.

Le Grimellec, C., Roinel, N., and Morel, F. 1974a. Simultaneous Mg, Ca, P, K, Na and Cl analysis in rat tubular fluid. III. During acute Ca plasma loading. Pfluegers Arch. 346:171.

Le Grimellec, C., Roinel, N., and Morel, F. 1974b. Simultaneous Mg, Ca, P, K and Cl analysis in rat tubular fluid. IV. During acute phosphate plasma loading. Pfluegers Arch. 346:189.

MacKenzie, A. P. 1965. Factors affecting the mechanism of transformation of ice into water vapor in the freeze-drying process. Ann. N.Y. Acad. Sci. 125:522.

MacKenzie, A. P. 1969. Apparatus for the partial freezing of liquid nitrogen for the rapid cooling of cells and tissues. Biodynamica 10:341.

Maynard, P. V., Elstein, M., and Chandler, J. A. 1975. The effect of copper on the distribution of elements in human spermatozoa. J. Reprod. Fert. 43:41.

Messing, R. A. 1976. Adsorption and inorganic bridge formations. In: *Methods in Enzymology*, Vol. 44. Immobilized Enzymes (Mosbach, K., ed.), p. 148. Academic Press, New York.

Moher, T., and Lechene, C. 1975. Automated electron probe analysis of biological samples. Biosci. Commun. 1:314.

Morel, F., and Roinel, N. 1969. Application de la microsonde electronique a l'analyse elementaire quantitative d'echantillons liquides d'un volume inferieur a 10^{-9}l. J. Chim. Phys. 66:1084.

Morel, F., Roinel, N., and Le Grimellec, C. 1969. Electron probe analysis of tubular fluid composition. Nephron 6:350.

Moseley, J. G. J. 1913. The high-frequency spectra of the elements. Phil. Mag. 26:1024.

Mroz, E. A., and Lechene, C. 1979. Fluorescence analysis of nanoliter volumes of liquid droplets. Fed. Proc. 38:1046.

Nei, T. 1962. Electron microscope study of microorganisms subjected to freezing and drying: cinematographic observations of yeast and coli cells. Exp. Cell Res. 28:560.

Osborn, D., and Hamilton, T. C. 1977. Electron microbeam analysis of calcium distribution in the ciliated protozoan, *Spirostonum ambiguum*. J. Cell. Physiol. 91:409.

Pastoriza-Munoz, E., Colindres, R. E., Lassiter, W. E., and Lechene, C. 1978. Effect of parathyroid hormone on phosphate reabsorption in rat distal convolution. Am. J. Physiol. 235:F321.

Peterson, S. K., Frishkopf, L. S., Lechene, C., Oman, C. M., and Weiss, T. F. 1978. Elemental composition of inner ear lymphs in cats, lizards, and skates by electron probe microanalysis of liquid samples. J. Comp. Physiol. 126:1.

Prayer, D. J., Bowan, R. L., and Vurek, G. G. 1965. Constant volume, self-filling nanoliter pipette: Construction and calibration. Science 147:606.

Quamme, G. A., Wong, N. L. M., Dirks, J. H., Roinel, N., Rouffignac, D. de, and Morel, F. 1978. Magnesium handling in the dog kidney: A micropuncture study. Pfluegers Arch. 377:95.

Quinton, P. M. 1975. Energy dispersive x-ray analysis of picoliter samples of physiological fluids. Proc. 10th Ann. Conf. of Microbeam Analysis Society, Las Vegas, p. 50A.

Quinton, P. M. 1976. Construction of picoliter-nanoliter self-filling volumetric pipettes, J. Appl. Physiol. 40:260.

Quinton, P. M. 1978a. Techniques for microdrop analysis of fluids (sweat, saliva, urine) with an energy-dispersive x-ray spectrometer on a scanning electron microscope. Am. J. Physiol. 234:F255.

Quinton, P. M. 1978b. Ultramicroanalysis of biological fluids with energy dispersive x-ray spectrometry. Micron 9:57.

Quinton, P. M. 1979. Energy dispersive x-ray analysis of biological microdroplets. In: *Microbeam Analysis in Biology* (Lechene, C., and Warner, R. R., eds.), pp. 327–346. Academic Press, New York.

Quinton, P. M., Warner, R. R., and Lechene, C. 1979. Minimum detectable concentrations with the liquid droplet technique. In: *Microbeam Analysis 1979* (Newbury, D. E., ed.), p. 73. San Francisco Press.

Rebhun, L. I. 1972. Freeze-substitution and freeze-drying. In: *Principles and Techniques of Electron Microscopy:* Biological Applications, Vol. 2 (Hayat, M. A., ed.). Van Nostrand Reinhold, New York.

Reed, S. J. B. 1975a. *Electron Microprobe Analysis.* Cambridge University Press, Cambridge.

Reed, S. J. B. 1975b. The shape of the continuous x-ray spectrum and background corrections for energy-dispersive electron microprobe analysis. X-ray Spectrometry 4:14.

Reeves, A. L. 1977. Beryllium in the environment. Clin. Toxicol. 10:37.

Rick, R., Horster, M., Dörge, A., and Thurau, K. 1977. Determination of elecolytes in small biological fluid samples using energy dispersive x-ray microanalysis. Pfluegers Arch. 369:95.

Roblero, L., Biggers, J. D., and Lechene, C. P. 1976. Electron probe analysis of the elemental microenvironment of oviductal mouse embryos. J. Reprod. Fert. 46:431.

Rochow, E. G. 1951. *An Introduction to the Chemistry of Silicones.* John Wiley and Sons, Inc., New York.

Roinel, N. 1975. Electron microprobe quantitative analysis of lyophilised 10^{-10}l volume samples. J. Microscopie Biol. Cell. 22:261.

Roinel, N., and Passow, H. 1974. A study of the applicability of the electron microprobe to a quantitative analysis of K and Na in single human red blood cells. FEBS Lett. 41:81.

Roomans, G. M., and Seveus, L. A. 1976. Subcellular localization of diffusible ions in the yeast *Saccharomyces cerevisiae:* quantitative microprobe analysis of thin freeze-dried sections. J. Cell. Sci. 21:119.

Rouffignac, C. de, Morel, F., Moss, N., and Roinel, N. 1973. Micropuncture study of water and electrolyte movements along the loop of Henle in *Psammomys* with special reference to magnesium, calcium and phosphorus. Pfluegers Arch. 344:309.

Saubermann, A. J., and Echlin, P. 1975. The preparation, examination and analysis of frozen hydrated tissue sections by scanning transmission electron microscopy and x-ray microanalysis. J. Microscopy 105:155.

Shipp, J. C., Hanenson, I. R., Windhager, E. E., Schatzman, H. J., Whittenburg, G., Yoshimura, H., and Solomon, A. K. 1958. Single proximal tubules of the *Necturus* kidney. Methods for micropuncture and microperfusion. Am. J. Physiol. 195:563.

Shirley, D. G., Poujeol, P., and Le Grimellec, C. 1976. Phosphate, calcium and magnesium fluxes into the lumen of the rat proximal convoluted tubule. Pfluegers Arch. 362:247.

Shuman, H., and Somlyo, A. P. 1976. Electron probe x-ray analysis of single ferritin molecules. Proc. Nat. Acad. Sci. 73:1193.

Siegel, B. M., and Beaman, D. R. 1975. *Physical Aspects of Electron Microscopy and Microbeam Analysis.* John Wiley and Sons, Inc., New York.

Sjöstrand, F. S., and Elfvin, L. -G. 1964. The granular structure of mitochondrial membranes and of cytomembranes as demonstrated in frozen-dried tissue. J. Ultrastruct. Res. 10:263.

Skadhauge, E., Lechene, C., and Maloiy, G. M. O. 1979. *Tilapia grahami:* Role of intestine in osmoregulation under conditions of extreme alkalinity. In: *Epithelial Transport in Lower Vertebrates* (Lahlou, B., ed.), pp. 133–142. Cambridge University Press, London. In press.

Southworth, D., Fisher, K., and Branton, D. 1975. Principles of freeze-fracturing and etching. In: *Techniques of Biochemical and Biophysical Morphology,* Vol. 2 (Glick, D., and Rosenbaum, R. M., eds.), p. 247. John Wiley and Sons, Inc., New York.

Stolte, H., Galaske, R. G., Eisenbach, G. M., Lechene, C., Schmidt-Nielson, B., and Boylan, J. W. 1977. Renal tubule ion transport and collecting duct function in the elasmobranch little skate, *Raja enrinacea.* J. Exp. Zool. 199:403.

Taylor, P. G., and Burgess, A. 1977. Cold stage for electron probe microanalyser. J. Microscopy 111:51.

Timourian, H., Jotz, M. M., and Clothier, G. E. 1974. Intracellular distribution of calcium and phosphorus during the first cell division of the sea urchin egg. Exp. Cell Res. 83:380.

Van Kampen, E. J., and Zijlstra, W. G. 1961. Standardization of hemoglobinometry II. The hemoglobin cyanide method. Clin. Chim. Acta 6:538.

Warner, R. R., and Lechene, C. 1976. Electron probe analysis of limiting transepithelial inorganic ion concentration differences across the rat proximal tubule. Proc. 11th Conf. of the Microbeam Analysis Soc., Miami, p. 59A.

Warner, R. R., and Lechene, C. 1979. Analysis of standing droplets in rat proximal tubules. I. Na, Cl, and raffinose limiting concentrations, solvent drag and active transport. J. Gen. Physiol. (Submitted for publication.)

Watson, R. E., and Perlman, M. L. 1978. Seeing with a new light: Synchrotron radiation. Science 199:1295.

Chapter 9

QUANTITATIVE X-RAY MICROANALYSIS OF BULK SPECIMENS

F. Duane Ingram and Mary Jo Ingram

The Cardiovascular Center, College of Medicine,
The University of Iowa,
Iowa City, Iowa

INTRODUCTION

Localized intracellular electrolyte concentrations are of interest in various aspects of study of biological function. Average intracellular electrolyte concentrations can be estimated from gross tissue specimens using standard techniques with extracellular space markers, such as radioinulin, and some extrapolated estimations can be obtained using intracellular ion activity determined with ion-selective electrodes. In addition, histological techniques are available for localizing certain elements by electron microscopy and by fluorescence microscopy. The only means, however, of combining analytical capabilities for nearly all the elements with the spatial localization of electron microscopy requires electron beam instruments employing x-ray detection devices. This chapter presents procedures for quantitatively measuring intracellular electrolyte concentrations using such equipment.

Fluids, soft tissues, such as cells and connective tissue, and hard tissues, such as bones and teeth, can be studied with the electron probe microanalyzer. Hard tissues present different problems for the analyst and are not dealt with here. Soft biological tissue samples for electron-probe analysis fall into three basic categories: 1) tissue sections thin or transparent to the electron beam; 2) thick tissue sections in which the electron beam is severely attenuated by the sample; or 3) solid samples infinitely thick or opaque to the electron beam. Characteristics, advantages, and methods of quantitation are different for each type sample. Other chapters in this book address themselves to analysis of fluids and to thin section analysis. The present chapter is devoted to a discussion of quantitation with embedded soft tissue samples infinitely thick to the electron beam, sometimes referred to as bulk samples. It relies heavily on analytical methods we have developed for use with an electron probe microanalyzer equipped with three wavelength spectrometers (Ingram and Ingram, 1975).

SPECIMEN PREPARATION

Since electron column instruments require a highly modified, dried, or otherwise fixed specimen that is incompatible with life, the treatment of sample preparation is central to any treatise on electron probe methodology. It is also the most controversial aspect of soft tissue electron probe work. This is particularly true for work involving measurements of soluble constituents.

Quantitative electrolyte measurements present some of the most difficult problems for an electron probe analyst. The solubility and mobility of the various electrolytes force stringent requirements on the work. Because each sample preparation procedure has particular advantages and each has special demands for equipment and technical expertise, the analyst must carefully examine the problem to be undertaken and select a preparation regimen that is within the realm of possibility with his equipment constraints, yet capable of providing tissue in the proper state for analysis.

Until analysis of hydrated frozen samples has been perfected to the point where it is practical on a routine basis, extracellular space measurements must be limited to those regions where fluid can be collected for microdrop analysis. For instances in which spatial resolution is of utmost importance, thin section analysis is mandated. Bulk sample analysis can be chosen for studies in which spatial resolution is of secondary importance to straightforward quantitative intracellular concentration measurements.

Besides the freeze-dried, plastic-embedded tissue preparation, other preparations that have been used for electrolyte analysis include frozen-hydrated tissue, unembedded freeze-dried tissue, and tissue prepared by freeze-substitution (see the chapters by Marshall in this book).

The principal advantage provided by the hydrated ice specimen lies in the promise of obtaining meaningful extracellular measurements of electrolyte content. The hydrated ice sample also appears at first glance to be the simplest preparation, in that it avoids both the drying and the plastic embedding common to most other bulk tissue preparations. It does, however, require a highly specialized cold stage for the electron probe. In addition, special precautions must be adopted in handling and analysis that will not allow the sample to freeze-etch in the vacuum or to be etched by the electron beam. The necessary work has not yet been done to determine what stability can be expected from the ice sample in the electron beam environment.

Imaging of the specimen in ice is difficult unless some freeze-etching is permitted to provide surface relief. This etching must be carefully controlled, because it is done at the expense of spatial resolution and complicates quantitation. Ideally, a method can be developed to allow for highly accurate measurements to be routinely made with reasonable spatial resolution.

At present, thin sections appear to offer the most promise of success for work with hydrated ice where modest spatial resolution is needed. No preparation regimen is without controversy, however, and serious questions remain unanswered about cryosectioned tissue. Tormey (1978) has presented evidence that cryosectioning can cause translocation of soluble substances. In his studies he has demonstrated a contamination of intracellular spaces with extracellular material. Presumably, material is wiped across the sample by the knife during cryosectioning. Saubermann (1979) presented data that reveal that more energy is required to cryosection a sample than should be expected to only cut through the sample. This excess energy is puzzling, since it is of sufficient magnitude to melt the entire specimen. Anyone embarking on cryosectioning is thus advised to carefully assess the preparation and to adopt criteria for ensuring that significant cutting artifacts do not compromise the preparation.

Specimens can be freeze-dried and analyzed in the dried state. This avoids plastic embedding, but because of sample nonuniformity and low overall density, spatial resolution is difficult to control and is about a factor of five poorer than with embedding. Concentration measurements are in terms of moles per unit dry mass, units that are sometimes difficult to relate back to wet weight concentrations. The most success with unembedded freeze-dried tissue has been realized with cryosectioned and freeze-dried thin sections. Procedures for thin-section quantitative analysis are handled by Roomans in this book.

Critical point drying has enjoyed wide acceptance as a preparation for scanning electron microscopy. It is not useful when measurements are to be made for electrolytes, however, because chemical fixation and the various aqueous steps are not compatible with maintenance of natural distributions of diffusible substances.

Freeze-substitution has also been used as an alternative to freeze-drying (Spurr, 1972). At low temperatures, tissue ice is replaced with a chemical, such as diethyl ether. The sample is subsequently warmed and embedded. This method has been used for plant tissues, which are freeze-dried only with extreme difficulty. Lechene and Warner (1977) tested freeze-substitution on microdrops and found that electrolytes were washed away with the substitution medium. This casts serious doubt on the advisability of using freeze-substitution as a routine preparation. For animal tissue preparations that can be treated adequately with freeze-drying there is no reason to use this questionable method (see also Marshall in this book).

The freeze-dried, plastic-embedded tissue sample involves what some investigators interpret as an unnecessary and compromising step—the introduction of an embedding medium. Embedding tissue imparts some highly desirable characteristics to the sample, however, and the plastic matrix does not appear to interfere with intracellular electrolyte measurements. Although extracellular material remains and does not appear to be translocated across cell boundaries, lack of a protein matrix in extracellular spaces of most tissue types results in nonhomogeneous distributions of constituents in these spaces. Thus, only in special cases are extracellular measurements permissible with freeze-dried, plastic-embedded tissue. This failure to preserve natural electrolyte distributions in extracellular spaces is used by some investigators as a reason to reject the embedded tissue preparation (Lechene and Warner, 1977). However, because cell boundaries do not appear to be violated (Coulter and Terracio, 1977), this is not sufficient reason for rejecting the preparation for intracellular measurements. The reader must be cautioned, though, that unless freeze-drying is carried out in a controlled manner at low temperatures, and unless tissue is embedded in an appropriate medium, the investigator may find an unnatural distribution of ions.

The principal advantages of using an embedding medium result from the increase in sample density, the increase in sample uniformity, and the variety of ways the embedded sample can be prepared. The method also provides some advantages for low-budget operations because of the relatively modest special equipment requirements for the preparation, the most expensive and critical non-standard item being the freeze-dry apparatus.

The increase in density provided by the plastic has the very desirable effect of limiting electron diffusion in the sample, thus increasing spatial resolution, and the uniform density imparted to the sample by the plastic simplifies quantitation and provides more predictable spatial resolution than would be the case if the thick sample were not embedded. The versatility of the embedded sample for options of sectioning and staining in study of morphology, and its options for mode of presentation to the electron probe, are among its more useful attributes. Thin sections can be mounted for optical microscopy or ultrathin sections can be collected for electron microscopy. If collected dry, the thin sections can be analyzed in the electron probe using techniques described elsewhere in this book. For studies in which spatial resolution is not as important as the collection of accurate quantitative data from a relatively large number of cells, it is most convenient to present the face of the sample block to the electron beam and analyze the sample as a thick sample. Plastic may be removed from the sample with sodium methoxylate for scanning electron microscopy of the block face (Erlandsen et al., 1973). Of course, the block is no longer useful for x-ray analysis after such treatment.

Before deciding upon the specimen preparation regimen to adopt for a given study, each investigator must examine his laboratory, his equipment, the type of biological problem he faces, and the type and amount of technical skill available. There are a number of directions that can be taken with the preparation, but they have not all been thoroughly evaluated. Unfortunately, this cannot be done in any one laboratory. Rather, the advocates of each of the various preparation procedures should be given an opportunity to prepare samples of the same specimen and have the results compared. Until controlled comparisons are made, the reader must always examine very carefully the critical remarks any given author makes about techniques that that author has not mastered in his own laboratory.

Tissue Collection

One of the most crucial aspects of tissue preparation is specimen collection itself. The precise protocol for tissue collection must be worked out for each type of tissue. There are, however, two general principles that should be observed: 1) specimens must contain complete, intact cells, and 2) excised tissue should be in the desired physiologic state when frozen.

It is not reasonable to expect that small fragments of muscle fibers or long nerve processes will faithfully retain natural electrolyte distributions representative of viable, intact cells. Kidney tissue that has been diced before shock freezing will contain collapsed tubules and be in an unnatural state when processed. For such unique problems, special methods of tissue collection must be devised. Surprising or unusual results that are attributed to the use of a controversial preparation technique may actually result from the choice of a poor tissue collection regimen.

Shock Freezing

Many of the currently available methods for tissue preparation require special skills, equipment, and procedures. Not all procedures appear to work as well for one investigator as they do for another, and consequently a procedure is sometimes discredited when it is not so much the fault of the procedure as it is the inability of a given investigator-equipment combination to make things work.

As an example, there is general agreement that the necessary first step to a preparation that will most faithfully preserve the natural distribution of soluble constituents is shock freezing the tissue. There is, however, substantial disagreement concerning the best methods for accomplishing this. Some groups prefer supercooled liquid nitrogen, some prefer chilled propane, some prefer chilled freon, and others insist that it is necessary to impact the tissue sample against the polished face of a cooled copper block.

To be useful, a freezing medium should have the following properties: 1) it should be colder than about $-130°C$, 2) it should have a high heat capacity, 3) it should have high thermal conductivity, 4) it should be nontoxic, and 5) it should be safe to handle.

In our laboratory we have found that although results are somewhat variable, quench freezing in either chilled liquid propane or supercooled liquid nitrogen can produce samples exhibiting excellent morphology observable by electron microscopy. Propane has better thermal conductivity than liquid nitrogen, and it has excellent wetting qualities. It must be treated with caution, however, since the propane vapor is highly flammable. Also, liquid oxygen can condense from room air and form an explosive mixture with the propane (Stephenson, 1954).

Liquid nitrogen cannot be used without supercooling, since a warm object will produce boiling of the nitrogen, forming an insulating layer of gas about the sample that impedes further heat transfer. Supercooled liquid nitrogen allows quench freezing of tissue without raising the liquid temperature to the boiling point. Liquid nitrogen does not have the thermal conductivity of copper or liquid propane, but it does have adequate conductivity to serve well as a quenching medium for electron probe bulk sample preparations.

Supercooling of liquid nitrogen is accomplished by pumping on a container of liquid nitrogen with a vacuum pump. Continued pumping can result in solidification of all or a substantial portion of the nitrogen. We find that about a liter of nitrogen in a 15 cm diameter styrofoam container serves as a convenient amount of fluid for supercooling. If too small a quantity, e.g., 100 ml, is used, it will probably all be pumped away before the temperature of the liquid is stabilized.

With the onset of vacuum pumping, a scum of solid nitrogen will form rather quickly on the surface, before the bulk of fluid has been cooled. This scum immediately disappears upon venting the chamber to atmosphere. With longer pumping times almost the entire bulk of liquid nitrogen can be solidified at $-210°C$. Such a bulk will remain at a temperature below $-205°C$ for 5 to 10 min, allowing ample time for freezing a few tissue specimens. Care must be taken that the fluid is well stirred, since solid nitrogen can remain in the bottom of the container after the top portion of fluid has heated to an unusable temperature, near the boiling point of liquid nitrogen, $-196°C$. Also, when liquid nitrogen is supercooled it adopts a cloudy appearance that is retained even after it is warmed to $-196°C$, an appearance that might be mistaken for evidence of a slush. Because it is difficult to know the temperature of liquid nitrogen by its appearance, it is generally considered safest to monitor the temperature with a suitable thermocouple.

The object of shock freezing is to obtain the highest possible freezing rates within tissue. Ideally, every portion of the specimen is cooled so rapidly that no ice crystals form and tissue water is transformed into a vitreous state. Shock freezing of tissue for preservation of morphology cannot be compared with freezing under conditions that allow tissue to remain viable when thawed. Shock freezing faithfully preserves tissue electrolytes as they were in the natural state before freezing, whereas for tissue to remain viable after thawing, it must contain a cryoprotectant and must be frozen slowly. In the frozen state, such tissue cannot be expected to contain electrolyte distributions and concentrations representative of functioning cells (Sjöstrand and Kretzer, 1975).

The rate of freezing is the most important factor determining quality of preservation of tissue morphology. For exterior portions of the specimen, the rate of freezing is dependent upon how effectively a given tissue type can be brought into intimate contact with the freezing medium and how rapidly that freezing medium can extract heat from the specimen. Since all heat from interior portions of the sample must be transferred through intervening tissue, the thermal conductivity of tissue is the limiting factor for determining freezing rates in interior regions when copper or liquid propane is used. Since the common shock-freezing media have adequate thermal capacity and adequately high thermal conductivity, quality of preservation in interior regions of tissue should be fairly independent of freezing medium.

Tissue samples should be fragments of the smallest practical size, ideally no larger than a fraction of a millimeter in one dimension, and some means should be devised for handling the sample and quickly bringing it into intimate contact with the freezing medium. We normally obtain good results by placing the tissue sample on a small strip of aluminum foil and plunging the foil into the freezing medium with a pair of forceps. The sample is maintained in constant motion in the fluid until freezing is accomplished. We sometimes observe the best preservation of morphology on the side of the specimen that had been in contact with the aluminum foil, indicating the importance of high thermal conductivity for the quenching medium–sample interface.

We find the cooled copper block to be more convenient, and it yields superior results in freezing large pieces of membranes, such as epithelium, from flux chamber experiments. Spreading the tissue on the end of a cork allows even pressure to be exerted over a large area, and tissue can be removed easily if the cork end is covered with aluminum foil. The copper block face must be polished with a fine abrasive before each use.

Freeze-Drying

Freeze-drying is performed in a vacuum, with the samples maintained at a higher temperature than a nearby cold surface, which serves as the condenser. The function of the vacuum is to provide a free path for water vapor to travel from sample to condenser. The distance between sample and condenser is then dictated by the vacuum capabilities of the system. We find a 2×10^{-5} torr vacuum to be adequate for a sample-to-condenser distance of about one centimeter. A representative list of works that contain critical examinations of freeze-dried tissue preparation procedures for electron microscopy would include: Hanzon and Hermodsson (1960), Staub and Storey (1962), and Coulter and Terracio (1977). In these studies spatial resolution requirements placed more stringent demands on preservation of tissue morphology than is necessary with thick-section electron probe analysis. Everything these authors present concerning apparatus design for electron microscope preparations holds for electron microprobe preparations as well. The only addition that is useful for electron microprobe preparations is providing some method for separately identifying a number of different samples in the freeze-dry apparatus. This is necessary because the electron probe preparations often involve comparison of samples of the same tissue that have been exposed to a variety of experimental conditions.

Tissue is manipulated and mounted in the freeze-dry apparatus under liquid nitrogen. Freeze-drying begins as the sample holder in the freeze-dry apparatus is allowed to warm to its equilibrium temperature overnight. With liquid nitrogen cooling the wall that serves as condenser, this equilibrium temperature is normally $-90°C$ to $-110°C$. The tissue is then warmed to $-80°C$ and held at this temperature for an hour to a day or more, depending upon sample size. Small samples that are 0.5 mm or less in one dimension and that will dry quickly are warmed in $10°C$ increments once each hour during the day, for complete drying in one day. They may in fact have been dry before the slow warming was commenced. Large tissue blocks may be held at low temperatures for a few days to ensure drying.

Acceptable preservation of tissue morphology, as revealed by light microscopy, is found when tissue samples are maintained at temperatures below $-70°C$ until drying is complete. Drying temperatures warmer than $-50°C$ result in clearly inferior preparations. Slow warmup after drying is adopted as a conservative measure to insure against tissue disruption resulting from too rapid release of any remaining water. It is not clear that this slow warmup is necessary, since there have been instances of equipment failure, with accompanying rapid temperature rise, and no sample damage. Premature warmup before drying is complete, however, is a disaster.

Freeze-dried tissue has a white, and in some places a very light pinkish, appearance. Dark coloring at this stage is normally indicative of a poor preparation.

Fixation

Osmium vapor fixation of freeze-dried tissue is considered necessary to ensure proper embedding of specimens to be used for electron microscopy (Hanzon and Hermodsson, 1960). For bulk electron microprobe specimens that are not examined as closely as an electron microscope preparation, we have found osmium fixation to be helpful, but not essential, for improving ease of sectioning. We find that gluteraldehyde vapor can be substituted for osmium tetroxide as a fixative, or fixation can be eliminated altogether.

Osmium fixation is desirable for electron probe preparations in general, however, because it provides the tissue with necessary contrast to allow poor but adequate imaging of tissue morphology during analysis. Osmicated tissue sections can be stained with azure blue for light microscopy, or with conventional agents for electron microscopy (Hayat, 1980).

In our laboratory we find it is convenient to place the dried tissue in a desiccator with a 0.25 g vial of osmium tetroxide crystals, to evacuate the desiccator, and to seal it off. Osmication appears to be complete in about 4 hr, with little change taking place after that. Because of hazards, primarily to the eyes, all work with osmium should be done under a fume hood.

Ideally, osmium vapor should be introduced to samples while they are still in the original vacuum of the freeze-dry apparatus. This is not possible with our equipment, and we must briefly expose dried tissue samples to room air while transferring from freeze-dry apparatus to desiccator and back. Care must be taken in this transfer, because dried samples can absorb water, thereby destroying the preparation.

Osmium tetroxide imparts a greyish cast to the tissue. Apparently, fixation will not be complete until after introduction of a fluid of some sort, at which time the tissue becomes black (Coulter and Terracio, 1977). The fact that tissue does not turn black until after infiltration with plastic is evidence that tissue manipulation subsequent to drying has not reintroduced water. If tissue turns dark before infiltration it should be rejected as a possibly compromised preparation, with probable redistribution of soluble constituents.

Embedding

An embedding medium should be selected that infiltrates tissue well, does not undergo large volume changes during curing, does not redistribute tissue electrolytes, contains little or no chlorine, has good cutting characteristics, and is stable under the electron beam. Epon 826 is the plastic that best meets these criteria in our hands. Although sections between 0.1 μm and 3 μm are relatively easy to prepare, this plastic does not lend itself well to ultrathin sectioning.

Epon 826 is very viscous, and hence vacuum embedding is necessary. Curing is done slowly in stages, starting at room temperature and progressing to 45°C, 52°C, and 65°C. The entire preparation process requires about a week for ideally small samples. Larger samples can require an additional week or more because of their longer drying times.

Sectioning

The sample block face is cut with either a steel or glass knife. The cutting serves two purposes: 1) it provides a flat, smooth surface that can be mounted in the electron probe perpendicular to the direction of incidence of the electron beam, and 2) it provides a microscope section for study of tissue morphology.

The typical preparation involves sectioning with a steel knife. The section adjacent to the block face, about 2 μm thick, is stained with azure blue and mounted conventionally for light microscopy. A high-quality light microscope is useful at this point for assessing quality of preparation, for orientation purposes, and to direct subsequent electron probe measurements.

Sections between 250 nm and 600 nm can also be cut dry from the block face for analysis with the electron probe. Since all sectioning must be done without water, mounting of these sections demands great patience and skill. An eyelash brush is used to mount the section on a grid that has been lightly brushed with paraffin. The mounted grid is then gently heated on a hotplate to drive off excess paraffin and to stick the section to the grid. Sections mounted in this fashion can be examined in an electron microscope or with STEM in the electron probe. Simple STEM detectors can be devised for most electron column instruments if they are not available from the manufacturer (Murphy and Metzger, 1968; Coleman et al., 1975).

For bulk section analysis, the cut blocks are mounted in suitable sample holders along with a small piece of quartz and a crystal each of KCl and NaCl. The small objects are all cemented to the sample holder with a conductive adhesive and vacuum coated with a thin layer of carbon for electrical and thermal conductivity.

The quartz is useful for focusing the electron beam and for establishing a reproducible electron beam current. The use of quartz is by no means necessary, but it serves as a convenient substance to use for setting up the instruments, and it provides a rapid reference for monitoring noise and background.

SELECTION OF ELECTRON BEAM OPERATING PARAMETERS

Appropriate selection of accelerating voltage and electron beam current is as important to electron probe analysis as is selection of a suitable sample preparation. The electron probe is a versatile piece of apparatus, but with that versatility comes a need for the analyst to make informed decisions concerning selection of accelerating voltage and electron beam current. The analyst must appreciate the relationships between spatial resolution, x-ray signal intensities, electron beam–induced specimen damage, and electron beam operating parameters. The discussion of these parameters in the following sections is applicable only to bulk sample analysis and is generally not relevant to a discussion of thin section analysis.

Accelerating Voltage

An empirical expression to predict electron beam penetration and hence size of x-ray excitation volume has been presented by Castaing (1951). A number of investigators have proposed values to be used for the coefficients in this expression, as outlined in the work of Beaman and Isasi (1972). The coefficients presented by Andersen and Hasler (1965) are possibly the most appropriate for biological applications. With their suggestions the relationship becomes:

$$R = \frac{0.064}{\rho} (E_o^{1.68} - E_c^{1.68}) \tag{1}$$

(where R is the electron spread or penetration in μm, ρ is sample density, E_o is electron beam accelerating voltage in kV, and E_c is energy corresponding to the critical absorption edge for the x-radiation of interest). Since the electron beam diffuses randomly in the plastic sample, the width of the excitation volume is about the same as the depth of penetration. With a beam diameter d, x-ray spatial resolution can be estimated by $R_x^2 = R^2 + d^2$. This typically overestimates the diameter of x-ray excitation volume, probably because of the small number of x-rays actually emanating from the extreme range, R. Experimental work reveals, actually, that 50% of the x-rays emanate from about the first 27% of R (Andersen, 1966, 1967).

As an example of the application of the formula for determining R, consider the case in which an analysis for sodium is performed with an 8 kV accelerating voltage. The maximum depth of excitation for Na Kα radiation will be 1.8 μm in the embedded sample, with density 1.15 g/cm^3. Excitation volumes for K Kα and Cl Kα will be somewhat smaller, with maximum penetration depths of 1.4 and 1.5 μm, respectively. Thus, a 2-μ thick section is called a bulk sample in common jargon.

Similar considerations reveal that a 30-kV electron beam would excite Na Kα radiation to a depth of 16.8 μm into the plastic sample. Thus, high accelerating voltages are not useful for thick sample analysis when spatial resolution is also important. For best spatial resolution on thick samples, the analyst must use the lowest possible accelerating voltage compatible with accurate x-ray counting.

The lower limit for practical accelerating voltages is determined by two factors. Most importantly, it has been determined with thin sections that the cross-section for x-ray production is highest for an electron beam accelerating voltage about a factor of three times the critical excitation voltage for a given characteristic line (Kyser and Geiss, 1977; Russ, 1977). This cross-section decreases rapidly to zero as accelerating voltage is lowered to the absorption edge. Also, at extremely low voltages, e.g., $\sim$ 5 kV, small electron beam diameters are more difficult to obtain and the electron beam diameter becomes the limiting factor for spatial resolution.

A compromise must be reached between adequate x-ray counting rates, which in general require higher voltages, and spatial resolution requirements, which are incompatible with high accelerating voltages. As a rule of thumb it is suggested that an electron beam accelerating voltage be adopted that is about a factor of three over the critical excitation voltage for the characteristic x-ray of interest. As a typical example, with three spectrometers tuned to Na Kα, Cl Kα, and K Kα, three times the overvoltage would be 3.2 kV, 8.5 kV, and 10.8 kV for the three radiations, respectively. Obviously it is not possible to simultaneously satisfy the overvoltage criterion for each of the three radiations.

The intracellular sodium and chlorine concentrations are quite low in the typical tissue sample, sometimes approaching the absolute detection limits for these elements. It thus becomes necessary to adopt operating parameters that are less compromising for Na Kα and Cl Kα radiations than for K Kα radiation when these three radiations are measured simultaneously. Although the signal-to-background ratio for Na Kα radiation peaks at about 15 kV, both signal intensities and signal-to-background ratios continue to improve with higher accelerating voltages for other electrolytes with thick electron probe samples. Hence the analyst is provided with a useful operating range of $\sim$ 8 to 15 kV from which to select a compromise accelerating voltage. For analyses involving calcium, where 2–3 μm spatial resolution is required on thick embedded tissue samples, the normally low intracellular concentrations of calcium dictate that accelerating voltages from 10 to 12 kV be used.

Thus there is no "correct" accelerating voltage to select; rather, there is a best compromise, which is dependent upon spectrometer geometry, crystal performance, electron optics design, source brightness, sample type, elemental atomic ratios, and objective of the analysis.

Electron Beam Current

Once an accelerating voltage has been chosen, it is necessary to select an operating current. It is unwise to attempt quantitative or even qualitative measurements on thick samples unless some means is provided for monitoring electron beam current at the sample. This is a serious limitation for many scanning microscopes; however, it is normally a relatively simple modification to add beam current monitoring to the system.

Ideally, a beam current should be selected that will produce sufficient x-ray intensities to allow the attainment of adequate counting statistics in a reasonable time without damaging the sample. It may appear at first glance that higher beam currents would impose compromises on x-ray spatial resolution, as the maximum current that can be confined in an electron beam goes as the 8/3 power of beam diameter (Castaing, 1960). From earlier discussion, however, we have seen that x-ray spatial resolution with bulk samples is determined primarily by electron diffusion within the sample, so there is little advantage in purposely limiting electron beam currents to achieve electron beam diameters less than about 0.5 μm. Electron beam current levels are thus limited principally by how much current the sample can withstand without suffering unacceptable damage.

Radiation Damage

Electron probe samples are modified or damaged to varying degrees by the electron beam. Samples that are poor thermal and electrical conductors containing large weight fractions of light elements appear to be most easily damaged. Neither qualitative nor quantitative electron probe studies should be attempted on any specimen unless the extent of accompanying radiation damage and its effect upon electron probe measurements are understood.

Although the precise mechanism of radiation damage to electron probe samples is not known, some general characteristics of the damage have emerged from research in radiochemistry and studies of radiation damage in electron microscopy (Kobayashi and Sakaoku, 1965; Reimer, 1965; Bahr et al., 1965; Stenn and Bahr, 1970; Isaccson, 1977). There appear to be two general components to the electron beam–induced radiation damage. One component is related to heating of the sample, and the second component, radiation damage, manifests itself even when great care is exercised to ensure that the sample experiences little or no temperature increase during irradiation. Radiation damage is often described as a dose-related component, since it is primarily a function of the total number of coulombs of charge delivered per square centimeter of sample bombarded by the electron beam.

This description of dose, often used by electron microscopists, is directly related to, but not the same as, the rad, the unit of radiation dose developed by nuclear physicists. The rad is strictly defined as the absorbed dose when 100 ergs per gram of absorbing material have been transferred to a sample by ionizing radiation. However, the simpler term used by electron microscopists is useful, and we shall refer to it in our present discussion. The reader should be aware that although the rad has been defined to be independent of radiation type and energy, there is some electron beam voltage dependence in radiation dose as we describe it.

It is possible and indeed very likely that the electron radiation dose-related damage can be greatly reduced or modified with the use of extremely low specimen temperatures (Hall and Gupta, 1974). An explanation for the anticipated reduction in radiation damage at low temperatures may be that the mobility of ions is so reduced at low temperatures that bonds ruptured by irradiation reform before the molecular fragments have had time to drift apart (Stenn and Bahr, 1970). More study is needed in this area, since the use of cold stages may provide a means of reducing or possibly even circumventing some of the electron irradiation-induced changes.

Sample chemistry changes are effected through scission and increased carbon binding, while some atoms and light molecules are ejected from the sample. Bahr et al. (1965) have presented data on the chemical changes that organic objects undergo as a result of electron irradiation. This work was done at accelerating voltages of 40 to 75 kV, which are high accelerating voltages for bulk sample analyses. It is anticipated that although similar irradiation damage would be experienced at the lower accelerating voltages typical of electron probe bulk sample analysis, there should be substantial reductions in extent of damage. Bahr et al. found that halogens, particularly fluorine, and light elements, such as oxygen and hydrogen, are often depleted from certain types of sample. Their studies of gelatin revealed that, as with plastics, of the major matrix components, oxygen was the most sensitive to radiation. With thin gelatin films exposed to 1.5×10^{-2} coul/cm^2 and an accelerating voltage of 75 kV, a relatively low exposure by electron probe standards, the gelatin was found to lose 25% of its total mass. Oxygen was reduced by 40%, hydrogen by nearly 30%, and carbon and nitrogen by about 10% each. Lineweaver (1963) described loss of oxygen from glass during electron bombardment and Hodson and Marshall (1971) have observed potassium migration in ultrathin electron probe sections.

Mass loss under the electron beam also results in shrinkage with some plastics. Bahr et al. (1965) found that with some of the less stable plastics, shrinkages of up to 30% in one dimension accompanied moderate electron radiation doses.

Even hard tissue samples are subject to damage from the ionizing energy of the electron beam. Edie and Glick (1976) reported 30% increases in Ca Kα x-ray count rates after teeth had been exposed to the electron beam. These increases were attributed to loss of light element components from the tooth matrix.

The ionizing radiation also has some desirable effects on the sample. Widespread cross-linking imparts brittle, hard characteristics that may also be accompanied by increased thermal conductivity (Reimer and Christenhusz, 1965). Not all materials are affected in the same way; while some plastics exhibit increased cross-linking, others experience scission. Thus it is necessary for the electron probe analyst to examine each different material used as an electron probe sample to determine how the inevitable radiation damage may affect analysis.

We have observed in our laboratory that there are important dose-related changes that occur in bulk samples of tissue embedded in Epon 826. These changes take place extremely rapidly, before less than 0.1% of the dose required for analysis has been delivered to the sample. As a result we find it necessary to "stabilize" fresh plastic samples with a large diffuse electron beam before analysis. An 8 or 10 kV electron beam about 200 μm across with a specimen current of 200 nA is routinely used for a period of 2 min to prepare the samples. During this stabilization period, K, Cl, and Na Kα x-ray counting rates are observed to increase by 5% to 10%. Counting rates thereafter are uniform as a function of time for accumulated large doses.

The 2-min stabilization period of gentle irradiation is better than a factor of two longer than necessary, since the radiation damage or transformation takes place rapidly. If the sample is observed with a light microscope during the stabilization process, a cathodoluminescence is observed that gradually decays. The sample appears to be adequately stabilized by the time it is no longer capable of supporting further cathodoluminescence. The electron beam has apparently transformed the plastic matrix into a different form that does not possess an available energy level for producing cathodoluminescence. The irradiated plastic also takes on a characteristic brown coloration that is indicative of unsaturation with accompanying increased C–C bonding (Stenn and Bahr, 1970). Although continued irradiation will probably result in further subtle transformations of the sample, subsequent changes in bulk samples are not accompanied by changes in sample appearance or changes in characteristic x-ray counting rates for tissue electrolytes in the embedded specimen.

Stabilization is also necessary for realizing the best preservation of tissue morphology. If a focused electron beam is allowed to raster across a fresh plastic sample that has not been stabilized, the raster lines are indelibly etched into the plastic. Figure 1 is a light micrograph of a section removed from a block of plastic that has been briefly exposed to a rapidly rastering 10 kV, 50 nA electron beam. The tracks appear as a series of bubbles in the plastic, supporting the concept of rapid liberation of gasses accompanying irradiation of fresh plastic. Such tracks are also evident in the backscattered electron image of tissue that has been exposed to a rastering beam before stabilization with a diffuse beam. Figure 2 is a backscattered electron image obtained on a sample of renal tubules in which only the left portion of the image had been stabilized before scanning. Although the imprints on the right do not affect subsequent x-ray count rates, they do interfere esthetically with sample imaging. We thus recommend treating the dose-related portion of sample radiation damage by establishing a stable, reproducible state for the sample.

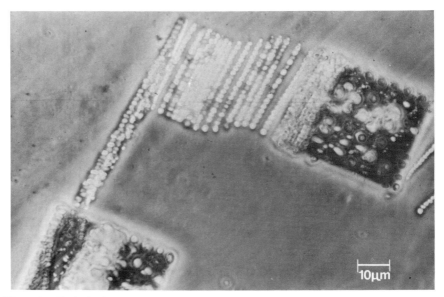

Figure 1.   Optical micrograph of section from a block of Epon 826 that has been exposed to a rapidly rastering 10 kV, 50-nA electron beam.

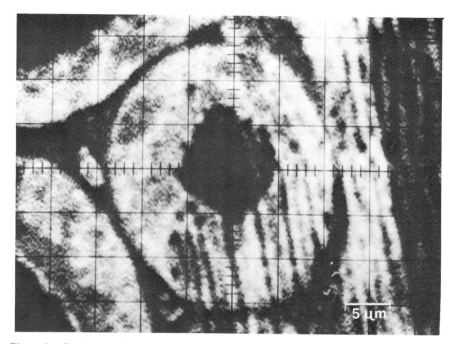

Figure 2. Backscattered electron oscillograph of mouse kidney. Left portion of sample had been stabilized with a large, diffuse electron beam before exposure to a rastering 10 kV, 50-nA electron beam. Right portion of sample had not been stabilized.

Thermal Damage

Subsequent analysis, while not producing further recognizable radiation damage, may be accompanied by thermally induced damage unless appropriate precautions are exercised during data collection. Thermal damage to an electron probe sample results when electron beam energy is deposited in the sample at a rate faster than it can be dissipated, with resultant temperature rise to damaging levels.

In his thesis, in which the electron probe microanalyzer was first described, Castaing (1951) provided computations that indicated that temperature rise in typical metallurgical electron probe samples would be insignificant. Organic samples, unfortunately, are not good conductors, and consequently it is necessary to examine more closely the effects of using a highly focused electron beam on bulk samples. Almasi et al. (1965) developed a different approach to computing the temperature rise in electron probe samples that allowed for evaporated layers of conductive coating materials on the sample. Data were presented to demonstrate that aluminum coating protected mercury telluride from thermal damage.

Friskney and Haworth (1967) also developed an expression for computing thermal heating when a nonconductor with thermal conductivity K is coated with a thin layer of conducting material having thermal conductivity K_f. The expression they developed was derived from Castaing's work and is in agreement with the results presented by Almasi et al. (1965). Because it is a simpler expression, it is more useful to the investigator. It is also capable of providing the temperature at a distance r from beam center. The expression is:

$$ T = \frac{3W_o \,(3 - r^2/r_o^2)\,(1 + 2 \ln R/r_o - r^2/r_o^2)}{4\pi\,[tK_f\,(13 - 3r^2/r_o^2) + 3Kr_o\,(1 + 2 \ln R/r_o - r^2/r_o^2)]} \tag{2} $$

(where W_o is total power absorbed, r_o is beam radius, R is specimen radius, and t is thickness of conductive coating). This expression is useful for estimating the relative effectiveness of available coating materials and for estimating the thickness of a given coating material required to restrict the sample temperature rise to a specific value. As an example a sample 0.5 cm across with thermal conductivity 0.002 W/cm^2/°C, subjected to a 1 μm diameter 8 kV, 50 nA electron beam, is found to experience a temperature rise of 955°C at the center of the electron beam. A 10-nm layer of aluminum or 18.5-nm layer of carbon will decrease this temperature rise to 145°C above ambient.

Accurate computations of sample heating are not possible with organic samples because of uncertainty in sample thermal conductivity. Although thermal conductivity values are available for virgin plastic, electron irradiation has the desirable effect of transforming the sample to a carbonized form, with attendant higher thermal conductivity than the original plastic. Thus, numbers generated by the formula can be considered upper limits on the temperature rise to be expected in plastic-embedded specimens.

Figure 3 was obtained from equation 2. From the figure it is obvious that the necessity for a conductive coating decreases as thermal conductivity of the sample increases. If thermal conductivity of a sample is known, it is then possible to determine the advantage that will result from the use of a conductive coating. For this purpose, carbon and aluminum are the most popular choices of conductive coating materials because they are pure substances, have low atomic numbers, and are relatively good thermal and electrical conductors. Echlin (1974) has presented a very useful chart listing the important physical properties of a number of possible coating materials.

Besides specimen coating, thermal damage can also be reduced in one or more of the following ways: 1) by lowering electron beam intensities to levels compatible with sample conductivity properties, 2) by providing a means of collecting data from any one point in a series of brief intervals that are too short to allow temperature excursions to reach damaging levels, and 3) by using a cold stage to maintain the sample at a low temperature during analysis.

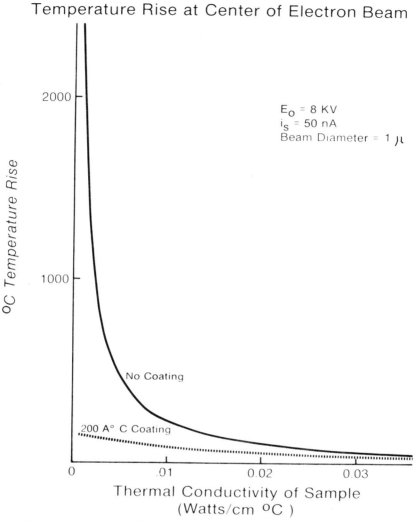

Figure 3. Demonstration of effectiveness of carbon coating for electron probe samples with low thermal conductivity. Curves were obtained using relationship developed by Friskney and Haworth (1967).

To capitalize on low beam currents it is normally necessary to use an energy detector system that can provide the required high solid angle of acceptance for x-rays. It is possible with most Si (Li) energy detector systems to optimize conditions for operating at low electron beam levels, where sample damage is not a problem.

Fortunately, damage can also be readily controlled with systems employing wavelength spectrometers. The procedures we have adopted for control of sample damage involve coating samples with a thin layer of carbon and using a computer to control electron beam positioning and beam dwell time at any given data collection point. Using these techniques, it is possible to obtain adequate x-ray signals and reasonable spatial resolution with no observable sample damage resulting from analysis.

BACKGROUND

The accurate determination of background is particularly important for intracellular electrolyte measurements on plastic-embedded tissue, since signal-to-background ratios are often low. Off-peak background is obtained by averaging the x-ray intensities obtained by detuning each spectrometer above peak and then the same number of wavelength units below peak. The average is used as an estimate of the Bremsstrahlung radiation, B_E, present under the spectral peak of element E.

Analysis for chlorine is further complicated in plastic-embedded tissue analysis by the unavoidable presence of chlorine in most embedding media. This problem is minimized with use of the low-chlorine plastic, Epon 826, which contains $\sim$10–20 mEq/kg Cl in the formulations we use. A chlorine-free embedding medium is available through the use of dibutyl phthalate as a substitute flexibilizer in low-viscosity Spurr plastic (Spurr, 1975).

Because it is essential to establish the concentration levels present in the embedding medium for all analyzed elements, it is necessary to obtain x-ray intensity measurements from the pure plastic when analyzing embedded tissue. The following expression describes the procedure for using the background information to convert a measurement of x-ray intensity, R_E, for a given element into a number, S_E, representing concentration:

$$S_E = C_E[(R_E - B_E) - (P_E - B_E')] \qquad (3)$$

In this expression, C_E is the calibration factor for element E, P_E is the on-peak counting rate on embedding medium, and B_E' is the Bremsstrahlung background from the embedding medium. All measurements must, of course, be obtained with the same spectrometer and be repeated for each spectrometer.

Normal Specimens

With most embedded soft tissues, off-peak background is higher on cells than on pure plastic, making it necessary to obtain off-peak background from tissue and not merely to estimate background from on-peak measurements on pure plastic. As a typical example, off-peak background counting rates on embedded albumin standards described later in this work were 3.10, 3.30, and 2.40 cps for K Kα, Cl Kα, and Na Kα, respectively, whereas the raw on-peak counting rates on pure plastic embedding medium were 2.46.

2.89, and 1.86 cps, respectively, for the three x-ray lines. Sample composition is normally sufficiently uniform, however, that it is not necessary to determine a different background for each point from which data are collected.

High-Contrast Specimens

Some tissues, such as the light receptors in retina, are particularly osmophilic, and high contrast is imparted to the specimen. With these tissues there is enough variation in average atomic number from one point in the specimen to another that Bremsstrahlung radiation will vary from point to point. Some alternative to osmication can be tried, or background can be measured at each point of analysis. Another means also presents itself if the backscattered electron signal is calibrated against off-peak background (Ingram and Ingram, 1974). Figure 4 shows such a calibration of the backscattered electron signal for K Kα off-peak background. The correlation coefficient for the linear regression is 0.89. Since the backscattered electron signal is routinely monitored for morphologic information by our computer data collection system, this information is readily available to us. The use of backscattered electron signal in this fashion nearly doubles the amount of data that can be collected in any given data-gathering session.

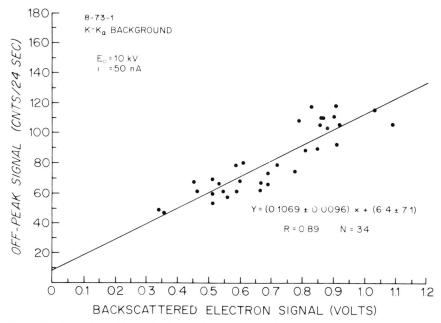

Figure 4.   Potassium off-peak background signal as function of backscattered electron signal.

Another procedure that has been used quite successfully involves detuning one of the three spectrometers, calibrating it against off-peak background for each of the other two spectrometers, and dedicating that spectrometer to continuous background monitoring.

ELECTRON PROBE SIGNAL

Calibration

The two major types of quantitative electron probe measurements commonly required are: 1) absolute concentration measurements, and 2) ratios of concentration of one element to another. Although both types of measurement place the same constraints on sample integrity and data collection technique, simplifications are realized in standards preparation for ratio measurements because the effects of many systematic errors cancel out.

The biologist desires concentrations described in millimoles per unit hydrated tissue mass. Tissue that has been dehydrated and analyzed dry will yield concentration measurements in millimoles per unit dry mass, a set of units that may require some interpretation for the biologist. The use of appropriate standards with an electron probe sample that is either hydrated or otherwise prepared with a stable medium replacing tissue water, however, makes it possible to relate electron probe measurements to wet weight concentrations.

A number of methods are available for transforming bulk sample measurements into numbers representing concentrations. These can be broken down into two general schemes: 1) the use of theoretical ZAF correction procedures to correct x-ray intensities for absorption and fluorescence effects, and 2) the use of prepared standards. The ZAF correction procedures have been worked out in great detail for metallurgical samples and are available in such forms as Magic (Colby, 1968) and FRAME (Yakowitz et al., 1973). The ZAF correction procedure has been adapted to biological applications by Warner and Coleman (1974). The use of such computer correction programs with plastic-embedded tissue is a bit hazardous, since correction factors for the important electrolytes are large, and accurate absorption coefficients for the plastic matrix are not known. Also, processed tissue has undergone gross distortions, resulting in tissue volume changes (Boyde et al., 1977; Ingram et al., 1976). The preponderance of evidence indicates that freeze-dried tissue preparations experience ~20% volume shrinkage. Since shrinkage of some magnitude accompanies all forms of tissue dehydration, a correction factor must be included to allow for preparation-induced volume changes if absolute concentration measurements are to be possible on processed tissue.

These problems are circumvented in large part if standard materials are selected that undergo the same preparation-induced distortions as the biological tissue, and if the prepared standards also interact with the electron beam in a manner qualitatively and quantitatively similar to tissue samples. Solutions of 20% bovine serum albumin in which known electrolyte concentrations are incorporated meet these criteria.

Primary Standards

There are two approaches to preparing primary electron probe standards. Either a set of standards samples all with the same concentration can be prepared from one well-characterized standards solution, or a series of different solutions can be used to prepare samples containing different concentrations. For the data presented in this publication, a set of standards samples was prepared from one 20% albumin solution. The albumin solution was analyzed with a flame photometer to contain 103.5 ± 0.1 mEq/kg K and 120.6 ± 2.1 mEq/kg Na. Chlorine concentration was 102.9 ± 0.1 mEq/kg as measured with a Buchler-Cotlove chloridometer. Confidence limits are the standard deviations of the mean of replicate measurements. Drops of solution were freeze-dried, vapor-fixed with osmium tetroxide, and embedded in Epon 826. The embedded standards blocks were faced with a steel knife and mounted in an electron probe specimen holder next to crystals of KCl and NaCl. A thin, approximately 20 nm, carbon layer was cast on the specimen holder in a vacuum evaporator. Table 1 presents counting information obtained from eleven standard blocks using a 10-kV electron beam with our three-spectrometer ARL-EMX electron probe microanalyzer

Table 1.   Probe counting on albumin standards blocks

Block No.	K – Kα		Cl – Kα	Na – Kα
72-1-6f	929* ±	45**	1405 ± 47	640 ± 40
72-1-6g	990 ±	57	1415 ± 50	686 ± 23
72-1-6a	992 ±	53	1263 ± 58	631 ± 30
72-1-6c	971 ±	48	1274 ± 41	626 ± 28
72-1-6q	964 ±	79	1311 ± 113	659 ± 39
72-1-6s	921 ±	69	1298 ± 70	623 ± 22
72-1-6h	964 ±	82	1326 ± 98	649 ± 55
72-1-6m	948 ±	102	1311 ± 95	637 ± 65
72-1-6k	977 ±	64	1295 ± 110	664 ± 54
72-1-6o	917 ±	54	1384 ± 45	666 ± 26
72-1-6j	916 ±	59	1362 ± 85	644 ± 48
Average	954 ±	29	1331 ± 52	648 ± 19

* Average of nine to eleven, 40 sec. counts
** S.D. of the mean
E_o = 10 kV
i = 52 nA (50 nA on quartz)

(Applied Research Laboratories, Sunland, California). A calibration coefficient, C_E, has been determined for each of the three spectrometers, with additional background information supplied in Table 2. Equation 3 was rearranged to provide C_E:

$$C_E = \frac{S_E}{(R_E - B_E) - (P_E - B_E)} \tag{4}$$

(where S_E is the concentration of element E in the standard).

With the alternative approach, a calibration coefficient would be obtained from the least-squares fit to the electron probe counting on standards made from a series of different solutions. The calibration coefficient, C_E, would be obtained from the slope of the regression line. This method is more laborious than the method we have used if the same careful characterization of the standards solution is repeated for each different standard solution. As work in our laboratory and elsewhere has demonstrated, linear relationship exists between concentration and x-ray intensity under given electron beam conditions, and this procedure is not necessary.

Table 2.   Calibration factors for converting x-ray intensities to numbers representing weight concentrations*

	Albumin standards					
	K		Cl		Na	
Concentration of Element (mEq/kg)	103.5	± 1.9	102.9	± 0.1	120.6	± 2.1
Counting on Standards Blocks (cnts/40 sec)	954	± 29	1331	± 52	648	± 19
Off-Peak Background (cnts/40 sec)	124	± 4	132	± 4	96	± 2
Corrected Counts on Pure Plastic (cnts/40 sec)	—		78	± 11	—	
Net Counting on Standards Blocks (cnts/40 sec)	830	± 29	1121	± 53	552	± 19
Calibration Factor (mEq/kg cnts/sec)	4.99 ±	.19	3.67 ±	.17	8.74 ±	.34
	Secondary Standards					
Reference Crystal	KCl		NaCl		NaCl	
Counting on Reference Crystal (cnts/sec)	4079	± 29	3143	± 46	2199	± 31

* All values are listed plus or minus the standard deviation of the mean. The confidence limits on the calibration factors are estimates of the standard deviation obtained from the square root of the sum of the squares of the standard deviations for each of the components. The counting on the reference crystals has been corrected for off-peak background and for a 2 μsec dead time.

E_o = 10 kV
i = 50 nA on quartz

Secondary Standards

During subsequent tissue analysis, stability is important in order to ensure that data are being collected under identical conditions to those used during calibration of the electron probe signal. Even small changes in operating conditions can change the calibration significantly, and with a system as complex as the electron probe, it is necessary to continually monitor instrument operation. Since x-ray counting rates on albumin standards are too low to serve as useful materials for periodic reference, it is advisable to identify substances to be used as secondary standards. These should be homogeneous, readily available, and contain high weight fractions of the elements of interest. Substances like NaCl and KCl serve well for sodium, chlorine, and potassium, and calcite is useful for calcium. The crystals of NaCl and KCl mounted next to the embedded tissue specimen receive the same layer of evaporated carbon as the sample, eliminating concerns about differences in absorption between sample and standard. The crystals that are used as secondary standards must be carefully calibrated by reference to the fabricated standards. Table 2 presents the potassium, chlorine, and sodium counting rates on KCl and NaCl crystals corresponding to the calibration coefficients obtained from the prepared standards.

These calibration factors can be used whenever a 10-kV electron beam is used on bulk samples. Periodically during data collection a series of 10-s counts are obtained on the crystals and are averaged at the end of the data collection session and referenced to Table 2 to establish the calibration factors for use in data reduction.

As a practical matter, counting errors may result from dead-time counting losses and counting rate–related pulse height shifts. The high count rates of pure crystals will normally place the detector systems in count rate regions where dead-time counting losses occur. Also, at the higher counting rates a sufficient reduction in signal pulse height may occur, resulting in counting losses when tight PHA windows are used (Bender and Rapperport, 1966). Because electronics are normally tuned with the high counting rates from crystals, pulse height shift losses are experienced on tissue counting. The solution is either to use wider PHA windows or to measure the shift and adjust amplifier fine gain accordingly for crystal or tissue samples. Increasing the detector bias voltages unnecessarily will increase the magnitude of pulse shifts at high count rates. In our laboratory, the adoption of crystals as secondary standards has probably been the single most important factor in developing a reliable, practical method for routine quantitative analysis for tissue electrolytes.

Data Collection

A cell or group of cells in the specimen under study is identified with the light microscope in the section adjacent to the electron probe sample. These same cells are then located in the electron probe sample by reference to either the backscattered electron image or the sample current image. Alternatively, cells may be selected at random from a given region of the sample block by reference to one of these images.

Figure 5 presents an example of what is possibly the most convenient and useful form for collecting data. A cursor on the electron probe display screen is positioned, using a joy stick, on the image of the cell of interest and a button is pushed that initiates a line scan profile through the cell. For the above examination and data collection, the electron beam is directed by a LINC-8 computer (Digital Equipment Corporation, Maynard, Mass.; see Figure 6). Data are collected at a series of points along the line scan. Blind scalers under computer control are used for x-ray counting, and timing is provided by a crystal-controlled clock. Data are collected in magnetic core memory and transferred to magnetic tape for permanent storage.

Figure 7 describes the interlacing data collection scheme that has been adopted to minimize thermal damage to the sample. The electron beam is positioned on a spot for a half-second data-counting interval, and then is moved to a different location 4 units away. This is repeated across a distance of up to 128 μm on the sample face. On the next pass the beam is

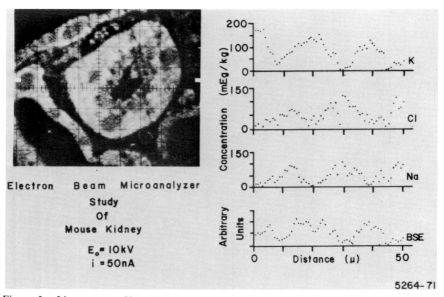

Figure 5. Line scan profiles of the potassium, chlorine, and sodium distributions across a proximal mouse kidney tubule depicted on the left. The picture is the oscillograph image of the electron backscatter signal (BSE). The tick marks on either side of the photograph delineate the location of each end of the line scan.

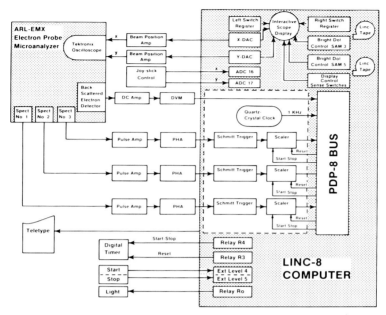

IOWA COMPUTERIZED ELECTRON MICROPROBE

Figure 6.   Schematic diagram of electron probe-computer system. The Digital Equipment Corporation LINC-8 computer is interfaced with an Applied Research Laboratories EMX to provide computer control of electron beam positioning and computer directed data collection. The interface has been greatly simplified in this drawing. Control circuits and level shifters have not been included.

positioned one unit to the right of the initial position and again a pass is made across the face of the sample in steps of 4 units. Four such passes are required to make the 64 points of a line scan profile. This requires a total of 32 s. The process is repeated enough times to build up the desired counting statistics.

With the short counting intervals, thermal loading of the sample is greatly reduced, since heat has an opportunity to dissipate before the electron beam again returns to deposit additional energy. The electron beam at any one sample location is pulsed at high power levels, to produce reasonable x-ray intensities, yet is of sufficiently short duration that energy delivery is maintained below damaging levels. Parameters are selected so that the pulsed 50-nA electron beam has about the same effect on the sample as a steady 3-nA electron beam.

It is convenient and is normally adequate to repeat the scanning twenty times, for an integrated time of 10 s per data point along the line scan and a total of 640 s per line scan set. All of the intervals, number of points per line

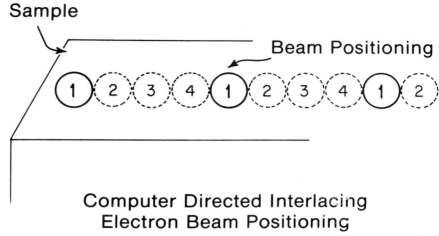

Computer Directed Interlacing
Electron Beam Positioning

Figure 7.   Schematic representation of interlacing electron beam positioning method used to decrease thermal loading of nonconducting electron microprobe sample. On the first pass across the sample, the electron beam impinges sequentially from the left onto spots labeled 1, stopping for a one-half second data counting interval at each location. On each successive pass the electron beam pauses at locations labeled 2, then 3, and finally 4. This entire process is repeated twenty times to integrate 10 s of counting for each point.

scan, and number of repetitions of scanning are software parameters that are readily changed. They have been arbitrarily optimized to provide convenience for the analyst, to provide adequate counting statistics, and to maximize the amount of data that can be collected in a normal day. Normally it is not possible to see where data have been collected on the sample face using these techniques, since beam-induced sample damage is virtually eliminated.

Data reduction is normally approached by examining the line scans on the computer's interactive display and integrating counts in a region of interest depicted by two operator-controlled cursors. These integrated counts are corrected for off-peak background and multiplied by a calibration coefficient from Table 2 to provide numbers representing the average concentration within the region of interest. It is most common to delineate cell boundaries with the two cursors so that the reduced data represent the intracellular concentrations in a particular cell. Table 3 presents an example of the use of this technique on frog muscle. Skeletal muscle samples were quench-frozen in cooled liquid propane, freeze-dried, and embedded in Epon 826. Calibration coefficients presented in Table 2 were used to convert x-ray counting intensities to numbers representing concentrations.

With computer-directed electron beam positioning, care must be exercised to ensure that the electron beam is not scanned off the Rowland circle. The spectrometers in our electron probe have flat focusing to better than 100 μm in the x direction of the electron beam movement. The light element spectrometer restricts excursion of the electron beam in the y direction to $\pm$ 25 μm from beam center. These distances are compatible with dimensions of normal cells.

Accuracy of Measurement

The electron probe is capable of giving measurements of high precision, provided necessary precautions have been exercised in tissue handling and data collection so that tissue is not destroyed before adequate counting statistics have been collected. Combining the uncertainty of measurement at each point in the calibration process leads to the precision estimates quoted in Table 2. They are 3.8% for potassium, 4.6% for chlorine, and 3.9% for sodium. Because data points on typical tissue have uncertainties of about 1% to 3%, an overall precision of less than 6% can normally be expected for a given data point. The accuracy of measurements, however, is not always as easy to determine, since inherent systematic errors are much more difficult to recognize and treat.

When relating measurements back to fabricated standards, accuracy of measurement can never be any better than the standards. For this reason, great care must be exercised in standards preparation and characterization. Less obvious are the subtle differences that may exist between standard and unknown due to preparation-induced distortions.

Our 20% albumin standards are appropriate only if the embedded, freeze-dried standards have properties of density and average atomic number similar to prepared tissue samples, and provided the standards

Table 3.   Electron probe measurements of electrolyte concentrations in frog skeletal muscle

Fiber number	Concentration (mEq/kg)		
	K	Cl	Na
1	108.4 ± 4.6 S.D	9.7 ± 1.0	5.1 ± 1.2
2	120.3 ± 5.0	8.4 ± 0.9	9.6 ± 1.2
3	106.4 ± 4.6	7.2 ± 0.9	3.3 ± 1.2
4	126.1 ± 5.3	7.4 ± 0.9	6.6 ± 1.2
5	104.9 ± 4.4	8.4 ± 0.9	6.8 ± 1.2
6	113.5 ± 4.9	11.5 ± 1.1	11.7 ± 1.5
7	105.9 ± 4.7	7.0 ± 1.0	7.3 ± 1.4
8	115.9 ± 4.8	6.4 ± 0.8	8.0 ± 1.2
9	104.6 ± 4.4	8.0 ± 0.9	7.6 ± 1.2
Average	111.8 ± 7.7	8.2 ± 1.6	7.3 ± 2.4

E_o = 10 kV
i = 50 nA

undergo the same preparation-induced distortions as tissue samples (Ingram et al., 1974).

Evidence collected in our laboratory and elsewhere indicates that all freeze-dried tissue undergoes about the same amount of volume shrinkage, 20%, during processing. If this is not true, then measurements will be inaccurate by whatever factor a given tissue shrinks more or less than the standards.

It is not necessary that fabricated standards similar to tissue be employed if correction is made for this known distortion. Spurr (1975) has developed a series of plastic standards in which alkali metals are incorporated into plastic. These standards are highly uniform and homogeneous. They can be applied to analysis of freeze-dried, plastic-embedded tissue, since the tissue samples typically compose only ~20% of a sample and the embedded tissue has an average atomic number and density little different from plastic standards. The only precaution is that the analyst apply the known correction for shrinkage of the tissue (Ingram et al., 1977).

When bulk sample analysis is attempted on tissue prepared by some method other than freeze-drying, the analyst must be careful either to select his standards appropriately or to measure the magnitude of preparation-induced distortions.

In the above discussion it has been assumed that no isolated, irregular imperfections exist in the prepared sample. Anyone familiar with freeze-dried tissue, is well aware of the lack of uniformity in quality of preparation that is particularly evident when large samples are shock frozen. As water freezes, large, pure ice crystals may form in some areas of the sample, during either shock freezing or the subsequent freeze-drying. The criterion we have adopted for judging tissue for adequate preparation is to insure that ice crystal artifact is present only to dimensions smaller than the resolution of the electron probe. Areas of tissue with obvious ice-crystal damage evident with light microscopy are not deemed acceptable for thick-sample analysis. It would be foolish to attempt electron probe analysis with spatial resolution better than the preparation could support.

Natural intracellular electrolyte distributions are readily changed by stresses to the cell before processing. Possibly the most serious compromise to the preparation lies in manipulations performed on tissue prior to shock freezing. Tissues mounted in flux chambers require 30 min or more to recover from trauma of dissection. Tissue mounted for intracellular electrical potential measurements must be allowed to recover and to stabilize for a period of 30 min or more before reproducible measurements are possible. Similarly, the investigator is well advised to incubate tissue to be analyzed for intracellular electrolytes in temperature-controlled, oxygenated bathing solution to allow recovery before shock freezing and subsequent analysis. Failure to do this will not necessarily have an adverse effect on accuracy of measurement, but the investigator will have measurements on altered tissue when he needs accurate measurements on viable, healthy cells.

Difficulties in relating electron probe electrolyte measurements to physiologically relevant measurements, such as electrical potentials, may result from the fact that the electron probe measures total electrolyte, whereas the physiologist often desires values describing the activity of a given ion species. The major fraction of cellular sodium, for instance, is believed to be bound. Because there is probably a stoichiometric relationship between the free and bound species of a given electrolyte under given circumstances, it may be necessary to allow adequate time for stabilization when tissue is subjected to experimental conditions before electron probe analysis.

CONCLUDING REMARKS

Techniques have been presented for making quantitative intracellular electrolyte distribution measurements on bulk biological tissue samples with an electron probe microanalyzer. The use of embedding media and low accelerating voltage permits spatial resolution better than 2 μm, and properly characterized secondary standards allow for accurate quantitative measurements on a routine basis. Computer-controlled data collection using a modified pulsed-beam technique has been recommended for efficient collection of data without sample damage, and it also allows for the presentation of data in one of the most useful formats for interpretation, the line scan profile.

Work upon which this chapter is based was supported by NIH Program Project Grant HL 14388.

LITERATURE CITED

Almasi, G. S., Glair, J., Ogilvie, R. E., and Schwartz, R. J. 1965. A heat-flow problem in electron-beam microprobe analysis. J. Appl. Phys. 36:1848.

Andersen, C. A. 1966. Electron probe microanalysis of thin layers and small particles with emphasis in light element determinations. In: *The Electron Microprobe* (McKinley, T. D., Heinrich, K. F. J., and Wittry, D. B., eds.), pp. 58–74. John Wiley and Sons, New York.

Andersen, C. A. 1967. An introduction to the electron probe microanalyzer and its application to biochemistry. In: *Methods of Biochemical Analysis*, Vol. XV (Glick, D., ed.), pp. 147–270. Interscience, New York.

Andersen, C. A., and Hasler, M. F. 1965. Extension of electron microprobe techniques to biochemistry by the use of long wavelength x-rays. Presented at the Congres International—L'Optique des Rayons X et al Microanalyse Orsay, Seine et Oise, France.

Bahr, G. F., Johnson, F. B., and Zeitler, E. 1965. The elementary composition of organic objects after electron irradiation. Lab. Invest. 14:1115.

Beaman, D. R., and Isasi, J. A. 1972. Electron beam microanalysis: the fundamentals and applications. ASTM STP 506, American Society for Testing and Materials, pp. 1–80.

Bender, S. L., and Rapperport, E. J. 1966. Nonproportional behavior of the flow proportional detector. In: *The Electron Microprobe* (McKinley, T. D., Heinrich, K. F. J., and Wittry, D. B., eds.), pp. 405–414. John Wiley and Sons, New York.

Boyde, A., Bailey, E., Jones, S. J., and Tamarin, A. 1977. Dimensional changes during specimen preparation for scanning electron microscopy. In: *SEM* (Johari, O., ed.), Vol. 1, pp. 507–518. IITRI, Chicago.

Castaing, R. 1951. Application des sondes electroniques a une methode d'analyse ponctuelle chimique et cristallographique. Thesis, Paris, 1951, O.N.E.R.A. Publ. No. 55. In translation: Rept. No. WAL 142/59-7, California Institute of Technology, Pasadena, California.

Castaing, R. 1960. Electron probe microanalysis. In: *Advances in Electronics and Electron Physics*, Vol. 13 (Marton, L., ed.), pp. 317–386. Academic Press, New York.

Colby, J. W. 1968. Quantitative microprobe analysis of thin insulating films. In: *Advances in X-ray Analysis*, Vol. 11 (Newkirk, J. W., Mallett, G., and Pfieffer, H., eds.), pp. 287–303. Plenum Press, New York.

Coleman, J. E., Davis, S., Halloran, B., and Moran, F. 1975. A simple transmitted electron detector (TED) for thin biological samples. Proc. Tenth Annual Conf. of Microbeam Analysis Society, pp. 45A–45G.

Coulter, H. D., and Terracio, L. 1977. Preparation of biological tissues for electron microscopy by freeze-drying. Anat. Rec. 187:477.

Echlin, P. 1974. Coating techniques for scanning electron microscopy. In: *SEM* (Johari, O., ed.), Vol. 1, pp. 1019–1028. IITRI, Chicago.

Edie, J. W., and Glick, P. L. 1976. Dynamic effects of quantitation in the electron probe analysis of mineralized tissues. Proc. Eleventh Ann. Conf. of Microbeam Analysis Society, pp. 65A–65F.

Erlandsen, S. L., Thomas, A., and Wendelschafer, G. 1973. A simple technique for correlating SEM with TEM on biological tissue originally embedded in epoxy resin for TEM. In: *SEM* (Johari, O., ed.), Vol. 1, pp. 349–356. IITRI, Chicago.

Friskney, B. A., and Haworth, C. W. 1967. Heat-flow problems in electron probe microanalysis. J. Appl. Phys. 38:3796.

Hall, T. A., and Gupta, B. L. 1974. Measurement of mass loss in biological specimens under an electron beam. In: *Microprobe Analysis as Applied to Cells and Tissues* (Hall, T., Echlin, P., and Kaufman, R., eds.), pp. 229–237. Academic Press, New York.

Hanzon, V., and Hermodsson, L. H. 1960. Freeze-drying of tissues for light and electron microscopy. J. Ultrastruct. Res. 4:332.

Hayat, M. A. 1980. *Principles and Techniques of Electron Microscopy: Biological Applications*, Vol. 1, 2nd Ed. University Park Press, Baltimore. In press.

Hodson, S., and Marshall, J. 1971. Migration of potassium out of electron microscopy specimens. J. Microscopy 93:49.

Ingram, F. D., and Ingram, M. J. 1974. Backscattered electron signal for background monitoring with biological tissue samples. Proc. Ninth Ann. Conf. Microbeam Society, pp. 11A–11C.

Ingram, F. D., and Ingram, M. J. 1975. Quantitative analysis with the freeze-dried, plastic embedded tissue specimen. J. Microscopie Biol. Cell. 22:193.

Ingram, F. D., Ingram, M. J., and Hogben, C. A. M. 1974. An analysis of the freeze-dried, plastic embedded electron probe specimen preparation. In: *Microprobe Analysis as Applied to Cells and Tissues* (Hall, T., Echlin, P., and Kaufmann, P., eds.), pp. 119–146. Academic Press, New York.

Ingram, F. D., Ingram, M. J., and Spurr, A. R. 1976. Volume changes in freeze-dried tissue. Proc. Eleventh Ann. Conf. Microbeam Analysis Society, pp. 63A–63B.

Ingram, F. D., Spurr, A. T., and Ingram, M. J. 1977. Comparison of two different types of Na standards for electron probe analysis of soft tissue. Analyst 102:515.

Isaccson, M. S. 1977. Specimen damage in the electron microscope. In: *Principles and Techniques of Electron Microscopy: Biological Applications*, Vol. 7 (Hayat, M. A., ed.) Van Nostrand Reinhold Co., New York.

Kobayashi, K., and Sakaoku, K. 1965. Irradiation changes in organic polymers at various accelerating voltages. Lab. Invest. 14:1097.

Kyser, D. F., and Geiss, R. J. 1977. Direct measurements of voltage-dependence of inner shell ionization cross-sections. Eighth International Conference on X-Ray Optics and Microanalysis and Twelfth Annual Conference of the Microbeam Analysis Society, Boston, pp. 31A–31C.

Lechene, C. P., and Warner, R. R. 1977. Ultramicroanalysis: x-ray spectrometry by electron probe excitation. Ann. Rev. Biophys. Bioeng. 6:57.

Lineweaver, J. L. 1963. Oxygen outgassing caused by electron bombardment of glass. J. Appl. Phys. 34:1786.

Murphy, A. P., and Metzger, C. A. 1968. Transmission electron microscopy with an ARL microprobe. Rev. Scient. Instrum. 39:1705.

Reimer, L. 1965. Irradiation changes in organic and inorganic objects. Lab. Invest. 14:1082.

Reimer, L., and Christenhusz, R. 1965. Determination of specimen temperature. Lab. Invest. 14:1158.

Russ, J. C. 1977. Selecting optimum kV for STEM microanalysis. In: *SEM* (Johari, O., ed.), Vol. 1, pp. 335–340. IITRI, Chicago.

Saubermann, A. J. 1979. General considerations of x-ray microanalysis of frozen hydrated tissue sections. In: SEM (Johari, O.), Vol. 2, p. 607.

Sjöstrand, F. S., and Kretzer, F. 1975. A new freeze-drying technique applied to the analysis of the mitochondrial and chloroplast membranes. J. Ultrastruct. Res. 53:1.

Spurr, A. R. 1972. Freeze-substitution additives for sodium and calcium retention in cells studied by x-ray analytical electron microscopy. Bot. Gaz. 133(3):263.

Spurr, A. R. 1975. Choice and preparation of standards for x-ray microanalysis of biological material with special reference to macrocyclic polyether complexes. J. Microscopie Biol. Cell. 22:287.

Staub, N. C., and Storey, W. F. 1962. Relation between morphological and physiological events in lung studied by rapid freezing. J. Appl. Phys. 17(3):381.

Stenn, K., and Bahr, G. F. 1970. Specimen damage caused by the beam of the transmission electron microscope, a correlative reconsideration. J. Ultrastruct. Res. 31:526.

Stephenson, J. L. 1954. Caution in the use of liquid propane for freezing biological specimens. Nature 174:235.

Tormey, J. M. 1978. Validation of methods for quantitative x-ray analysis of electrolytes using frozen sections of erythrocytes. In: *SEM* (Johari, O., ed.), Vol. 2, pp. 259–266.

Warner, R. R., and Coleman, J. R. 1974. Quantitative analysis of biological material using computer correction of x-ray intensities. In: *Microprobe Analysis as Applied to Cells and Tissues* (Hall, T., Echlin, P., and Kaufman, R., eds.), pp. 249–268. Academic Press, New York.

Yakowitz, H., Myklebust, R. L., and Heinrich, K. F. J. 1973. FRAME—An on-line correction procedure for quantitative electron probe microanalysis. NBS Technical Note 796.

Chapter 10

QUANTITATIVE X-RAY MICROANALYSIS OF THIN SECTIONS

Godfried M. Roomans

Wenner-Gren Institute,
University of Stockholm,
Stockholm, Sweden

INTRODUCTION

During recent years, the need for quantitation in biological x-ray microanalysis has been increasingly recognized. In x-ray microanalytical studies carried out so far on biological specimens, localization and identification of a xenobiotic or otherwise "unusual" element have often been the primary goals. In such cases, qualitative analysis was usually sufficient. The potential of biological x-ray microanalysis can, however, be greatly expanded by employing quantitative techniques.

Quantitation may allow correlation of results obtained by x-ray microanalysis with results obtained by other techniques, such as chemical elemental analysis by flame spectrophotometry or atomic absorption spectrometry, isotope studies, and microelectrode studies. Valuable information may be obtained from such a combined approach, for instance in studies of ion transport in cells and tissues.

Quantitation also allows independence of the results from the actual instrument on which the measurements were made. In a pathological examination one may wish to compare a "suspected" specimen with a range of "normal" values. It is of great practical importance when base-line values can be established separately, and the investigation of a "suspected" specimen would not have to involve the simultaneous analysis of controls every time. In addition, data from several laboratories can be compared.

Quantitative x-ray microanalysis of biological specimens has mostly been carried out on thin sections. In addition to the advantage of a superior spatial resolution of analysis in thin sections, the process of quantitation is relatively simple as compared to bulk specimens (see Ingram and Ingram in this volume) and the quantitative techniques are now well established. It is useful to define what is meant by a thin section. For the purpose of this chapter, two criteria are used: 1) the electrons reach the "far" surface of the section without losing an appreciable fraction of their initial energy, and 2) the x-rays generated in the section are able to reach the surface of the section without being absorbed to an appreciable extent.

It should be realized that in these two criteria the thickness as such is of no importance, but the mass through which electrons and x-rays have to pass is significant. Therefore, a section of freeze-dried biological tissue may be much thicker than a metal foil, while still being "thin" according to the above definition. In addition, the elemental composition of the specimen and the initial energy of the electrons are important parameters. Hence, a section may be "thin" when analyzed at 100 kV, but "thick" at 10 kV.

Approximately 1–2 μm thick sections of biological materials are usually still considered thin enough to comply with the above criteria. The higher limit would be appropriate for freeze-dried sections, whereas the lower limit would be appropriate for frozen-hydrated sections or for material embedded in resin.

Reliable quantitation in x-ray microanalysis is not simple, and a number of difficulties are typical for biological specimens. Several requirements have to be fulfilled for accurate quantitation:

1. *Separation of peaks from background and deconvolution of overlapping peaks.* Since the accuracy of quantitation depends to a great extent on the accuracy with which the number of characteristic counts in a peak can be determined, this first step in the quantitation procedure is extremely critical.

2. *Standards.* Unlike metallurgical specimens, in which all measurable elements add up to 100% and standardless quantitation can be carried out, biological specimens require the use of a homogeneous standard containing the element(s) of interest in a known concentration for analysis. Preparation of such standards has been one of the major problems in quantitation of biological specimens.

3. *A quantitative model* to compare the unknown specimen with the standard. The theory for quantitation of x-ray analysis of biological thin sections was developed and presented by Hall (1971), and the "Hall method" is now generally applied in biological x-ray analysis.

4. *Optimal instrumental configuration.* The specimen holder, the grid, and parts of the electron microscope surrounding the specimen may contribute to the spectrum. The accuracy of quantitation in biological microanalysis is positively affected by a reduction of this contribution, which is obtained by optimizing the instrumental configuration. In addition, problems like contamination of the specimen are partly dependent on instrument performance.

The newer x-ray analytical systems can be equipped with a minicomputer, and software for quantitative analysis is being provided by the manufacturer. Hence, it may often seem that quantitation is only a matter of pushing one or two buttons on the keyboard. In fact, few, if any, of the programs marketed at present are perfect in the sense that they perform accurately under all conditions and on any specimen. Usually, programs contain "hidden assumptions" that may make them invalid or inaccurate for some kinds of problems. Since much software has been developed for use with metallurgical or mineral specimens, the biologist should be extremely careful in applying these programs to biological specimens. A thorough understanding of the principles of quantitation is still needed, and cannot be replaced by the belief that the machine will do everything.

This chapter, then, is aimed at giving the reader an understanding of how quantitative x-ray microanalysis of biological specimens should be carried out. After discussion of the x-ray spectra of biological thin films in more detail, attention is given to the first step of the quantitative procedure, namely the separation of peaks and background to determine the net and relative peak intensity. In a following section, preparation and use of standards for biological microanalysis are discussed. Theory and practice of the continuum method of quantitation in thin films are considered in detail. One section is devoted to various experimental difficulties such as contamination, mass loss, and extraneous background. In addition, the determination of the limits of detectability of an element by x-ray microanalysis is discussed. The chapter is concluded by a review of important applications of quantitation in biological x-ray microanalysis.

THIN FILM SPECTRA

The interaction of the electron beam with the specimen gives rise to the generation of both characteristic and continuum x-rays. Since the detector cannot distinguish between the two "kinds," the thin film spectrum shows both. In the following section, the generation of characteristic and continuum radiation is considered, and the way in which the collection of the x-rays in energy-dispersive analysis affects the generated x-ray spectrum is discussed.

Characteristic Radiation

Excitation of characteristic x-rays occurs as a consequence of two subsequent phenomena: 1) *inner shell ionization*, the creation of a vacancy in one of the inner shells under the impact of the electron beam, and 2) a *radiative transition*, the filling of the vacancy by an electron from a higher shell and the liberation of energy in the form of an x-ray photon. Not all incident electrons (in fact only very few) give rise to inner shell ionizations, and only part of these cause a radiative transition. Thus, the ultimate probability of x-ray excitation is equal to the product of the probabilities of inner shell ionization and radiative transition.

In a thin film of a pure element, the number of ionizations (n) caused by one incident electron during its travel through a film of thickness Δx is given by:

$$n = Q \frac{N\rho}{A} \Delta x \tag{1}$$

(where Q is the *ionization cross-section*, N is Avogadro's number, ρ is the density, and A the atomic weight of the element; $N\rho/A$ equals the number of atoms per unit volume).

The ionization cross-section Q is given by

$$Q = \frac{7.92 \times 10^{-14}}{E_o E_c} \ln \frac{E_o}{E_c} \tag{2}$$

(where E_o is the intial energy of the electron and E_c the critical excitation energy, i.e., the energy required to raise an electron to the first vacant energy level). Values of E_c can be found in tables (see Reed, 1975).

The probability of a radiative transition is called the *fluorescence yield* (ω); the fluorescence yield depends strongly upon the atomic number of the element. Values of ω can be found in tables (see Reed, 1975).

Hence, the following expression for the characteristic intensity per incident electron, I_A (in counts per second), may now be obtained:

$$I_A = \omega Q \frac{N\rho}{A} \Delta x \tag{3}$$

A salient feature of this equation is that the intensity of the characteristic radiation is directly related to the number of atoms in the irradiated volume. In thin sections, there is a linear relationship between section thickness and characteristic intensity.

If we define N_x as the number of atoms x in the irradiated volume, then, since, $N_x = N\rho\Delta x/A$, we may simply write:

$$I_A = \omega Q N_x \tag{4}$$

Continuum Radiation

The continuum radiation or background radiation consists of photons emitted by electrons suffering energy loss in collisions with atoms. In most cases this energy loss is quite small, typically below 1 keV, and, even in thin films, part of these very low energy photons may already be absorbed in the specimen. The theory of continuum intensity is rather complicated, and an approximative theory is generally used. A short and simplified outline of the theory is given below.

Let us consider a small part, dE, of the spectrum between energies E and dE. The continuum energy in this part of the spectrum ($E_{cont}dE$) can be assumed to be a fraction dE/E_o of the total continuum energy (E_{cont}), which ranges up to the limit E_o, the accelerating voltage. The continuum energy will be dependent on the energy loss ΔE suffered by the electrons in the thin film. In addition, we have to take an efficiency factor f into account, to correct for absorption of low-energy photons in the film. The continuum intensity, or the number of continuum x-rays, in the fraction dE of the spectrum ($I_c dE$) is equal to the continuum energy divided by the energy of the continuum x-rays, and hence:

$$I_c dE = \frac{E_{cont}dE}{E} = a_1 f \frac{dE}{E_o} \frac{1}{E} \Delta E \tag{5}$$

(where a_1 is a constant). The efficiency factor f is fiven by

$$f = a_2 Z E_o \tag{6}$$

and the energy loss ΔE is given by the approximative equation:

$$\Delta E = a_3 \frac{1}{E_o} \frac{Z}{A} \rho\Delta x \tag{7}$$

(where a_2 and a_3 are constants). Combining Equations 5–7, we obtain:

$$I_c dE = a \frac{Z^2}{A} \frac{1}{E_o E} \rho \Delta x \, dE \qquad (8)$$

As was the case with the characteristic intensity, the continuum intensity is directly dependent on section thickness. Comparing Equations 3 and 8, it can be seen that the ratio of characteristic to continuum intensity is independent of section thickness.

With the number of atoms in the irradiated volume defined as $N_c = N\rho \, \Delta x / A$, Equation 8 turns into

$$I_c dE = a \, Z^2 \frac{1}{E_o E} N_c \, dE \qquad (9)$$

(where a is a constant). If we have a thin film consisting of several elements, each element contributes to the continuum radiation, and the total continuum radiation, and the total continuum radiation is the sum of these contributions. Hence

$$I_c dE = \frac{a}{E_o E} \sum_c (N_c Z_c^2) dE \qquad (10)$$

Equation 10 will be used later in the description of the continuum method of quantitation in thin films. A salient point is that, whereas the characteristic intensity of element x is directly related to the number of atoms x (N_x) in the irradiated volume of the section, the continuum intensity is related to the total number of atoms $\sum_c N_c$ in the irradiated volume. The ratio of characteristic intensity to continuum intensity, called the peak-to-background ratio (P/B ratio), is hence an (approximate) measure for the concentration of element x in the irradiated volume. As stated above, the P/B ratio is independent of section thickness.

In contrast with the characteristic radiation, which is isotropic, the continuum radiation is strongly anisotropic, with a bias to a "forward" direction (the direction of the electron beam). The position of the x-ray detector relative to the specimen is therefore very important, and high x-ray take-off angles minimize continuum intensity and maximize the P/B ratio.

Energy-Dispersive Analysis

Due to peculiarities of the Si (Li) semiconductor detector used in energy-dispersive x-ray analysis, the spectrum appearing on the monitor is not identical to the original spectrum generated in the thin film. The main factors affecting the x-ray spectrum are the energy resolution of the detector, the detector efficiency, and the occurrence of escape peaks.

Energy Resolution   The use of the Si (Li) semiconductor for detection of x-rays is based on the absorption of x-rays in the material of the sensitive region, and the subsequent ionization of the detector material. The mean number of ionizations produced by an x-ray of energy E is given by E/ϵ, where ϵ is the energy required for one ionization (about 3.8 eV in the silicon detector). The number of ionizations produced is, however, subject to statistical fluctuations; theoretically their distribution is very close to a Gaussian function. The energy resolution, usually defined as the full width of a peak at half its maximum height (FWHM), equals 2.355 times the standard deviation of the Gaussian.

In a Si (Li) detector, the energy resolution is not purely dependent on ionization statistics, but also on variations in the efficiency of charge collection. In addition, thermal noise in the field effector transistor and fluctuations in the leakage current in the detector influence the energy resolution. Since the energy resolution varies with x-ray energy, a reference energy is used to define the energy resolution in a particular system. Usually, the energy of 5.9 keV (Mn $K\alpha$ line) is selected for this purpose.

Instead of the very narrow energy band of monochromatic x-rays generated in the specimen, the spectrum now shows peaks of a certain width. Since the continuum intensity is virtually unaffected by the detector resolution, spreading of the characteristic x-ray counts over a wider energy band results in a markedly lowered P/B ratio. Due to its inferior energy resolution (about 145 eV as compared to 20 eV for a wavelength-dispersive detector), the energy-dispersive detector gives a P/B ratio that is lower than that of a wavelength spectrometer by a factor of 5. This affects the accuracy of quantitation of trace elements adversely. In addition, the increased incidence of peak overlaps may cause serious difficulties in the quantitative procedure.

The theoretical Gaussian shape of the peaks may be somewhat affected by incomplete charge collection due to energy loss in the "dead layer" of the detector and at the detector sides. This will lead to a higher proportion of low-energy counts, and the resulting deviation is called a "low-energy tail."

Detector Efficiency   In order to maintain the necessary purity of the detector material, the detector is normally mounted in an evacuated encapsulation. The vacuum is maintained by a beryllium entrance window. This means that, before entering the detector, the x-rays have to pass through a beryllium layer typically about 7–8 μm thick. Low-energy x-rays are strongly absorbed in this layer and the efficiency of detection declines steeply at about 1 keV. In addition, attenuation takes place in the gold layer (<20 nm) covering the detector surface and in the "dead layer" (about 200 nm) of the detector. The absorption in the thin gold layer can usually be neglected. The combined effect of the absorbing layers on characteristic and continuum radiation is shown in Figures 1 and 2. Elements lighter than sodium cannot normally be detected by an energy-dispersive detector.

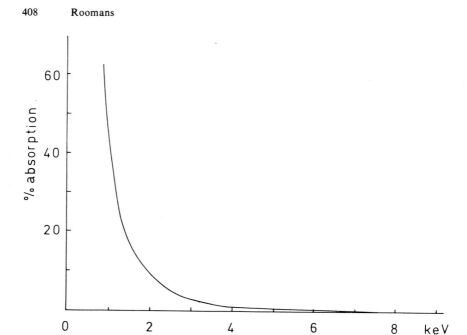

Figure 1.   Effect of the beryllium window on characteristic radiation. Na Kα rays (1.04 keV) are absorbed to about 50% by a typical beryllium window.

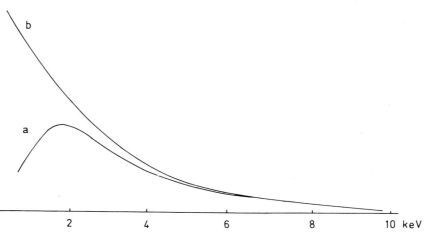

Figure 2.   A typical continuum shape from a low-atomic number target is drawn in curve a. If no absorption of x-rays by the beryllium window would occur, the continuum would have the shape indicated in curve b.

Escape Peaks   The energy transfer in the detector occurs partly through photoelectric absorption, which gives rise to the generation of Si $K\alpha$ rays. A small portion of these escape from the detector, and the corresponding amount of energy (1.74 keV) is thus lost. Relative to the full energy peak E, the energy deposited in the detector is (E-1.74) keV. As a consequence, a small satellite peak will appear in the spectrum at 1.74 keV below the main peak. Of course, only x-rays with an energy higher than the Si K absorption edge produce these satellite peaks, which are called "silicon escape peaks." The ratio between the intensity of the silicon escape peak and the main peak is 0.0141 for phosphorus, and decreases to less than 0.002 for cobalt and heavier elements. Hence, in biological microanalysis, silicon escape peaks will hardly ever be significant.

SEPARATION OF PEAKS FROM CONTINUUM

Since the P/B ratio gives a first estimate of the concentration of a particular element in the analyzed spot, the first step in quantitative analysis should obviously be the accurate determination of the net peak intensity and the background intensity. Due to the non-linearity of the continuum intensity, this presents some difficulties, and much attention has been given to the development of reliable methods for the separation of peaks from the background (often called "background subtraction" or "background suppression"). The existing methods all require some kind of computer facilities. Depending on the complexity of the calculation to be carried out (which is, ideally, related to the ultimate accuracy), more or less memory capacity and computing time are needed for background subtraction. An extra complication is introduced if peak overlaps occur: in this case, a deconvolution procedure also has to be included.

In biological thin sections several factors contribute to an aggravation of an already complicated problem: 1) Most elements of biological interest have characteristic lines in the region of 1–4 keV, where the curvature of the background is strongest, and the background shape most difficult to calculate. In addition, the approximate theory given in the previous section has been tested mainly at high x-ray energies, and there is reason to assume some deviations from the theory in the energy range of interest to the biologist. 2) The count rate in biological thin sections is often low, so that the statistical noise superimposed on the spectrum may be significant, and this affects the accuracy of the background subtraction. 3) A considerable part of the continuum does not originate from the specimen itself, but from extraneous origins: the grid, the specimen holder, and the parts of the microscope immediately surrounding the specimen. Combined with the previous factor, this will further decrease the accuracy of the background subtraction. 4) The P/B rate is often low, which aggravates both the statistical and the methodical errors.

The existing methods for separation of peaks from background are continuously being implemented. It should be realized, though, that even a perfect method will still be dependent on counting statistics, and this may well remain a limiting factor in biological microanalysis.

In this section, the principles of the most commonly used methods for background subtraction and deconvolution of peak overlaps are discussed. Copies of the actual computer programs used are supplied by most manufacturers on request.

Linear Interpolation

The principle of linear interpolation is that the height of the background on either side of the peak is determined, and a straight line is drawn to connect these points (Figure 3). The advantage of this method is its relative simplicity: it requires little computer memory and time, and is thus cheap and fast. The disadvantage is that the linear interpolation does not accurately represent the shape of the non-linear continuum. This deviation is most important in the low-energy region of 1–4 keV, where the curvature is very strong. In addition, the determination of the background level (which often is at the discretion of the operator) may be inaccurate in the case of peak overlaps, and when the statistical noise is appreciable. Linear interpolation may be considered as an alternative for other methods of background subtraction only at energies higher than 4 keV, and only in the case of isolated peaks with high P/B ratio. Recently, methods for non-linear interpolation have become available. A fit is then made to a second-order curve. This method obviously offers advantages over simple linear interpolation.

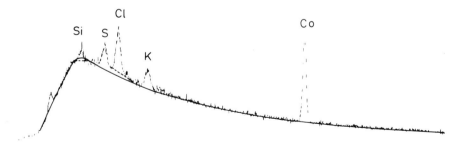

Figure 3.   Linear interpolation as method of background subtraction may lead to considerable errors in the determination of peak intensity, especially in the case of a small peak in the low-energy region (silicon) or in the case of partially overlapping peaks (sulfur and chlorine). The method of linear interpolation may be permissible with relatively high peaks at higher energies (cobalt). The calculated background level is given by the drawn line, linear interpolations by a broken line.

Frequency Filter

The frequency filter method of background suppression makes use of the assumption that peaks have a certain shape (a moderately slow rate of rise and fall) that does not occur in the background, and can be distinguished from it. The background is considered to be a smooth function of energy, with superimposed high-frequency statistical noise. There is some variation in the actual algorithms and computer programs applied in the frequency filter methods. The following simplified account illustrates the principle of the method.

A narrow energy band comprising only a few channels is compared with the neighboring bands. If its average height is not significantly higher than that of one of the neighboring bands, it is considered not to be a peak, and it is used to determine the background level. Energy bands rising significantly above the background are recognized as peaks, "clipped off," and removed, leaving only the background. By averaging a number of channels, instead of using a single channel, statistical noise is filtered out. The complete spectrum is scanned one or more times in this way.

The frequency filter method has several advantages. It gives a more truthful representation of the actual background than, for example, the linear interpolation method. In contrast to some of the methods to be discussed later, it does not require previous knowledge about the composition of the specimen—the program locates the peaks using objective criteria. As far as computer programming is concerned, the method is more complicated than linear interpolation, but less demanding than some of the other methods. Difficulties associated with this method are that it may be "fooled" by peak overlaps, giving rise to a single broad peak, and that it has difficulties in handling peaks occurring in the 1–3 keV region, the strongly curved part of the continuum.

Calculation of Background

As discussed earlier, the continuum intensity may be described by a smooth function of energy. It should therefore be, at least theoretically, possible to calculate the expected background intensity and to subtract this from the spectrum (Fiori et al., 1976). This method of background subtraction, also called the Reed-Fiori method, assumes that the shape of the background is dependent only on energy and not on the precise elemental composition of the specimen—the elemental composition affects only the height of the background. This assumption is in accordance with Equation 9 and also with more complicated models of continuum generation. In practice, background subtraction, according to this method, is carried out by selecting a number of channels that contain only background counts, and by fitting an equation of the form of Equation 9 to these points. Among the more complicated equations used (Reed, 1975; Fiori et al., 1976) are:

$$I_c(E) = k \, T_E \left(a \, \frac{E_o - E}{E} \right) \tag{11}$$

$$I_c(E) = k \, T_E \left(a \, \frac{E_o - E}{E} + b \, \frac{(E_o - E)^2}{E} \right) \tag{12}$$

$$I_c(E) = k \, T_E \, E \, [a(E_o - E)^2 + b(E_o - E) + c] \tag{13}$$

(where k is a constant, depending on the (mean) atomic number of the specimen; a, b, c are fitting constants; and T_E is the detector efficiency, which decreases sharply below 3 keV, due to absorption of x-rays in the beryllium window and the silicon dead layer of the detector). Ways of determining T_E are discussed below.

The drawback of this approach is the uncertainty of the theory in the energy range below 3 keV. Reed (1975) argues, however, that the various deviations from the theory probably cancel out; this might be responsible for the relative success of the method.

Multiple Least-Squares Fitting of Peaks and Background

The most complicated method of determining net peak and continuum intensity is the multiple least-squares fitting routine (Shuman et al., 1976). A set of standard spectra is collected and stored for fitting of the individual peaks and background to the unknown spectrum. These standard spectra may be obtained from thin films of the pure element, or thin crystals for which only one characteristic line is present (e.g., LiCl). Alternatively, generated Gaussian peaks may be used. A continuum standard may be obtained from a film of organic material or calculated, as seen in the preceding section.

A thin section spectrum composed from these standards may be described as:

$$N_j = a_o N_{oj} + \sum_{i=1}^{m} a_i N_{ij} \tag{14}$$

if m elements with characteristic peaks are present in the section. N_j is the number of counts in the jth channel. N_{oj} is the background intensity, and N_{ij} the characteristic intensity; a_o and a_i are, respectively, the magnitudes of the background and of the characteristic lines. If M_j is the number of counts in the jth channel of the unknown spectrum, the coefficients a_o and a_i can be obtained by the least-squares fit of N_j to M_j, by minimizing χ^2:

$$\chi^2 = \sum_{j=1}^{n} (M_j - N_j)^2 / \sigma_j^2 \tag{15}$$

(where σj is the standard deviation in M_j in the jth channel and N_j is assumed to be noise-free); in total the spectrum has n channels. The mathematical procedure of the least-squares fitting is performed according to standard methods.

In addition to separating peaks and background, the least-squares fitting method also accomplished a simultaneous deconvolution of overlapping peaks. The least-squares fitting method may be "embedded" in a complete quantitative routine according to the continuum method described in a subsequent section.

The least-squares fitting method requires extensive minicomputer facilities and the availability of stored standard spectra, and it is rather time-consuming. On the other hand, it is, in principle, the most accurate method for the separation of peaks and background. The method is sensitive for the quality of the stored standard spectra—generally these should be acquired under exactly the same circumstances as are used during analysis of the unknown. In addition, the method is sensitive for miscalibration of the energy scale (which is discussed in more detail later).

Curve-fitting has become one of the dominant methods employed in the analysis of energy-dispersive spectra, and will doubtlessly continue to be that.

Deconvolution of Peak Overlaps

Due to the inferior energy resolution, peak overlaps occur much more frequently in energy-dispersive analysis than in wavelength-dispersive analysis. Peak overlaps should be avoided as much as possible, for instance by omitting heavy metal fixation (osmium), and heavy metal staining (lead, uranium, tungsten) during specimen preparation. Some peak overlaps may be unavoidable; of particular interest for biologists is the K $K\beta$–Ca $K\alpha$ overlap in sections of frozen or freeze-dried tissue. In most tissues, the concentration of potassium is several orders of magnitude higher than that of calcium, so that even the K $K\beta$ line may be relatively intense, compared to the Ca $K\alpha$ line.

The principal methods of peak stripping are 1) the use of overlap factors, and 2) least-squares fitting of library spectra or generated peak shapes.

The first method is primarily used in cases where a major elemental line is overlapped by a minor line of another element; an example of this is the K $K\beta$–Ca $K\alpha$ overlap. The relative magnitude of the K $K\alpha$ and $K\beta$ peaks is then determined from a thin standard that does not contain calcium. In the unknown spectrum, the K $K\alpha$ peak and the convoluted K $K\beta$–Ca $K\alpha$ peak are now determined, and the known ratio between the K $K\alpha$ and K $K\beta$ peak is used to strip the convoluted peak from the K $K\alpha$ and K $K\beta$, leaving the Ca $K\alpha$ intensity.

In thin sections, where no absorption occurs, the ratio between, for instance, a $K\alpha$ and a $K\beta$ line of the same element may be assumed to be constant and independent of specimen composition. This assumption may not be valid in thick specimens.

Least-squares fitting of library spectra or generated peak shapes is carried out according to the principles described in the previous sections. When generated peak shapes are used, especially in the case of the more complex L or M lines, serious errors may be made unless all minor lines are included in the fit. A rather extensive library of reference spectra is needed for this method. In thick specimens an elemental spectrum may be influenced by the presence of other elements in the specimen; in thin sections this complication is absent.

Fitting methods are extremely sensitive to miscalibration of the energy scale. Changes in energy calibration may occur, for example, because of temperature changes, or when the pulse height analyzer is temporarily switched off. Russ (1976) and Shuman et al. (1976) show examples of the effects of miscalibration on the quantitative results; even errors of 5 eV (less than a channel) may have devastating effects. Hence, the operator should make sure that the energy calibration of the standard spectrum is exactly the same as that of the unknown. Since a calibration that is sufficiently accurate for quantitation purposes is hard to achieve for the operator using his eyes only, several manufacturers now provide computer routines for calibrating the energy scale to the required degree of accuracy.

STANDARDS

One of the requirements for reliable quantitation is a good standard. Preparing standards for analysis of biological specimens has presented a number of difficulties, and this is one of the factors that has delayed quantitation in biological microanalysis. In recent years, it has been attempted to overcome some of the practical difficulties encountered in the early stages of quantitative biological microanalysis. Existing methods developed for a limited group of elements were extended to include more elements. Various types of standards, virtually for all elements, have now become available.

The ideal standard should be chemically well defined and homogeneous at the level of resolution used. These requirements are self-evident and need no further discussion. In addition, the standard should resemble the specimen in its physical and chemical properties. The reasons for this requirement are related to mass loss during irradiation by the electron beam, and to the quantitation model.

During irradiation of the specimen by the electron beam, chemical bonds within the specimen are broken, which leads to the formation of small organic molecules and loss of the elements carbon, hydrogen, nitrogen, and oxygen from the specimen. Thus, the concentration of elements detectable by energy-dispersive x-ray microanalysis is increased during analysis. For various reasons, it is as yet difficult to correct for this effect; it is generally assumed that, if specimen and standard resemble each other closely in their chemical and physical properties, mass loss will occur to the same extent in specimen and standard, and corrections for mass loss will cancel.

A consequence of the continuum model for quantitation in thin sections (this model is discussed in detail in the following section) is that the P/B ratio is not linearly related to the concentration of the element, since the background intensity is dependent upon the composition of the specimen. A correction factor has to be included to account for this effect. If the composition of the specimen differs markedly from that of the standard, the correction factors may become unproportionally important in the calculation. Since the theory of the continuum model is only an approximative theory, this is generally not desirable.

Therefore, since the standard should resemble the specimen, the choice of the standard is determined by the kind of specimen under analysis. Consequently, this section treats the main types of thin biological specimens, namely, thin sections of embedded material, cryosections, and microdroplets, individually.

Sections of Embedded Tissue

For quantitative analysis of sections of tissue embedded in epoxy resin, a standard dissolved in resin seems a logical choice. Mineral salts cannot be dissolved homogeneously into epoxy resins, only dispersed; elements to be dissolved in the resin must form part of a molecule with organic properties. To qualify as a suitable standard, the compound to be dissolved in the resin should fulfill the following requirements (Roomans, 1979a):

It should contain the element(s) of interest, preferably in a high mass fraction.
It should be easily soluble in the epoxy resin.
It should be stable during analysis under the electron beam; the element(s) of interest should not be volatilized.
It should be non-reactive toward the resin components (especially the accelerator), and neither should it be sensitive to oxygen (air) or water.
It should have a sufficiently high boiling point (and a low vapor pressure) to avoid loss during polymerization at elevated temperatures.

Sometimes these requirements are in conflict; solubility is increased if the molecule has a large organic part, but on the other hand this reduces the mass fraction of the element of interest. In addition, good solubility is often associated with high volatility. The final choice of the compound then has to be a compromise.

In principle, it is possible to prepare standards for (nearly) all elements using suitable organic or organometallic compounds and dissolving these in epoxy resin (Table 1). The survey in Table 1 is not limiting; for a number of elements alternatives can easily be found.

Table 1. Standards for quantitative analysis of sections of resin-embedded material

Elements	Compound	Reference	Comments
Sodium, potassium	Crown ethers	Spurr (1974; 1975)	Can also be used for calcium (low concentrations)
Zinc	Crown ethers	Chandler (1976)	
Magnesium, iron	Pentadienyl-derivatives	Roomans and Van Gaal (1977)	
Nitrogen	Acetyl-acetonates	Roomans and Van Gaal (1977)	
Copper, zinc, cadmium, mercury, lead, bismuth	Dithiocarbamates	Roomans and Van Gaal (1977)	Analogues can be used for all other transition elements
Phosphorus, arsenic, antimony	Triphenylphosphine and analogues	Roomans and Van Gaal (1977)	
Sulfur	DY 041	Jessen et al. (1974)	Flexibilizer (Araldite)
Sulfur	Crown ether-thiocyanide complexes	Spurr (1974)	
Sulfur	Dithiocarbamates	Roomans and Van Gaal (1977)	
Chlorine	Epoxy resin	Pallaghy (1973)	Spurr's resin may contain about 15 g/l chlorine as contaminant
Chlorine	Dichloro-phenoxyacetic acid	Roomans (1979b)	
Bromine	Bromophenols	Roomans (1979b)	
Iodine	Iodobenzoic acid	Roomans (1979b)	

Alkali and earth-alkaline metals may form complexes with, usually, rather large organic molecules. The use of macrocyclic polyethers, or crown ethers (Pedersen, 1967), has been suggested by Spurr (1974, 1975). Sodium and potassium complexes of one of these crown ethers, dicyclohexyl-18-crown-6, have repeatedly been used to prepare standards for these elements. Up to 200 mEq sodium or potassium per kg, the standards are completely homogeneous, but at higher concentrations problems are encountered (Shuman et al., 1976). This type of standard is prepared as follows: equimolar amounts of the crown ether and a suitable sodium or potassium salt are dissolved in methanol, where complex formation takes place. The methanol is removed by heating the solution to 70°C; the complex, a honey-colored viscous mass, crystallizes out and is dried at 70°C. This dried complex can then be dissolved into the epoxy resin by stirring. The mixing procedure can be made easier by gently heating the resin mixture (from which the accelerator has been omitted) to make it less viscous. The accelerator is then added later (Roomans, 1979a). Polymerization of the final mixture is reported to be slower than usual, depending on the concentration of the crown ether complex (Spurr, 1974, 1975). The anion of the sodium or potassium salt forms part of the complex. However, not all salts give stable complexes: cyanides and thiocyanides give the best results. Iodides and bromides can also be used, but other salts (cyanoferrates, permanganates, chlorides, and phosphates) should not be used.

Less well-tested alternatives for the crown ethers are cryptates: complexes of alkali or earth alkaline metals with 4,7,13,16,21,24-hexaoxa-1,10-diazobicyclo-(8,8,8)-hexacosan, marketed as Kryptofix (Merck, Darmstadt, German Federal Republic), or with related compounds (Boekestein, personal communication). Recently, standards containing up to 1 mol/kg calcium were prepared with calcium-naphtenate (Ornberg, personal communication).

Transition metals have a rich coordination chemistry, providing numerous alternatives for the preparation of resin-soluble standards (Roomans and Van Gaal, 1977). Among the possibilities are cyclopentadienyl-derivatives, acetylacetonates, and dithiocarbamates. Dithiocarbamates are complexes of metals with the general formula $Me (R_2 dtc)_m$, where Me is the metal, R is an alkyl group (C_nH_{2n+1}), dtc is dithiocarbamate, and m varies from 1 to 5. Nearly all transition metals have been shown to form dithiocarbamate complexes (Willemse et al., 1976). The complexes are not available commercially, but are not very difficult to make.

The preparation is carried out in two steps: 1) Sodium di-N-n-butyldithiocarbamate is made by adding, under cooling with ice, 1 mol CS_2 to a stirred aqueous solution of NaOH and the secondary di-n-butylamine (1 mol of each). This solution can be used as it is for the next step. 2) The required stoichiometric amount of the metal chloride or sulfate (dissolved in water) is added, and the complex precipitates and is isolated by filtration. Recrystallization can be carried out from various solvents (acetone, ethanol) and the complex is dried in a vacuum above P_2O_5. (Note: Na- or K-dithiocarbamates are soluble in water but not in resin, and cannot be used to make standards for these elements in resin.) Sodium di-N-n-ethyldithiocarbamate is available commercially, so that the complex can be made in one step. Generally, however, the longer the alkyl chain, the better the solubility of the complex in epoxy resin; for this reason the butyl analogues are preferred to ethyl analogues (Roomans and Van Gaal, 1977).

Standards with these compounds can be easily prepared. Some compounds dissolve directly into epoxy resin, others dissolve rapidly under gentle heating, for instance, in a water bath. In the latter case, the accelerator was added after the complex had been dissolved and the resin had been cooled to room temperature. Polymerization of the blocks can be carried out overnight.

Not only transition metals, but also metals of groups IVb and Vb of the periodic system, such as silicon, lead, or bismuth, may form dithiocarbamate complexes that are suitable for the preparation of standards (Roomans and Van Gaal, 1977). Standards for phosphorus can be prepared by dissolving triphenylphosphine in epoxy resin; the analogous arsenic and antimony compounds can be used to prepare standards for these elements (Roomans and Van Gaal, 1977).

The dithiocarbamate standards described above can also serve as a standard for sulfur. An alternative is the use of the sulfur-containing flexibilizer DY 041 in Araldite as a standard (Jessen et al., 1974). However, the number of suitable sulfur-containing compounds is so large that the selection of alternative possibilities should not be a problem.

Also for the halogens, theoretically many suitable compounds are available. Among the halogen-substituted benzenes and benzene derivatives, there are many compounds that can easily be dissolved into the resin. Especially in the case of chlorine-substituted benzene derivatives, a substantial side chain on the benzene ring is needed to decrease the volatility. With bromine- or iodine-substituted aromatic compounds, the side chains may be shorter (Roomans, 1979b). Table 1 lists a number of compounds that were found satisfactory; structural formulae of some of these are given in Figure 4.

It is easier to prepare standards in the low-viscosity Spurr resin than in the more viscous Epon (Roomans and Van Gaal, 1977; Roomans, 1979a). Standards have also been prepared in Araldite (Jessen et al., 1974) and in Vestopal W (Roomans, 1979b).

A problem associated with this type of standard is that it actually resembles only part of the specimen in its chemical and physical properties,

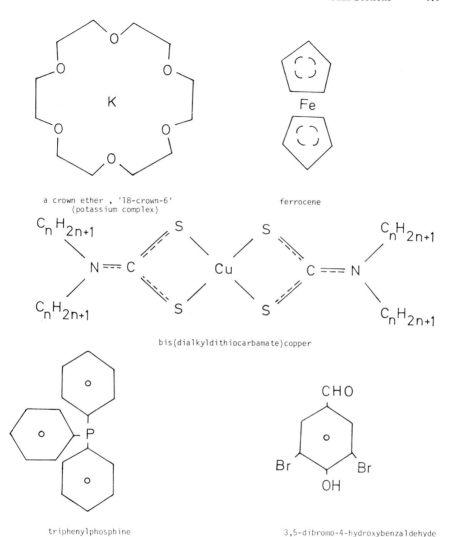

a crown ether , '18-crown-6'
(potassium complex)

ferrocene

bis(dialkyldithiocarbamate)copper

triphenylphosphine

3,5-dibromo-4-hydroxybenzaldehyde

Figure 4. Structural formulae of compounds used as standards for quantitative analysis of resin-embedded specimens.

namely the embedding medium, but not the tissue itself. A type of standard that would, at least theoretically, overcome this difficulty was suggested by Ingram et al. (1974). Drops of a 20% albumin solution containing the elements of interest in a known concentration were shock-frozen, freeze-dried, osmium-fixed, and embedded. These blocks cannot be sectioned onto distilled water as a trough liquid, because of loss of diffusible elements. The amount of experience with this technique is as yet insufficient to allow a comparison with other methods of standard preparation.

Cryosections

Important elements, such as sodium, potassium, chlorine, magnesium, and calcium, are lost partly or completely during the conventional preparation procedure. Especially for the study of diffusible elements, cryotechniques have therefore become the method of choice. The tissue is shock-frozen and sectioned at very low temperature in a cryoultramicrotome; no trough liquid is used. The sections are either transferred to the microscope in the frozen-hydrated state and kept in this state during examination, or they are freeze-dried. In accordance with the notion that the standard should resemble the specimen as much as possible, generally a standard is used that consists of mineral salts in an organic matrix.

It is not difficult to select a suitable salt for this type of standard. In principle, all water-soluble salts can be used, as long as they are stable under the electron beam under the conditions of analysis. Just as in the case of preparation of any other specimen by cryomethods, the main problem is to avoid ice crystallization during freezing, and recrystallization and diffusion artifacts during sectioning and drying (see chapters by Marshall and Morgan in this volume).

Various organic compounds have been used as matrix; gelatin and albumin are most commonly used (Lehrer and Berkeley, 1972; Dörge et al., 1974; Roomans and Sevéus, 1976; 1977). Concentrations of 20%–30% are applied, which resemble the conditions in biological tissue. Commercially available gelatin and albumin may contain impurities, which may be removed by dialization against distilled water. Sometimes cryoprotectants are also added to minimize ice-crystal formation. The standard is shock-frozen and sectioned under the same conditions as the specimen. Roomans and Sevéus (1977) reported good results with a standard consisting of 20% gelatin to which 5% glycerol was added as a cryoprotectant; in addition, mineral salts, in known concentration, were added.

Standards for various elements have been prepared by these methods. Naturally, sodium, potassium, and chlorine have been the elements most frequently introduced in these kinds of standards, but the method is equally well suited for other elements like phosphorous and calcium as well as heavier elements like rubidium and cesium.

Instead of gelatin and albumin, tissue homogenates have also been used as matrix (Hall and Peters, 1974); although these standards may resemble the specimen more truthfully, their inferior homogeneity is a serious disadvantage.

In some cases, specimen and standard are joined. Gupta et al. (1976) dissect the specimen in saline solution containing 10%–20% dextran (which serves as a cryoprotectant). This solution, which fills the intercellular spaces in the specimen (e.g., the lumen in salivary glands) is then used as a standard. Dörge et al. (1974, 1978) cover the specimen with a layer of albumin standard solution containing various elements in a known concentration before freezing. The albumin layer is preserved during freezing and freeze-drying and is sectioned with the specimen. Standards of this kind (also called "peripheral" standards) offer the advantage that specimen and standard can be analyzed under exactly the same instrumental conditions, without the need for specimen exchange. In addition, one is sure that specimen and standard have been subjected to exactly the same preparative procedure.

Freezing artifacts may be circumvented by using the protein standards suggested by Shuman et al. (1976). Bovine serum albumin and phosvitin contain, respectively, covalently bound sulfur or phosphorus in known concentrations. Standards may be prepared either by dusting bare copper grids with the crystals or by placing a drop of 3% aqueous protein solution on a carbon film covered grid and then drying the grid at 60°C for 1 hr.

For the quantitative analysis of erythrocytes, Kirk et al. (1974, 1978a) devised a special method of standard preparation, by loading erythrocytes with known amounts of sodium and potassium, using the nystatin method. Standard cells and unknowns were sprayed onto a support and air-dried.

Microdroplets

Analysis of the elemental composition of extremely small volumes of fluid, e.g., in renal physiology, has become a major field of application of x-ray microanalysis. The technique of sample preparation varies from one laboratory to another, but is generally carried out according to the following principle (Garland et al., 1978).

The samples are withdrawn with the help of a specially designed micropipette precisely controlled by a micromanipulator. Subsequently, the samples are kept under water-equilibrated mineral oil to prevent evaporation (since the volume of these microdroplets is extremely small, in the order of magnitude of 0.01–0.5 nl). The droplets are deposited, still under oil, on a substrate. Usually, thick substrates are used: quartz slides (Ingram and Hogben, 1967), beryllium (Morel and Roinel, 1969; Lechene, 1970), aluminum (Garland et al., 1973), or carbon (Moss, 1976). In principle, however, carbon films on grids may also be used. The droplets are air-dried or freeze-dried after removal of the mineral oil by an organic solvent like xylene. Deposits formed from the microdroplets comprise a regular distribution of 1–2 μm crystals over an area 80–150 μm in diameter. These deposits are generally analyzed with a defocused beam, so that the whole deposit is included. The characteristic signal is then linearly related to the amount of element in the deposit. The micropipette can be accurately calibrated (for instance by using a radioactive solution) and the concentration can then be calculated. The standards are simple salt solutions; mostly four or five concentrations in a relevant range are used. Organic substances (propylene glycol, urea, albumin) may be used to increase the viscosity of the droplets and to produce better droplets. The need for calibration of the pipette may be circumvented by adding a known amount of an element that does not otherwise occur in the samples, to both standards and unknowns, so that the analyzed quantity may be compared (Davies and Morgan, 1976).

To avoid absorption of x-rays, microcrystals should, indeed, be very small. This is especially the case with the light elements sodium, magnesium, and aluminum. To conform to the definition of a thin specimen, no appreciable absorption may take place within the crystals. If, for technical reasons, it is impossible to prepare crystals that are thin enough, a correction for absorption has to be made.

A simple but quite satisfactory way of doing this is to assume that all x-rays are generated in the middle of the crystal and travel an equal length through the crystal. Then Beer's law for x-ray transmission may be used:

$$\frac{I}{I_o} = e^{-\mu\rho x} \tag{16}$$

(where I/I_o is the fractional transmission, ρ the density in g/cm^3, x the thickness in cm, and μ is the mass absorption coefficient). In the (usual) case that the crystal consists of more than one element, μ is to be replaced by $\sum c_i\mu_i$, where c_i is the mass fraction of the ith element. Data on mass absorption coefficients of elements of biological importance are given in Table 2. (Complete tables of μ have been published by Heinrich, 1966.)

Table 2.   Absorption coefficients for Kα radiation

Absorber	Emitter								
	Na	Mg	Al	Si	P	S	Cl	K	Ca
C	1534	905	557	357	235	160	111	57	42
N	2450	1448	893	573	378	258	179	93	68
O	3698	2190	1353	869	575	392	274	141	104
Na	571	5409	3359	2168	1441	986	690	359	266
Mg	770	464	4377	2825	1877	1285	899	468	346
Al	1021	615	386	3493	2325	1593	1117	583	432
Si	1333	802	503	328	2840	1949	1368	716	531
P	1696	1021	641	417	280	2371	1664	870	645
S	2103	1266	794	518	347	239	1966	1031	765
Cl	2578	1552	974	635	425	294	207	1210	898
K	3729	2245	1409	918	615	425	300	158	1190
Ca	4413	2657	1667	1086	728	502	354	187	139

After Heinrich (1966) and Reed (1975).

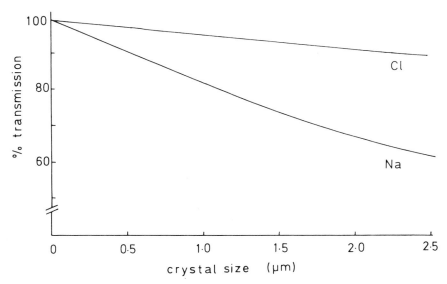

Figure 5.   Absorption of Na and Cl Kα x-rays in NaCl crystals; it is assumed that the x-rays originate from the center of the crystal (see text).

From these data, the absorption of x-rays in crystals of a certain size can easily be calculated. From Figure 5 it can be seen that for sodium, absorption effects may become appreciable already at crystal sizes of 0.6 μm.

Semi-Quantitative Standards: The Ratio Model

If one is less interested in the absolute amount or concentration in which an element occurs, but only in the relative amounts in which the elements occur, the need for standards may be circumvented by using the so-called "ratio model" (Russ, 1974). The ratio model is also very useful if one desires absolute quantitation but does not have access to standards for all elements present in the specimen. The intensity ratio of two elements (I_1/I_2) in a thin specimen in which no absorption occurs may be converted to the concentration ratio (C_1/C_2) by

$$C_1/C_2 = p\,(I_1/I_2) \tag{17}$$

(where p is a proportionality constant). There are basically two approaches to determine p, one theoretical and one experimental approach. The theoretical approach is based on Equation 4 and is given by

$$p = \frac{\omega_2 Q_2 T_{E2} L_2}{\omega_1 Q_1 T_{E1} L_1} \tag{18}$$

(where ω is the fluorescence yield, Q the ionization cross-section, T_E the detector efficiency, and L a correction factor that is to be applied if different lines are compared, e.g., Lα or Kβ with Kα). The detector efficiency depends on the transmission of x-rays in the beryllium window and the silicon dead layer:

$$T_E = \exp - [(\mu\rho x)_{Be} + (\mu\rho x)_{Si}]$$ (19)

If the data required to determine T_E from Equation 19 are not provided by the manufacturer, or are not sufficiently accurate, T_E has to be determined experimentally. In that case it is just as easy, or even easier, to determine the whole product $\omega QT_E L$ experimentally. The experimental approach to determine p is therefore more common.

The microdroplet standards described previously are an excellent help in the determination of p. Assuming that no absorption occurs, p can be simply determined from the intensity ratio between two elements in the microdroplet of which the concentration ratio is known.

$$p = \frac{C_1/C_2}{I_1/I_2}$$ (20)

A curve giving the sensitivity of the analysis as a function of the atomic number can be prepared by 1) including a number of salts in carefully weighed amounts in the microdroplets (Davies and Morgan, 1976) or 2) analyzing a number of microdroplets, each containing only one salt (two or three elements in a known atomic ratio), taking care that a sufficient number of elements occurs at least twice (Chandler, 1976). Such a curve is shown in Figure 6.

Alternatively, other types of standards may be used, such as the thin cryosectioned standards. Standards made up of a solution of dithiocarbamates in epoxy resin contain sulfur and one other element in a known atomic ratio and may hence be used for the construction of a sensitivity curve. These alternatives are admittedly more tedious, but avoid the problems connected with absorption corrections, which may occur in microcrystals.

If one is interested in obtaining separate values for T_E this may be done by correcting the sensitivity curve for ω and Q of the various elements. The practical determination of p has to be repeated if the accelerating voltage is changed, since Q, and hence p, is dependent on this parameter.

If the concentration of one of the elements occurring in the analyzed volume can be accurately ascertained by other methods of analysis, this element can serve as an internal standard, and the determination of the relative concentration of the other elements using the ratio method gives at the same time their absolute concentration.

Theoretically, the use of the ratio method should not affect the accuracy of the quantitation. It is, however, recommended to restrict the use

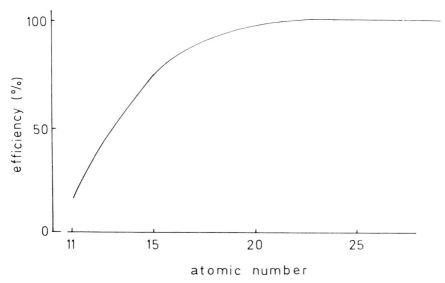

Figure 6.   Curve giving the overall efficiency (sensitivity) of analysis for the range of elements of particular interest for biological x-ray microanalysis. The curve was prepared using the microdroplet method, at relatively high accelerating voltage. The shape of the curve is sensitive to the experimental conditions (window thickness, accelerating voltage).

of the ratio model to elements that do not differ too much in atomic number. In addition, especially if microcrystals are used, the ratio model may be less accurate in the case of low-energy x-rays.

QUANTITATION IN THIN FILMS: THE CONTINUUM METHOD

To determine the absolute concentration of an element in the analyzed volume, the spectra recorded from the unknown specimen and the standard have to be compared. In a previous section, it was shown that the P/B ratio of an element was related to its concentration in the analyzed area of the specimen. As a first approximation, the P/B ratios in standard and specimen may be compared. Since, according to Equation 10, the continuum intensity is also dependent on the total composition of the sample, the P/B ratio of an element is not only related to its concentration, but is also influenced by the concentrations of other elements in the sample. The commonly used method for quantitation in biological thin sections is the "continuum method" (Hall, 1971; Hall et al., 1973). In this section, the theory of the method is presented and we show how to apply the continuum method to biological thin sections.

Theory

Equations 4 and 10 can be rewritten in a more simple form:

$$I_x = f_1 N_x \qquad (21)$$

$$I_c = f_2 \sum_c N_c Z^2{}_c \qquad (22)$$

(where f_1 and f_2 are constants). The mass of element x in the analyzed volume equals the number of atoms N_x multiplied by their atomic weight A_x. Similarly, the total mass of the analyzed volume is given by $\sum_c N_c A_c$. The mass fraction C_x of element x is then given by

$$C_x = \frac{N_x A_x}{\sum_c N_c A_c} \qquad (23)$$

After some simple algebraic manipulation, Equation 22 can be written as:

$$I_c = f_2 \sum_c N_c A_c (\overline{Z^2/A}) \qquad (24)$$

where $\overline{Z^2/A}$ is the weighted sum of Z^2/A for all elements of the specimen:

$$\overline{Z^2/A} = \sum_c [(Z_c{}^2/A_c) \cdot C_c] \qquad (25)$$

By combining Equations 21, 23, and 24 we get:

$$C_x = f_3 A_x \frac{I_x}{I_c} \overline{Z^2/A} \qquad (26)$$

(where f_3 is a constant). The peak-to-background ratio I_x/I_c can be written as R_x, the relative peak intensity of element x. Equation 26 is, of course, valid for both specimen and standard. By introducing the appropriate subscripts, sp and st, and dividing the resulting equations, we finally obtain:

$$C_{x,sp} = C_{x,st} \cdot \frac{R_{x,sp}}{R_{x,st}} \cdot \frac{\overline{Z^2/A}_{sp}}{\overline{Z^2/A}_{st}} \qquad (27)$$

since $f_3 A_x$ occurs in both equations and cancels out. Equation 27 can be used directly to calculate the concentration of element x in the specimen. The notation of Equation 27 is slightly different from the notation used by Hall (1971, 1973):

$$C_{x,sp} = A_x \cdot \frac{R_{x,sp}}{R_{x,st}} \cdot (N_x/\sum NZ^2)_{st} \cdot \overline{Z^2/A}_{sp} \qquad (28)$$

but the mathematical equivalence of Equations 27 and 28 is easily seen by evoking Equations 23 and 25.

Practical Procedures for Quantitation

After characteristic peaks and background have been separated according to one of the methods discussed in a previous section, the net intensity of the characteristic peak and the contiuum intensity can be determined, so that values for $R_{x,sp}$ and $R_{x,st}$ are obtained. The background radiation does not come only from the specimen; there may also be a considerable extraneous component from the specimen holder, grid, and supporting film. Since we are only interested in the background radiation coming from the specimen, the extraneous background has to be subtracted from the total radiation. Most of the practical problems associated with quantitation have to do with the accurate determination of the background. The total accuracy of quantitation depends heavily on the accuracy of the continuum determination.

The relative peak intensity R_x can be defined as

$$R_x = \frac{(P - b)_x}{W - W_b} \tag{29}$$

where $(P - b)_x$ is the net peak intensity (total peak minus background under the peak), W is the total continuum radiation ("white" radiation), and W_b is the white radiation from extraneous sources. The practical procedure for quantitation involves the determination of $(P - b)_x$, W, and W_b. Attention should be given to the following points:

1.   Net peak intensities of *all* elements present in the specimen should be obtained. Omission of one or more elements may lead to inaccuracies in the calculation (since $\overline{Z^2/A}_{sp}$ will be incorrectly calculated). The inaccuracies may be serious if the element is present in high concentration or has a high atomic number and thus contributes appreciably to $\overline{Z^2/A}_{sp}$.

2.   There is some disagreement in the literature on where to determine the background intensity. The main requirement is that the part of the spectrum that is used for continuum determination is free from all characteristic lines. In addition, the band chosen should contain sufficient counts that the statistical error is relatively low. Hall and coworkers argue that the continuum band should be taken at high energies, although well below the column voltage. For accelerating voltages of 60 kV or greater, a band of 20–40 keV is suggested. It is argued that, in comparison with the spectrum from a thin specimen, the spectrum of continuum radiation from the thick surrounding objects (grid, specimen holder) is biased toward low energies. Hence, if a high-energy band is used, the contribution of extraneous sources to the continuum is relatively low, which would increase the accuracy of the determination of $W - W_b$ (Hall, 1971; Hall et al., 1973; Gupta et al., 1977).

On the other hand, Shuman et al. (1976) argue that the grid-generated continuum is relatively low in the low-energy region, due to absorption of x-rays in this solid source. Since the continuum production is concentrated toward the ends of the electron trajectories (Reed, 1975), most of the extraneous continuum produced by grids and specimen holder originates several micrometers below the surface. Shuman et al. (1976) recommend the region of 1.34–1.64 keV (the Al Kα region) for the determination of continuum intensity, provided that no characteristic lines (Al Kα, As, Se, Br, and Rb Lα) are present. Alternatively, the Ar Kα region (2.81–3.11 keV) can be chosen; interfering characteristic lines are Ag and Cd Lα and U Mα. Recently, Roomans and Kuypers (1980) showed experimentally that the contribution of the extraneous background from a copper or titanium grid increases with increasing x-ray energy, and that the relative contribution of the specimen to the continuum is largest at low energies.

The accuracy of the determination of the continuum intensity of the specimen is considerably increased if the contribution of extraneous sources to the total contiuum can be minimized. Observations should therefore not be carried out near the edge of a grid or near a grid bar and, as discussed in the next section of this chapter, low atomic number materials for grids and specimen holders might be preferred. If the extraneous continuum is only a small part of the total contiuum intensity, the choice of the energy region is probably not critical, and the most convenient region can be chosen.

3.   Hall et al. (1973) suggested that the contribution of the specimen to the total continuum should be determined by measuring the continuum on an "empty" part of the grid, that is, a part of the grid not covered by a section, and subtracting this value from the total continuum intensity. In this way, the contribution of the specimen holder and surrounding parts of the microscope, the grid, the supporting film covering the grid, and, if present, the carbon layer(s), are subtracted, and only the continuum coming from the specimen is left. The measurement on the empty part of the grid should be carried out at the same relative position to the grid bars.

A complication arises, however, due to lateral scattering of electrons. Part of the x-rays originating from the grid are generated by electrons that have been scattered by the specimen. Since the amount of electrons scattered is dependent on the thickness of the layer through which they pass, more electrons are scattered by the section together with the supporting film alone. It is easily observed that, with a specimen placed on a copper grid, the copper signal obtained when the beam is placed on the specimen is more intense than the copper signal from the part of the grid covered by the support film only. In the case of thick sections, or sections of material with a high density, considerable errors may be made (Janossy and Neumann, 1976).

To correct for this effect, the following implementation was suggested (Gupta and Hall, 1979; Roomans and Kuypers, 1980): A measurement is made *on* a grid bar, or if the grid is of low atomic weight material, on the rim of the specimen holder, and the reciprocal value r of the P/B ratio of the metal is determined:

$$r = W_{me}/(P - b)_{me,gr} \qquad (30)$$

(where $(P - b)_{me,gr}$ is the number of characteristic metal counts measured on the grid bar, and W_{me} is the continuum intensity). From a measurement on a part of the grid not covered by section or support film one determines the intensity due to uncollimated radiation, $(P - b)_{me,u}$ and W_u. From a measurement on the support film, one determines:

$$W_f = W_{t,f} - r\,[(P - b)_{me,f} - (P - b)_{me,u}] - W_u \qquad (31)$$

(where $(P - b)_{me,f}$ is the number of metal counts measured on the supporting film due to lateral scattering and uncollimated radiation, $W_{t,f}$ is the continuum intensity measured on the supporting film, which includes the contribution of x-rays from the grid, and W_f is the *net* continuum intensity from the film). Now the extraneous background W_b is given by

$$W_b = W_f + r\,[(P - b)_{me,s} - (P - b)_{me,u}] + W_u \qquad (32)$$

(where $(P - b)_{me,s}$ is the number of metal counts measured on the sections due to lateral scattering and uncollimated radiation). The second term in Equations 31 and 32 gives the contribution of the laterally scattered electrons to the continuum.

A complete quantitative sample measurement thus includes:

1. A measurement on the section to determine $(P - b)_x$ for *all* elements present, $(P - b)_{me,s}$ and W.
2. A measurement on a part of the grid covered by the support film only to determine $W_{t,f}$ and $(P - b)_{me,f}$.
3. A measurement on the grid bar or the rim of the specimen holder to determine r.
4. A measurement on a completely empty part of the grid to determine W_u and $(P - b)_{me,u}$.
5. Calculation of W_f from Equation 31.
6. Calculation of W_b from Equation 32.
7. Calculation of $R_x = (P - b)_x/(W - W_b)$.

This should be done for both specimen and standard.

Calculations

In Equation 27, $R_{x,sp}$ and $R_{x,st}$ have been determined by recording the spectra of specimen and standard; $C_{x,st}$ and $\overline{Z^2/A}_{st}$ are known, since the composition of the standard is known. However, there are some problems in determining $\overline{Z^2/A}_{sp}$, since this parameter depends on the (yet unknown) concentrations of elements in the specimen. This apparent difficulty can be solved as follows.

It is assumed that the specimen consists of one element x with an atomic number of 11 or higher, of which the characteristic radiation can be measured, in an organic matrix. If the mass fraction of x equals C_x, the mass fraction of the matrix is equal to $1 - C_x$. If we define Z^2/A_m as the mean weighted value of Z^2/A for the organic matrix elements, then:

$$\overline{Z^2/A}_{sp} = C_x \cdot Z^2/A_x + (1 - C_x) \cdot Z^2/A_m \qquad (33)$$

Now, Equation 33 can be combined with Equation 27 and $C_{x,sp}$ can be written explicitly as:

$$C_{x,sp} = \frac{C_{x,st} \cdot \dfrac{R_{x,sp}}{R_{x,st}} \cdot \dfrac{Z^2/A_m}{\overline{Z^2/A}_{st}}}{1 - C_{x,st} \cdot \dfrac{R_{x,sp}}{R_{x,st}} \cdot \dfrac{Z^2/A_x - Z^2/A_m}{\overline{Z^2/A}_{st}}} \qquad (34)$$

This method is quite well feasible for one element x, and, as we see in one of the next sections, in the case that x has a relatively low atomic number, one may use the simplifying assumption that $\overline{Z^2/A}_{sp}$ can be represented by Z^2/A_m, thereby avoiding the rather bulky Equation 34. In principle, the above method can be extended to include the case of more than one element $(Z \geq 11)$ (Hall, 1971). An alternative method, the calculation of C_x and $\overline{Z^2/A}_{sp}$ by an iterative procedure, is, however, more attractive. This method is outlined and illustrated below.

The calculation is carried out in a series of steps:

1. As a first approximation, a value for $\overline{Z^2/A}_{sp}$ is assumed, e.g., $\overline{Z^2/A}_{sp} = Z^2/A_m$ or $\overline{Z^2/A}_{sp} = \overline{Z^2/A}_{st}$.
2. Using this assumed value of $\overline{Z^2/A}_{sp}$, the values of C_x for all elements with characteristic radiation are calculated.
3. Using the calculated values of C_x, a new value for $\overline{Z^2/A}_{sp}$ is calculated according to Equation 25.
4. Using the new value of $\overline{Z^2/A}_{sp}$, new values for C_x are calculated.
5. Steps 3 and 4 may be repeated: it will then be seen that each time the new calculated values of C_x and $\overline{Z^2/A}_{sp}$ will differ less from those of the preceding iteration. After a certain number of iterations, the values of C_x and $\overline{Z^2/A}_{sp}$ will not change appreciably. A criterion for accuracy may be set, and if this is fulfilled no new iteration is started.
6. The final values of C_x and $\overline{Z^2/A}_{sp}$ are recorded.

There is some uncertainty in the value of Z^2/A_m if the composition of the matrix is not exactly known. Representative values for Z^2/A_m for various kinds of matrix are given in Table 3; small variations in composition have only little effect.

Table 3.   Values of Z^2/A_m for some biological specimens*

Specimen	Z^2/A_m Value
Gelatin	3.31
Albumin	3.20
Frozen-dried tissue	
Liver	3.19
Heart	3.14
Spleen	3.15
Frozen-hydrated tissue	
Liver	3.53
Heart	3.54
Spleen	3.56
Epoxy resin (Spurr)	3.06

* For determination of Z^2/A_m, only elements not detectable by energy-dispersive x-ray microanalysis (hydrogen, carbon, nitrogen, oxygen) have been taken into account.

To make the reader familiar with procedure, the following calculated example is given. We assume the following data:

Standard Element	C_x	R_x	Z^2/A_x	Specimen R_x
Cl	0.038	1.58	8.15	0.84
K	0.014	0.65	9.23	0.17
Rb	0.030	0.70	16.02	0.00
Cs	0.047	1.24	22.76	1.95
Matrix				
Gelatin	0.871		3.31	"Cytoplasm" 3.25

$$\overline{Z^2/A}_{st} = 4.87$$

Before starting the calculation, it is necessary to define the original assumption of $\overline{Z^2/A}_{sp}$ and to define the convergence criterion. We assume that, as a first approximation, $\overline{Z^2/A}_{sp} = \overline{Z^2/A}_{st}$. It is convenient to define the convergence criterion as a maximal relative change in $\overline{Z^2/A}_{sp}$, e.g., of 0.1%. The first iteration will give the following results:

$$C_{Cl,sp} = 0.038 \times \frac{0.84}{1.58} \times \frac{4.87}{4.87} = 0.0202$$

$$C_{K,sp} = = 0.0037$$

$$C_{Rb,\,sp} = = 0.0$$

$$C_{Cs,sp} = = 0.0739$$

$$\text{Matrix} = 1 - 0.0978 = 0.9022$$

which gives $\overline{Z^2/A}_{sp} = 4.812$; the relative change in $\overline{Z^2/A}_{sp}$ is (4.87 −

$4.812)/4.812 = 0.012$, or 1.2%. This means that a second iteration has to be carried out:

$$C_{Cl,sp} = 0.038 \times \frac{0.84}{1.58} \times \frac{4.812}{4.87} = 0.0200$$

$$C_{K,sp} = \qquad\qquad = 0.0036$$

$$C_{Rb,sp} = \qquad\qquad = 0.0$$

$$C_{Cs,sp} = \qquad\qquad = 0.0730$$

$$\text{Matrix} = 1 - 0.0966 \qquad = 0.9034$$

and the new $\overline{Z^2/A}_{sp} = 4.794$; the relative change in $\overline{Z^2/A}_{sp}$ has now been reduced to 0.4%. The third iteration gives:

$$C_{Cl,sp} = 0.038 \times \frac{0.84}{1.58} \times \frac{4.794}{4.87} = 0.0199$$

$$C_{K,sp} = \qquad\qquad = 0.0036$$

$$C_{Rb,sp} = \qquad\qquad = 0.0$$

$$C_{Cs,sp} = \qquad\qquad = 0.0728$$

$$\text{Matrix} = 1 - 0.0963 \qquad = 0.9037$$

Now $\overline{Z^2/A}_{sp} = 4.789$ and the relative change in this parameter is only 0.1%. The final values are:

$$C_{Cl,sp} = 0.038 \times \frac{0.84}{1.58} \times \frac{4.789}{4.87} = 0.0199 \ (1.99\%)$$

$$C_{K,sp} = \qquad\qquad = 0.0036 \ (0.36\%)$$

$$C_{Rb,sp} = \qquad\qquad = 0.0$$

$$C_{Cs,sp} = \qquad\qquad = 0.0727 \ (7.27\%)$$

During the last iterations, only the last, unimportant, decimal was subject to change. A stricter limit than a relative change of 0.1% in $\overline{Z^2/A}_{sp}$ is thus not needed. A relative change of $n\%$ in $\overline{Z^2/A}_{sp}$ also implies a relative change of $n\%$ in each of the computed concentrations.

The relatively low number of iterations is, in part, due to the fortunate choice of the first approximation $\overline{Z^2/A}_{sp} = \overline{Z^2/A}_{st}$. Had we chosen to set $\overline{Z^2/A}_{sp} = Z^2/A_m$ instead, then 8–9 iterations would have been needed. However, seeing that both specimen and standard contained relatively large amounts of heavy elements, we could have expected that the ultimate value of $\overline{Z^2/A}_{sp}$ would be closer to $\overline{Z^2/A}_{st}$ than to Z^2/A_m; thus, the choice of the first approximation in the above situation was the most logical.

In the case discussed here, all elements present in the specimen were also present in the standard. This will not always be the case. If one or more elements occurring in the specimen are missing in the standard, absolute quantitation can still be carried out, provided that the intensity ratio between the "missing" elements and at least one suitable element in the specimen or standard is known. Then, the ratio model can be inserted into the continuum method, as in the example below.

Let it be assumed that in the specimen, in addition to the peaks for potassium, chlorine, and cesium also a peak for sulfur, with a relative intensity of 1.09, had been observed, and that it had been determined that equal amounts of potassium and sulfur would give, under the experimental conditions applied, characteristic intensities with a ratio of $1:0.8$. During the first iteration, the following result will be obtained:

$$C_{Cl,sp} = 0.0202$$
$$C_{K,sp} = 0.0037$$
$$C_{Rb,sp} = 0.0$$
$$C_{Cs,sp} = 0.0739$$

Then, according to Equation 20:

$$C_{S,sp} = \frac{1.09}{0.17} \times \frac{1.0}{0.8} \times 0.0037 = 0.0297$$

and it follows that the mass fraction of the matrix is 0.8725, which gives $\overline{Z^2/A}_{sp} = 4.954$; the other iterations are carried out according to the same general scheme.

Application of the ratio model within the calculation of absolute concentrations with the continuum method has the same limitations as the ratio model in general.

Computer Programs

Iterative calculations as described in the previous section are tedious to carry out by hand, but generally are very well suited for calculation by a computer. A number of manufacturers provide a continuum-method type correction procedure with their software. The correction procedure may be joined with a multiple least-squares method of separation of peak and background.

A separate computer program for quantitative analysis according to the continuum model is generally not difficult. A flow-chart of such a program is shown in Figure 7 and the calculations are those described above. The program will converge within a few iterations, if $\overline{Z^2/A}_{sp}$ is estimated well. It is also possible to include a convergence accelerator, e.g., the method of Wegstein (see Reed, 1975) in the program, but this should not be generally necessary.

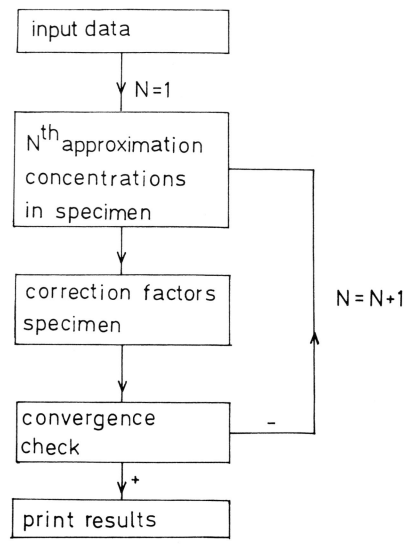

Figure 7.   General flow chart of a computer program for iterative calculation of elemental concentrations by x-ray microanalysis.

Interpretation of Quantitative Data

In interpreting the results of quantitative analysis, one should be aware that the microprobe measures the local weight fraction of the specimen *as it is under analysis*. If we disregard mass loss in the specimen during analysis, or assume that this is completely corrected for by the use of an appropriate standard, we may interpret the data as local weight fractions of the specimen *as it is after preparation*. Evidently, what one is interested in are data that refer to the tissue *in vivo*; also, for electrolytes, the most significant way to express the results is as mmol per 1 cell water rather than as mmol per kg dry or wet weight. For a biologically significant interpretation of the quantitative data, corrections of the original data have to be applied.

By far the most accurate method is that employed by Gupta, Hall and coworkers, which is based on the comparison of x-ray signals from the selected region of the (frozen) specimen before and after dehydration. Since the continuum originating from the specimen $(W - W_b)$ is proportional to the mass of the analyzed area, the dry mass fraction, f_d, can be calculated from

$$f_d = \frac{dry\ mass}{total\ mass} = \frac{R_{x,hydr}}{R_{x,dehydr}} = \frac{(W - W_b)_{dehydr}}{(W - W_b)_{hydr}} \tag{35}$$

while the aqueous fraction, f_h, is given by $f_h = 1 - f_d$. Knowledge of f_h and f_d allows the results to be expressed in terms of mmol per kg or 1 cell water (Gupta et al., 1977).

A difficulty arises when no analysis on hydrated sections can be carried out. In the presence of a "peripheral" standard—made by surrounding the tissue with, for example, albumin or dextran containing mineral salts in known concentration (Gupta et al., 1976; Dörge et al., 1974, 1978), the following procedure may be used.

This procedure is based on the assumptions that the mass per unit area is the same in the peripheral standard and the analyzed region of the specimen in the hydrated state (i.e., when the section is cut), and that the relative masses are not affected by shrinking during the drying process. In that case:

$$\frac{f_{d,sp}}{f_{d,st}} = \frac{(W - W_b)_{sp}}{(W - W_b)_{st}} \tag{36}$$

(Jones et al., 1979).

If an external standard is used, this simplifying approach is no longer valid and the dehydrated mass/hydrated mass ratio has to be determined by independent means to allow adequate correction of the data.

Even more problems arise when quantitative analysis of resin-embedded sections is to be carried out. In this case, a correction has to be applied that accounts for the replacement of water by resin, which has a higher density ($1.17\ g/cm^3$ for the Spurr low viscosity resin). This type of correction is usually difficult to make, and the accuracy of the results may be seriously impaired.

In interpreting quantitative results, attention should also be given to the spatial resolution of analysis, and to the fact that the incident beam usually is at an angle of 45°–60° to the specimen. Especially in the somewhat thicker sections, structures may collapse on top of each other during freeze-drying (Appleton, 1978). Another artifact described in the literature is ascribed to the scattering of electrons in lateral directions within the specimen. If the structure measured contains a low concentration of a certain element, but is surrounded by a matrix with a high concentration of this element, the low concentration area may show overly high values, since the matrix will also contribute to the spectrum (Tormey, 1979).

Approximations

Because of the work involved in the calculation of mass fractions according to the continuum method described above, it is tempting to consider an approximative procedure. One possible approximation is to neglect the cor-

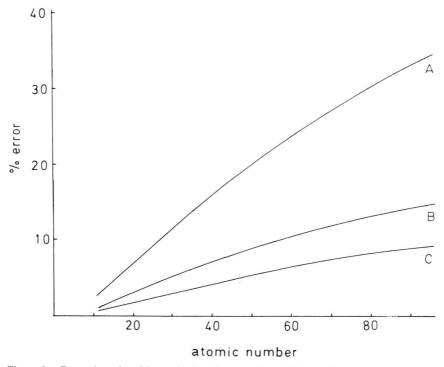

Figure 8. Errors introduced by neglecting the correction of the background intensity for the effect of Z^2/A, as a function of the atomic number (this is equivalent to the assumption that $\overline{Z^2/A}_{sp}$ equals $\overline{Z^2/A}_{st}$). In the calculation an organic matrix was simulated that contained one element with atomic number higher than 11. In case A the standard contained 4% of this element and the specimen 0.5%; in case B the standard contained 2% and the specimen 0.5%; in case C the standard contained 2% and the specimen 1%. The magnitude of the error depends 1) on the atomic number of the element concerned and 2) on the concentration difference between standard and specimen.

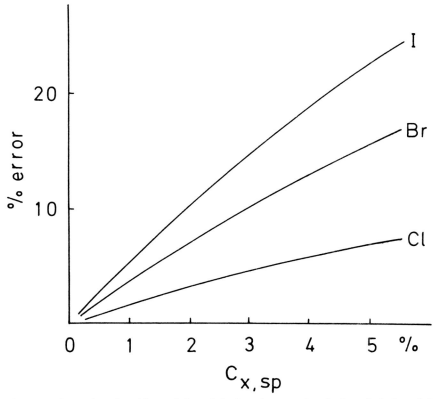

Figure 9.   Errors introduced by omission of the iterative procedure in the calculation of elemental concentrations (equivalent to the assumption that $\overline{Z^2/A_{sp}}$ equals Z^2/A_m). In this calculated example, $C_{x,st}$ is taken to be 5%. The magnitude of the error depends 1) on the atomic number of the element concerned and 2) on its concentration in the specimen.

rection of the background intensity for atomic-number effects, and assume a linear relationship between P/B ratio and elemental concentration for comparison of an unknown with a standard. This approximation amounts to the assumption that $\overline{Z^2/A_{sp}}$ = $\overline{Z^2/A_{st}}$ in Equation 27. Considerable errors may be introduced, depending on the concentration difference between specimen and standard, and on the atomic number of the element under consideration (Roomans, 1979a), as shown in Figure 8. Alternatively, one may assume that $\overline{Z^2/A_{sp}}$ = Z^2/A_m and avoid the iterative procedure. This assumption is permissible if the specimen contains only elements with a low atomic number at moderate or low concentrations (Hall et al., 1973). As shown in Figure 9, at high Z or high concentrations unacceptable errors can be made.

EXPERIMENTAL DIFFICULTIES

In quantitative x-ray microanalysis, a number of experimental difficulties are encountered that may seriously affect the accuracy of quantitation. The three main problems, which are now discussed in more detail, are: the extraneous background, specimen damage and mass loss during irradiation, and contamination of the specimen.

Extraneous Background

As pointed out previously, part of the background in the x-ray spectrum is not due to the specimen, but to extraneous sources; this presents a serious problem in quantitative analysis. Since the determination of mass fractions in thin sections rests on an accurate determination of the background due to the specimen, the correction of the total background radiation for the contribution of extraneous sources is of vital importance. Spurious characteristic and continuum x-rays can originate from the grid, the specimen holder, and parts of the specimen chamber, either due to uncollimated radiation (electron tails and high-energy x-rays) striking these targets or to electrons scattered by the specimen. Although the amount of uncollimated radiation and scattered electrons is relatively small compared to the total radiation striking the specimen, the mass of the grid and specimen holder is several orders of magnitude more than that of the specimen itself, so that the extraneous radiation can have the same or even greater intensity than the radiation from the specimen.

The uncollimated radiation can be monitored by placing the beam in the middle of a grid square (the grid should not be covered by a supporting film). The signal thus produced, the so-called "hole count," is due to uncollimated radiation. The principal cause of the hole count is still a matter of controversy, but it may also depend on the type of instrument used. The major contributions seem, however, to arise from high-energy x-rays generated in the condenser system, and from stray electrons not eliminated by the condenser apertures (Williams and Goldstein, 1977). Examination of the design of the illumination system ought to give a clue about whether this stray radiation can occur (Bentley et al., 1979). The fact that x-ray fluorescence gives a higher P/B ratio than generation of x-rays by electrons may be used in identifying the species responsible for the hole count (Bentley et al., 1979).

The major source of the high-energy x-rays is the second condenser (C_2) aperture, which, although it is thick enough to stop 100 kV electrons, will transmit a considerable part of the characteristic and continuum x-rays produced in the C_2 aperture at higher beam voltages. This radiation can be reduced by placing a thick (0.5–1 mm) heavy metal (Pt, Pb) insert directly below the aperture holder, or by inserting a non–beam-defining collimator system between the C_2 aperture and the specimen (Williams and Goldstein, 1977; Bentley et al., 1979). The insert will also prevent any stray electrons passing around the C_2 aperture from reaching the specimen. The actual solution chosen for the reduction of uncollimated radiation will depend on the type of microscope used.

The scattering of electrons by the specimen cannot be prevented, but the contribution of this phenomenon to the total spectrum can be minimized by using low atomic number grids and specimen holders. Grids made of beryllium, graphite-polymer, and nylon are commercially available. Care should be exercised in the handling of beryllium grids: although beryllium metal itself is not toxic, the grids are brittle and break easily, and the metal will powder if crushed under the foot. Inhalation of particulate beryllium is extremely dangerous. Likewise, the grids should not be cleaned with acid, to avoid formation of toxic beryllium salts (Panessa et al., 1978). Graphite polymer grids were recently developed expressly for x-ray microanalysis and give even less extraneous radiation than a beryllium grid, because they are less dense (Panessa et al., 1978). Woven nylon grids contain some titanium and a variable Ti $K\alpha$ peak is observed when analysis is carried out on these specimens. The grids are not flat and one must be aware that a grid bar could block x-rays emerging from the specimen. This artifact can easily be recognized: the maximum of the continuum curve, which is at about 2 keV for biological specimens, is shifted toward higher energies if low-energy x-rays are blocked.

Specimen holders of beryllium or graphite (Liljesvan and Roomans, 1976) have been developed for several types of instruments and may be commercially available. The graphite holders have to be rather thick, so that some risk for blocking of x-rays is present if analysis is carried out on an eccentric part of the grid.

Although these low atomic number grids and holders have considerable advantages in reducing the extraneous background, their use raises the problem that it is now more difficult to determine the exact contribution of grid and holder to the total spectrum, since no characteristic signal from these sources is obtained (Shuman et al., 1976). This information is thus missing in the quantitative procedure described in the previous section (Equations 30–32). It is not yet clear how this affects the total accuracy of the determination and further study on this point is needed.

Specimen Damage and Mass Loss

Electron irradiation produces chemical changes in all biological specimens. Ionizations of molecules in the specimen leads to scission and subsequent vaporization of small fragments. Loss of mass, due to loss of the elements carbon, hydrogen, nitrogen, and oxygen from the specimen (presumably in the form of small organic molecules like CO_2, CH_4, and C_2H_6, etc.), occurs almost instantaneously.

Specimen damage due to irradiation has been a cause of concern from the beginning of electron microscopy, and a number of studies have been carried out to determine its extent and its origin. Depending on the composition of the specimen and the experimental conditions, losses of 10%–90% of organic material have been reported; for biological thin sections losses of 30%–40% appear typical (Hall and Gupta, 1974). Hydrogen, nitrogen, and oxygen are lost to a greater extent than carbon, but are still present in the final product (Bahr et al., 1965). The effects depend in part on the radiation dose rate, and appear to be in the first place a radiation damage effect rather than temperature damage (Stenn and Bahr, 1970a, b). However, especially in x-ray microanalysis, where, in contrast to conventional electron microscopy, the electron beam is kept focused on a small spot on the specimen for a considerable time, temperature effects may also be of importance. Increasing specimen conductivity may help to reduce thermal effects (Chandler, 1977). Mass loss and chemical changes occur almost instantaneously and level off after some time (Stenn and Bahr, 1970a). The loss of carbon, hydrogen, nitrogen, and oxygen affects the remaining elements in the specimen. The mass fraction of these elements is increased during irradiation. This has serious consequences for quantitative analysis, which is, in fact, being carried out on a specimen of which the composition changes during analysis. Although determination of the mass loss during irradiation can be carried out, it is unfeasible to do this every time quantitative x-ray analysis is performed. One approach to solving the problem is to have the standard resemble the specimen, so that mass loss occurs in both specimen and standard to the same extent, and corrections cancel. The assumption of comparable mass loss in specimen and standard seems reasonable, and appears to work well in practice, but it has to be admitted that a fundamental uncertainty is introduced in the quantitation.

The rate of mass loss can be reduced considerably by lowering the temperature (Hall and Gupta, 1974; Shuman et al., 1976). Although the initial radiation damage is not temperature sensitive, the vaporization of the small fragments can be arrested at the temperatures of $-110°$ to $-130°$ used in these studies. As the technical difficulties of including a cold stage in transmission electron microscopes are now gradually being solved, the low-temperature approach appears to be extremely promising for the future.

Not only are the light elements constituting the organic matrix lost, but also the elements detectable by energy-dispersive analysis. The halogen elements are notoriously unstable (Bahr et al., 1965; Chandler, 1976; Hobbs, 1978; Roomans, 1980), especially when they occur in crystalline form. The main instrumental factor determining the extent of halogen loss is the beam current (Chandler, 1976). Also, other elements, e.g., bismuth, may be lost from the specimen (Fowler and Goyer, 1975).

Contamination

Hydrocarbons of the residual gas in the column condense on solid surfaces within the vacuum. Under the impact of the electron beam, ionization takes place, and a polymer is formed by cross-linking of the hydrocarbon radicals. After further irradiation, only a carbon skeleton remains deposited on the specimen. Under some conditions a contamination layer may build up on the specimen at a rate of 0.1 nm/s or more. With the commonly used analyzing times, this implies the possibility of the contamination layer approaching a thickness in the same order of magnitude as the thin section examined. The phenomenon of mass loss may be partly or even completely obscured by the addition of contaminating material to the specimen.

In addition to degradation of image quality, contamination has two effects on quantitative analysis: x-rays emerging from the specimen may be absorbed in the contamination layer, and the mass fraction of elements of interest is decreased.

To attack the problem of contamination, one must first identify its source. There are three main sources for contamination (Hren, 1978; Isaacson, 1977; Isaacson et al., 1979): 1) the electron microscope, 2) the specimen itself, and 3) specimen handling during preparation. With the improvement of vacuum techniques, the contribution of the electron microscope is being considerably reduced. Biological specimens are semi-infinite sources of contaminant (the organic molecules lost from the specimen may in part "reappear" as contamination) and this poses a fundamental limitation. Considerable contamination may be introduced during the preparation of the specimen for analysis. If the specimen comes into contact with contaminated surfaces on its way to the microscope, contamination of the specimen itself is unavoidable. Obviously, every effort should be made to ensure that the specimen (including grid and specimen holder) enters the microscope as clean as possible, although this is certainly not easy to attain. In addition, the anti-contamination cold trap should always be used during analysis.

LIMITS OF DETECTION

Because x-ray microanalysis is a localized analysis, it is not possible to link it with a more sensitive method that would allow the detection of even smaller quantities. The limit of detection is therefore of great importance in x-ray microanalysis.

The minimal detectable concentration of an element (under given experimental conditions) is conventionally determined in the following way: the number of net counts in a peak is defined as $(P - b)$ and the background under the peak as b. According to statistics, the standard deviation of $(P - b)$ can be given as

$$\sigma_{(P-b)} = (P + b)^{1/2} \qquad (37)$$

At the very low concentrations at which the limits of detection are approached, the number of net peak counts is very low; by approximation, P equals b, so that

$$\sigma_{(P-b)} = (2b)^{1/2} \qquad (38)$$

For an element to be identified with 95% certainty, the number of net counts should exceed two times the standard deviation. If a higher degree of certainty is required, the detection limit can be set by three or even six times the standard deviation.

If a standard with concentration C_{st} of the element gives a net peak count of $(P - b)_{st}$, then a specimen with the minimal detectable concentration C_{min} will give a net peak count of $(P - b)_{min}$, according to

$$\frac{(P - b)_{min}}{C_{min}} = \frac{(P - b)_{st}}{C_{st}} \qquad (39)$$

Since $(P - b)_{min}$ should equal at least twice the standard deviation in $(P - b)_{min}$, according to Equation 38,

$$(P - b)_{min} > 2(2b_{min})^{1/2} \qquad (40)$$

and, on combining Equations 39 and 40, the minimal detectable concentration can be given as

$$C_{min} = \frac{2C_{st}(2b_{min})^{1/2}}{(P - b)_{st}} \qquad (41)$$

Although the above approach to calculating the minimal detectable concentration is the most commonly used (Chandler, 1977), it is a simplification. Alternative methods with a greater validity have been suggested by Shuman et al. (1976) and Kotrba (1977).

As shown in Equation 41, the minimal detectable concentration is dependent on the background intensity. A considerable part of the background comes from extraneous sources, and the sensitivity of analysis can be greatly improved by minimizing the extraneous background. Ways of doing this have been discussed in the previous section of this chapter. Under favorable conditions, concentrations as low as 0.005% might be detected (Shuman et al., 1976).

The minimal detectable mass can be determined theoretically as follows: from Equation 4 the characteristic intensity per incident electron obtained from one atom can be calculated; this figure should be corrected for detector efficiency and the solid angle of detection. If one knows the current density, the count rate per atom can be calculated. Shuman et al.

predicted a minimal detectable mass of about 10^{-19} g; this prediction was tested for the iron core of single ferritin molecules and good experimental agreement with the theory was shown (Shuman and Somlyo, 1976).

It is obvious that an element can be detected only if *both* its minimal detectable amount and its minimal detectable concentration are exceeded in the analyzed volume.

APPLICATIONS

The biological applications of quantitative x-ray microanalysis range from studies on microorganisms to studies on human material. Especially in the fields of ion- and fluid-transporting epithelia and muscle physiology, x-ray microanalysis has made important contributions to our knowledge and understanding of biological systems. A salient result is also that the older theory of the nucleus as a repository for excess cell sodium can no longer be held. Studies on several cell types show that nuclear ionic concentrations follow the average cytoplasmic concentrations and that the dominant nuclear cation, measured in situ, is potassium rather than sodium (Gupta and Hall, 1979; Jones et al., 1979; Roomans and Sevéus, 1976; Somlyo et al., 1979).

Ion-Transporting Epithelia

Studies on ion distribution in the tubular fluid-secreting insects were carried out by Gupta and coworkers (Gupta et al., 1976, 1977; Gupta and Hall, 1979) to elucidate the mechanism of the iso-osmotic fluid secretion. Local sodium, potassium, and chlorine concentrations were measured in *Rhodnius* Malpighian tubules and in *Calliphora* salivary glands. The use of frozen sections, which were first analyzed in the hydrated and then in the dehydrated state, permitted the results to be expressed as mmol per liter of fluid, thus allowing comparison of the microprobe results with microelectrode measurements. Good agreement was found when the potassium ions occurred free in solution. In *Rhodnius* Malpighian tubules, measurements were carried out in the microvillate brush border and in the cytoplasm. Although the ion concentrations were highest in the brush border, as was expected, it was surprising to find that the concentrations of sodium, potassium, and chlorine increased from the base to the tip of the brush border. This contradicted an earlier model of fluid transport, and pointed to the possibility of a significant water flow through the leaky septate junctions. In *Calliphora* salivary glands, the apical membrane is deeply invaginated to form a pair of blind secretory canaliculi. These canaliculi were found to be hypertonic with regard to the other tissue compartments. The concentration profile in the canaliculi showed a decrease in ionic concentration from base to tip; this decrease was not gradual, but occurred suddenly in the middle of the canaliculi. This concentration profile is attributed to the existence of both a transcellular and a paracellular water flux.

Electrolyte concentrations in frog skin were determined on 1 μm thick freeze-dried cryosections, to elucidate the mechanism of transepithelial sodium transport (Rick et al., 1978). Granular, spiny and germinal cells showed low sodium and high potassium concentration; the sodium to potassium ratio could be reversed by addition of ouabain to the medium. The cornified cells, on the other hand, had an ionic composition similar to that of the external medium.

Measurements of ion concentration profiles in epithelial cells and inter-cellular spaces of rabbit ileum revealed the existence of a narrow zone of peripheral cytoplasm immediately adjacent to the cell membrane, which has much less dry mass than the cytoplasm in the cell core, but has a much higher average concentration of potassium (Gupta et al., 1977; Gupta and Hall, 1979). It is speculated that this zone acts as a channel for rapid con-duction of ions. The lateral intercellular spaces were found to be hypertonic with respect to the bathing solution and the capillary fluid. A detailed study of the concentration profile along the intercellular space showed a maximum in the ion concentration about halfway along the channel.

Electron probe x-ray microanalysis has frequently been applied to study the handling of ions by the renal tubular epithelium (reviewed by Gar-land et al., 1978), using the microdroplet technique. Comparative investiga-tions showed the x-ray analytical methods to give similar values to more conventional techniques and, because of the broader spectrum of analysis and the extremely small amounts of fluid required, more extensive informa-tion is gained. Most of the studies have been on mammalian kidney func-tion, but work has also begun on amphibians and fish. Sodium, magnesium, potassium, calcium, chlorine, and phosphate fluxes have been studied under a variety of experimental conditions. Transtubular water movements have been studied using ferrocyanide (Fe) and vitamin B_{12} (Co) as a tracer; these compounds appear to be handled in a similar manner to inulin, which has no suitable marker element for x-ray microanalysis.

Muscle

The distribution of various elements in frog striated muscle was determined on thin, dried, cryosections of rapidly frozen muscles (Somlyo et al., 1977a). No chlorine was sequestered in the terminal cisternae of resting muscles, and mitochondria partially excluded chlorine. The calcium concentration of the cisternae was much higher than that of the cytoplasm. Hypertonicity produced swollen vacuoles adjacent to the Z-lines, which contained high concentrations of sodium and chlorine, but no calcium; granules containing calcium, magnesium, and phosphorus were found in the sarcoplasmic reticulum.

A study on elemental distribution in the swimbladder muscle of the toadfish showed that neither chlorine nor sodium was sequestered in the sarcoplasmic reticulum, in contrast to calcium (Somlyo et al., 1977b).

Analysis on vascular smooth muscle of the rabbit (Somlyo et al., 1979) showed that the mitochondrial calcium content of normal fibers was as low as 0.8 ± 0.5 mmol per kg. In damaged fibers, massive mitochondrial accumulation of up to 2 mol of calcium per kg dry weight could be demonstrated. In addition it could be shown that nuclei did not contain more calcium than the cytoplasm.

Other Applications

Roomans and Sevéus (1976) carried out quantitative x-ray microanalysis of thin, dried cryosections of the yeast *Saccharomyces cerevisiae*. The intracellular potassium was partially replaced by rubidium and cesium and it was found that these ions had the same intracellular distribution as potassium, suggesting that the major intracellular compartments are equally accessible to these monovalent cations. In a subsequent study (Roomans, 1980) quantitative analysis of calcium- and phosphate-rich cytoplasmic granules in yeast was carried out.

Red algae are very rich in brominated phenols; one of these occurs in the red pigment floridorubin, which has been isolated from several red algae. A series of quantitative investigations of the bromine content of chloroplasts and cuticles was carried out using epoxy-resin embedded sections (Pedersén et al., 1980a,b; Roomans, 1979b). In *Polysiphonia nigrescens* the cuticle contained about 1–2% bromine, depending on age, and the chloroplasts 2.5–4.0%. In *Chondrus crispus*, the cuticle contained 2.2–2.3% bromine, and in the chloroplasts concentrations of 6%–7% bromine were measured (all values in percent of the fresh weight). Pallaghy (1973) investigated the distribution of chlorine and potassium in freeze-substituted leaves of *Zea mays*. Light-induced shifts in the ionic concentrations in cytoplasm, vacuole, and chloroplasts were studied.

Calcium binding in the contractile spasmoneme of the ciliate *Zoothamnium* was studied by Routledge et al. (1975). A quantitative determination of the calcium content of contracted and extended spasmonemes allowed a confirmation of the model for contraction suggested by the authors.

The subcellular distribution of sodium, potassium, and chlorine in chick red blood cells throughout embryonic development was studied by Jones et al. (1979). A salient finding was that the nuclear sodium concentration resembled that of the cytoplasm, rather than that of the medium. It could be shown that high nuclear sodium levels were preparative artifacts. Kirk et al. (1978b) investigated changes in elemental composition in dog red blood cells during maturation.

Jessen et al. (1974, 1976) studied the composition of keratohyalin granules from the epidermis and from the lingual and esophageal epithelium of the rat. The results allowed the distinction of two chemically different types of granules, one type (the "single" granule) containing only a sulfur-rich component (2.5%–3.6% S), whereas composite granules also contained low-sulfur keratohyalin (0.6%–0.9% S).

Several cases of "green hair" were examined by Forslind et al. (1979). High concentrations of copper (due to elevated concentrations of copper in tap water) were found in these hairs. Measurement of the concentration profile of copper in cross-sections of the hair showed that the copper had completely penetrated the hair, although a concentration gradient (from 0.5% to 0.3%) was still discernible. Concentration profiles of calcium, phosphores, and sulfur were measured in sections of predentine of young rat incisors (Nicholson et al., 1977); zones of calcium enrichment could be demonstrated.

Several quantitative studies have been carried out with the microdroplet technique, such as the determination of the inner ear lymph composition in various animals (Peterson et al., 1978) and of the elemental composition of mouse blastocele fluid (Borland et al., 1977).

CONCLUDING REMARKS

This chapter has dealt with the quantitative analysis of thin sections of biological material. Relatively few quantitative studies have been done on bulk material of biological origin, because of the inferior spatial resolution of bulk analysis. For some problems, however, analysis with a spatial resolution at the cellular level gives results of sufficient interest. The amount of time and work saved by using a simpler preparation method may then become a factor of importance.

The theory of quantitative x-ray microanalysis is more complicated with bulk specimens than with thin sections. With a bulk specimen (defined as a specimen that is infinitely thick with respect to the electron beam, so that no electrons reach the far surface of the specimen), the observed intensity of a characteristic line depends on three factors: 1) the primary generated intensity, 2) the fraction of primary generated intensity absorbed in the specimen, and 3) the secondary generated intensity. These three factors should be taken into account when specimen and standard are compared.

1. *The primary generated intensity.* As shown in Equation 1, the number of ionizations caused by an incident electron is dependent on the ionization cross-section, Q, and according to Equation 2, Q is in turn dependent on the energy of the electron. In the definition of a thin section

we assumed that the electrons reached the far surface of the section without losing an appreciable fraction of their initial energy, i.e., that their energy was approximately constant and equal to E_o. This assumption does not hold for a bulk specimen: here, the energy of the incident electron decreases along its path through the sample until it has given off all its energy. The rate of energy loss, called the stopping power (S), is dependent on the initial energy of the electron and on the composition of the specimen. To calculate the primary generated radiation, these factors have to be taken into account. In addition, a correction has to be made for the fraction of electrons that is backscattered from the specimen. This fraction depends on the accelerating voltage and on the average atomic number of the sample; in biological specimens it is usually low. Since the backscatter factor (R) and the stopping power are dependent on the same parameters, these corrections are taken together as one correction (R/S) called the "atomic number correction," Z.

2. *Absorption of primary intensity.* The second criterion in the definition of a thin section, namely that the x-rays generated in the section are able to reach the surface of the section without being appreciably absorbed, cannot be applied to bulk specimens. The extent to which primary generated x-rays are absorbed depends, as shown in Equation 16, on the energy of the x-ray, the composition of the sample (μ, ρ), and the path length of the x-ray in the sample (x). This path length is not the same for all generated x-rays, because x-rays are produced at various depths. The absorption correction (A) is hence also dependent on the depth-distribution of x-ray production, which in turn is a function of the accelerating voltage and the composition of the specimen.

3. *The secondary generated intensity.* Secondary fluorescence may be a consequence of absorption of x-rays in the sample. It occurs only if both the element-emitting x-rays and the elements-absorbing x-rays are present in high concentrations, and in biological specimens the correction for secondary fluorescence (F) can usually be neglected.

The complete correction procedure is called the ZAF correction and is carried out using an iterative procedure of greater complexity than in the case of quantitative analysis of thin sections. The established ZAF-correction programs (for a review see Martin and Poole, 1971, and Reed, 1975) have mainly been developed for metallurgical problems and contain several assumptions that may not apply to biological specimens. Whereas in metallurgical specimens the sum of the concentrations of the elements detectable by energy-dispersive analysis adds up to 100% (or, alternatively, the elements are present as their oxide with a known ratio of metal to oxygen), a typical biological specimen may contain only a few percent of elements with an atomic number higher than 10. On the other hand, if the specimen contains only elements of relatively low atomic number in low or moderate concentrations, the ZAF-correction factors may be so small that omission of the correction procedure may cause only small errors (Zs. Nagy et al., 1977; Sumner, 1978). For a more general approach, the existing ZAF routines have to be adopted for use with biological specimens (Boekestein et al., 1980).

Applications of quantitative x-ray microanalysis of bulk specimens include studies on potassium distribution in hyphae of the mold *Neurospora crassa* (Roomans and Boekestein, 1978), distribution of electrolytes in rat liver and brain cells (Pieri et al., 1977), and analysis of sulfur and phosphorus in different types of human muscle fibers (Wróblewski et al., 1978a). The application of thick cryosections in pathology (Wróblewski et al., 1978b; Trump et al., 1979) could be a field of future developments. Quantitative x-ray analysis of biological bulk specimens may need and deserve a more extensive description in future reviews.

A somewhat different case is presented by sections that are too thick to conform to the criteria of a thin section, but which on the other hand are not infinitely thick with respect to the electron beam, so that part of the incident electrons hit the substrate on which the specimen is placed. The quantitative procedure may include parts of the ZAF correction, and in addition a determination of sample thickness. Although the subject has been treated theoretically (Hall, 1975; Warner and Coleman, 1974), this type of analysis has not been widely practiced.

Reviewing the present state of quantitative x-ray microanalysis of thin sections, one may conclude that considerable progress has been made during the last few years. The standards developed for biological microanalysis are not yet perfect, in the sense that they do not affect the total accuracy of the quantitative determination at all, but neither can it be said that the preparation of standards is a major bottleneck in quantitative analysis. Accurate separation of peaks from the background, especially in low-energy region that is of interest to biologists, may still require some work. Major problems still reside in the poorly controlled mass loss and contamination during analysis, although the low-temperature approach appears promising (Shuman et al., 1976). Under optimal conditions, the variations in measured elemental concentrations in a biological sample can be within the range of counting statistics, and the variance within the sample is less than the variance between samples from different animals (Somlyo et al., 1977a); an accuracy of about 10% should be obtainable (Shuman et al., 1976). Application of quantitative procedures in x-ray microanalysis of biological specimens has expanded the scope of this technique and provided valuable results and will doubtlessly be of even more importance in the future.

Thanks are due to Ms. Ulla-Britt Åkerblom for assistance in the preparation of this manuscript.

This work was supported in part by a grant from the Swedish Natural Science Research Council.

LITERATURE CITED

Appleton, T. C. 1978. The contribution of cryoultramicrotomy to x-ray microanalysis in biology. In: *Electron Probe Microanalysis in Biology* (Erasmus, D. A., ed.), p. 148. Chapman and Hall, London.

Bahr, G. F., Johnson, F. B., and Zeitler, E. 1965. The elementary composition of organic objects are electron irradiation. Lab. Invest. 14:1115.

Bentley, J., Zaluzec, N. Y., Kenik, E. A., and Carpenter, R. W. 1979. Optimization of an analytical electron microscope for x-ray microanalysis. Scanning Electron Microscopy II:581.

Boekestein, A., Roomans, G. M., Stols, A. L. H., and Stadhouders, A. M. 1980. A ZAF-correction procedure for electron probe x-ray microanalysis of biological bulk specimens. A theoretical analysis of some inaccuracies involved. (In preparation.)

Borland, R. M., Biggers, D., and Lechene, C. P. 1977. Studies on the composition and formation of mouse blastocoele fluid using electron probe microanalysis. Dev. Biol. 55:1.

Chandler, J. A. 1976. A method for preparing absolute standards for quantitative calibration and measurement of action thickness with x-ray microanalysis of biological ultrathin specimens in EMMA. J. Microscopy 106:291.

Chandler, J. A. 1977. X-ray microanalysis in the electron microscope. In: *Practical Methods in Electron Microscopy* (Glauert, A. M., ed.), Vol. 5. North-Holland Publishing Company, Amsterdam.

Davies, T. W., and Morgan, A. J. 1976. The application of x-ray analysis in the transmission electron analytical microscope (TEAM) to the quantitative bulk analysis of biological microsamples. J. Microscopy 107:47.

Dörge, A., Gehring, K., Nagel, W., and Thurau, K. 1974. Intracellular Na^+-K^+ concentrations of frog skin at different states of Na-transport. In: *Microprobe Analysis as Applied to Cells and Tissues* (Hall, T., Echlin, P., and Kaufmann, R., eds.), p. 337. Academic Press, London.

Dörge, A., Rick, R., Gehring, K., and Thurau, K. 1978. Preparation of freeze-dried cryosections for quantitative x-ray microanalysis of electrolytes in biological soft tissues. Pflügers Arch. 373:85.

Fiori, C. E., Myklebust, R. L., Heinrich, K. F. J., and Yakowitz, H. 1976. Prediction of continuum intensity in energy-dispersive x-ray microanalysis. Anal. Chem. 48:172.

Forslind, B., Afzelius, B. A., Liljesvan, B., and Roomans, G. M. 1979. Green hairs—elementary x-ray analysis in the electron microscope. Proc. First International Hair Congress, Hamburg.

Fowler, B. A., and Goyer, R. A. 1975. Bismuth localization within nuclear inclusions by x-ray microanalysis. Effects of accelerating voltage. J. Histochem. Cytochem. 23:722.

Garland, H. O., Hopkins, T. C. Henderson, I. W., Haworth, C. W., and Chester Jones, I. 1973. The application of quantitative electron probe microanalysis to renal micropuncture studies in amphibians. Micron 4:164.

Garland, H. O., Brown, J. A., and Henderson, I. W. 1978. X-ray analysis applied to the study of renal tubular fluid samples. In: *Electron Probe Microanalysis of Biology* (Erasmus, D. A., ed.), p. 212. Chapman and Hall, London.

Gupta, B. L., and Hall, T. A. 1979. Quantitative electron probe x-ray microanalysis of electrolyte elements within epithelial tissue compartments. Fed. Proc. 38:144.

Gupta, B. L., Hall, T. A., Maddrell, S. H. P., and Moreton, R. B. 1976. Distribution of ions in a fluid transporting epithelium determined by electron probe x-ray microanalysis. Nature 264:284.

Gupta, B. L., Hall, T. A., and Moreton, R. B. 1977. Electron probe x-ray microanalysis. In: *Transport of Ions and Water in Animals* (Gupta, B. L., Moreton, R. B., Oschman, J. L., and Wall, B. J., eds.), p. 83. Academic Press, London.

Hall, T. A. 1971. The microprobe assay of chemical elements. In: *Physical Techniques in Biochemical Research* (Oster, G., ed.), Vol. 1A, p. 157. Academic Press, New York.

Hall, T. A. 1975. Methods of quantitative analysis. J. Microscopie Biol. Cell. 22:271.

Hall, T. A., and Gupta, B. L. 1974. Measurement of mass loss in biological specimens under an electron microbeam. In: *Microprobe Analysis as Applied to Cells and Tissues* (Hall, T., Echlin, P., and Kaufmann, R., eds.), p. 147. Academic Press, London.

Hall, T. A., and Peters, P. D. 1974. Quantitative analysis of thin sections and the choice of standard. In: *Microprobe Analysis as Applied to Cells and Tissues* (Hall, T., Echlin, P., and Kaufmann, R., eds.), p. 239. Academic Press, London.

Hall, T. A., Clarke Anderson, H., and Appleton, T. 1973. The use of thin specimens for x-ray microanalysis in biology. J. Microscopy 99:177.

Heinrich, K. F. J. 1966. X-ray absorption uncertainty. In: *The Electron Microprobe*, p. 296. Wiley, New York.

Hobbs, L. W. 1978. Radiation damage in electron microscopy of inorganic solids. Ultramicroscopy 3:381.

Hren, J. J. 1978. Specimen contamination in analytical electron microscopy: sources and solutions. Ultramicroscopy 3:375.

Ingram, M. J., and Hogben, C. A. M. 1967. Electrolyte analysis of biological fluids with the electron microprobe. Anal. Biochem. 18:54.

Ingram, F. D., Ingram, M. J., and Hogben, C. A. M. 1974. An analysis of the freeze-dried, plastic embedded electron probe specimen preparation. In: *Microprobe Analysis as Applied to Cells and Tissues* (Hall, T., Echlin, P., and Kaufmann, R., eds.), p. 119. Academic Press, London.

Isaacson, M. 1977. Specimen damage in the electron microscope. In: *Principles and Techniques of Electron Microscopy: Biological Applications*, Vol. 7 (Hayat, M. A., ed.). Van Nostrand Reinhold Co., New York.

Isaacson, M., Kopf, D., Ohtsuki, M., and Utlaut, M. 1979. Contamination as a psychological problem. Ultramicroscopy 4:97.

Janossy, A. G. S., and Neumann, D. 1976. Quantitative x-ray microanalysis: microcrystal standards and excessive background. Micron 7:225.

Jessen, H., Peters, P. D., and Hall, T. A. 1974. Sulphur in different types of keratohyalin granules: a quantitative assay by x-ray microanalysis. J. Cell Sci. 15:359.

Jessen, H., Peters, P. D., and Hall, T. A. 1976. Sulphur in epidermal keratohyalin granules: a quantitative assay by x-ray microanalysis. J. Cell Sci. 22:161.

Jones, R. T., Johnson, R. T., Gupta, B. L., and Hall, T. A. 1979. The quantitative measurement of electrolyte elements in nuclei of maturing erythrocytes of chick embryo using electron probe x-ray microanalysis. J. Cell Sci. 35:67.

Kirk, R. G., Crenshaw, M. A., and Tosteson, D. C. 1974. Potassium content of single human red cells measured with an electron probe. J. Cell Physiol. 84:29.

Kirk, R. G., Bronner, C., Barba, W., and Tosteson, D. C. 1978a. Electron probe microanalysis of red blood cells. I. Methods and evaluation. Am. J. Physiol. 235:C245.

Kirk, R. G., Lee, P., and Tosteson, D. C. 1978b. Electron probe microanalysis of red blood cells II. Cation changes during maturation. Am. J. Physiol. 235:C251.

Kotrba, Z. 1977. The limit of detectability in x-ray electron probe microanalysis. Mikrochim. Acta 1977/II:97.

Lechene, C. 1970. The use of the electron microprobe to analyse very minute amounts of liquid samples. Proc. Fifth National Conference Electron Probe Microanalysis, 32A.

Lehrer, G. M., and Berkeley, C. 1972. Standards for electron probe microanalysis of biological specimens. J. Histochem. Cytochem. 20:710.

Liljesvan, B., and Roomans, G. M. 1976. Use of pure carbon specimen holders for analytical electron microscopy of thin sections. Ultramicroscopy 2:105.

Martin, P. M., and Poole, D. M. 1971. Electron probe microanalysis: the relation between intensity ratio and concentration. Metall. Rev. 150:19.

Morel, F., and Roinel, N. 1969. Application de la microsonde électronique à l'analyse élémentaire quantitative d'échantillons liquides d'un volume inférieur à 10^{-9} l. J. Chim. Phys. 66:1084.

Moss, N. G. 1976. Micropuncture studies on renal tubular function in selected vertebrates. Ph.D. Thesis, University College of Wales.

Nicholson, W. A. P., Ashton, B. A., Höhling, H. J., Quint, P., Schreiber, J., Ashton, S. K., and Boyde, A. 1977. Electron microprobe investigations into the process of hard tissue formation. Cell Tiss. Res. 177:331.

Pallaghy, C. K. 1973. Electron probe microanalysis of potassium and chloride in freeze-substituted leave sections of Zea mays. Aust. J. Biol. Sci. 26:1015.

Panessa, B. J., Warren, J. B., Hren, J. J., Zadunaisky, J. A., and Kundrath, M. R. 1978. Beryllium and graphite polymer substrates for reduction of spurious x-ray signal. Scanning Electron Microscopy II:1055.

Pedersen, C. J. 1967. Cyclic polyethers and their complexes with metal salts. J. Am. Chem. Soc. 89:7017.

Pedersén, M., Roomans, G. M., and Hofsten, A. V. 1980a. Blue irridescence and bromine in the cuticle of the red alga Chondrus crispus Stackh. Bot. Mar. (In press.)

Pedersén, M., Roomans, G. M., and Hofsten, A. V. 1980b. The cuticle of the red alga Polysiphonia nigrescens. J. Phycol. (In press.)

Peterson, S. K., Frishkopf, L. S., Lechene, C., Oman, C. M., and Weiss, T. F. 1978. Element composition of inner ear lymphs in cats, lizards and skates determined by electron probe microanalysis of liquid samples. J. Comp. Physiol. A. 126:1.

Pieri, C., Zs. Nagy, I., Zs. Nagy, V., Giuli, C., and Bertoni-Freddari, C. 1977. Energy-dispersive x-ray microanalysis of the electrolytes in biological bulk specimen. II. Age-dependent alterations in the monovalent ion contents of cell nucleus and cytoplasm in rat liver and brain cells. J. Ultrastruct. Res. 59:320.

Reed, S. J. B. 1975. Electron microprobe analysis. Cambridge University Press, Cambridge.

Rick, R., Dörge, A., Von Arnim, E., and Thurau, K. 1978. Electron microprobe analysis of frog skin epithelium: evidence for a syncytial sodium transport compartment. J. Membrane Biol. 39:313.

Roomans, G. M. 1979a. Standards for x-ray microanalysis of biological specimens. Scanning Electron Microscopy II:649.

Roomans, G. M. 1979b. Quantitative x-ray microanalysis of halogen elements in biological specimens. Histochemistry 65:49

Roomans, G. M. 1980. Localization of divalent cations in phosphate-rich cytoplasmic granules in yeast. Physiol. Plant. 48:47.

Roomans, G. M., and Boekestein, A. 1978. Distribution of ions in *Neurospora crassa* determined by electron microprobe analysis. Protoplasma 95:385.

Roomans, G. M., and Kuypers, G. A. J. 1980. Background determination in x-ray microanalysis of biological thin sections. Ultramicroscopy. (In press.)

Roomans, G. M., and Sevéus, L. A. 1976. Subcellular localization of diffusible ions in the yeast *Saccharomyces cerevisiae*: quantitative microprobe analysis of thin freeze-dried sections. J. Cell Sci. 21:119.

Roomans, G. M., and Sevéus, L. A. 1977. Preparation of thin cryosectioned standards for quantitative microprobe analysis. J. Submicrosc. Cytol. 9:31.

Roomans, G. M., and Van Gaal, H. L. M. 1977. Organometallic and organometalloid compounds as standards for microprobe analysis of epoxy resin embedded tissue. J. Microscopy 109:235.

Routledge, L. M., Amos, W. B., Gupta, B. L., Hall, T. A., and Weis-Fogh, T. 1975. Microprobe measurements of calcium binding in the contractile spasmoneme of a vorticellid. J. Cell Sci. 19:195.

Russ, J. C. 1974. The direct element ratio model for quantiative analysis of thin sections. In: *Microprobe Analysis as Applied to Cells and Tissues* (Hall, T., Echlin, P., and Kaufmann, R., eds.), p. 269. Academic Press, London.

Russ, J. C. 1976. Processing of energy-dispersive x-ray spectra. EDAX Editor 6:4.

Shuman, H., and Somlyo, A. P. 1976. Electron probe x-ray analysis of single ferritin molecules. Proc. Natl. Acad. Sci. 73:1193.

Shuman, H., Somlyo, A. V., and Somlyo, A. P. 1976. Quantitative electron probe microanalysis of biological thin sections: methods and validity. Ultramicroscopy 1:317.

Somlyo, A. P., Somlyo, A. V., and Shuman, H. 1979. Electron probe analysis of vascular smooth muscle. Composition of mitochondria, nuclei and cytoplasm. J. Cell Biol. 81:316.

Somlyo, A. V., Shuman, H., and Somlyo, A. P. 1977a. Elemental distribution in striated muscle and the effect of hypertonicity. Electron probe analysis of cryosections. J. Cell Biol. 74:828.

Somlyo, A. V., Shuman, H., and Somlyo, A. P. 1977b. Composition of sarcoplasmic reticulum *in situ* by electron probe x-ray microanalysis. Nature 268–556.

Spurr, A. R. 1974. Macrocyclic polyether complexes with alkali elements in epoxy resin as standards for x-ray analysis of biological tissue. In: *Microprobe Analysis as Applied to Cells and Tissues* (Hall, T., Echlin, P., and Kaufmann, R. eds.), p. 213. Academic Press, London.

Spurr, A. R. 1975. Choice and preparation of standards for x-ray microanalysis of biological materials with special reference to macrocyclic polyether complexes. J. Microscopie Biol. Cell 22:287.

Stenn, K. S., and Bahr, G. F. 1970a. A study of mass loss and product formation after irradiation of some dry amino acids, peptides, polypeptides and proteins with an electron beam of low current density. J. Histochem. Cytochem. 18:574.

Stenn, K. S., and Bahr, G. F. 1970b. Specimen damage caused by the beam of the transmission electron microscope: a correlative reconsideration. J. Ultrastruct. Res. 31:526.

Sumner, A. T. 1978. Quantitation in biological x-ray microanalysis with particular reference to histochemistry. J. Microscopy 114:19.

Tormey, J. M. 1979. Origin of artifactual quantitation of electrolytes in frozen sections of erythrocytes. Scanning Electron Microscopy II:627.

Trump, B. F., Berezesky, I. K., Chang, S. H., Pendergrass, R. E., and Mergner, W. J. 1979. The role of ion shifts in cell injury. Scanning Electron Microscopy III:1.

Warner, R. R., and Coleman, J. R. 1974. Quantitative analysis of biological material using computer correction of x-ray intensities. In: *Microprobe Analysis as Applied to Cells and Tissues* (Hall, T., Echlin, P., and Kaufmann, R., eds.), p. 249. Academic Press, London.

Willemse, J., Cras, J. A., Steggerda, J. J., and Keyzers, C. P. 1976. Dithiocarbamates of transition group elements in unusual oxidation states. Structure and Bonding 10:83.

Williams, D. B., and Goldstein, J. I. 1977. A study of spurious x-ray production in a Philips EM 300 TEM/STEM. Proc. Eighth International Conference on X-Ray Optics and Microanalysis, 113A.

Wróblewski, R., Roomans, G. M., Jansson, E., and Edström, L. 1978a. Electron probe x-ray microanalysis of human muscle biopsies. Histochemistry 55:281.

Wróblewski, R., Gremski, W., Nordemar, R., and Edström, L. 1978b. Electron probe x-ray microanalysis of human skeletal muscle involved in rheumatoid arthritis. Histochemistry 57:1.

Zs. Nagy, I., Pieri, C., Giuli, C., Bertoni-Freddari, C., and Zs. Nagy, V. 1977. Energy-dispersive x-ray microanalysis of the electrolytes in biological bulk specimen. In: Specimen preparation, beam penetration and quantitative analysis. J. Ultrastruct. Res. 58:22.

AUTHOR INDEX

Subject Index

Absorption, of x-rays
  by beryllium window, effect on
      quantitative microanalysis,
      407, 408
  caused by protein in liquid droplet
      specimens, 330–332
  effect on detector efficiency, 21–24
  effect on x-ray production in bulk
      specimens, 7
  in microdroplet standard crystals,
      correction for, 422
  in quantitative microanalysis of
      plant tissues, 250–254
  relation to atomic mass, 7–8
  resulting from specimen topog-
      raphy, 243, 250–254
Absorption coefficients, for ions in
      microdroplet standard crys-
      tals, 422
Absorption edge, 8
Accelerating voltage
  as consideration in freeze-drying, 122
  in control of beam current, 36, 37
  as factor in excitation volume, 13
  measurement of, 43
  in microanalysis of bulk specimens,
      43, 377–379
      calculation of, 377–378
  in microanalysis of thin sections, 43
  relation to spatial resolution, 15–16
Acetone
  use in freeze-substitution
      as organic solvent, 110, 111
      as substitution fluid, 111
  use in preparing and cleaning micro-
      pipettes, 314, 319
Acrolein, use in freeze-substitution,
      211–212
Agar blocks, ion loss in, during freeze-
      substitution, 113, 115

Air drying, 99–105
  chemical redistribution in specimen
      during, 102, 104–105
  contamination of specimen during,
      100, 102, 103, 105
Algae, quantitative microanalysis of
      brominated phenols in, 445
Alkali metals, standards for quantita-
      tive microanalysis of, in plas-
      tic-embedded thin sections,
      416, 417
Alkaline earth metals, standards for
      quantitative microanalysis of,
      416, 417
Aluminum
  absorption coefficient for, 422
  as coating in quantitative microanal-
      ysis, 48
      for bulk specimens, to prevent
          thermal damage, 384
      for frozen-hydrated bulk speci-
          mens, 173
      for frozen-hydrated thin sections,
          202
Amplifier, see Pulse processor
Analog-to-digital converter (ADC), in
      multichannel analyzer, 34
Analytical resolution
  in freeze-substituted specimens,
      234–235
  in frozen-hydrated thin sections,
      197–198
Angle of incidence
  effect of specimen topography on,
      243, 244–250
  relation to interpretation of quanti-
      tative microanalysis data, 435
Anti-roll plate, use in dry cutting of
      frozen specimens, 293–295